Vancouver's
BRAVEST

120 YEARS OF FIREFIGHTING HISTORY

hancock
house

Alex Matches

ISBN-10: 0-88839-615-5
ISBN-13: 978-0-88839-615-0

Cataloging in Publication Data

Matches, Alex
    Vancouver's bravest : a firefighting history, 1886-2006 / Alex Matches.

ISBN 0-88839-615-5 (pbk.)

    1. Vancouver (B.C.). Fire and Rescue Services--History. 2. Fire extinction—
British Columbia—Vancouver—History. 3. Fire fighters—British Columbia—
Vancouver—History. I. Title.

TH9507.V3M383 2006   363.37'80971133   C2006-905501-7

Editing: Theresa Laviolette, Nancy Miller
Production: Rick Groenheyde
Photo editing: Laura Michaels
Cover design: Rick Groenheyde
Cover photo: Peter Battistoni - *Vancouver Sun* - Hi-To Fish Co. 3rd alarm, Dec 5, 1990

Printed in China — SINOBLE

*We acknowledge the financial support of the Government of Canada through the
Book Publishing Industry Development Program (BPIDP) for our publishing activities.*

Published simultaneously in Canada and the United States by

HANCOCK HOUSE PUBLISHERS LTD.
19313 Zero Avenue, Surrey, B.C. Canada V3S 9R9
(604) 538-1114   Fax (604) 538-2262
HANCOCK HOUSE PUBLISHERS
1431 Harrison Avenue, Blaine, WA USA 98230-5005
(800) 938-1114   Fax (800) 983-2262
Website: www.hancockhouse.com   *Email:* sales@hancockhouse.com

CONTENTS

CHAPTER 3   The Roarin' 20s, Amalgamation & Jubilee Year 1920-1936

CHAPTER 4   New Chiefs, the War and Major Fires 1937-1945 . . . .

Foreword

How important are firefighters? Just imagine the city without them.

In fact, we don't have to imagine. On June 13, 1886 — just two months after the little city of Vancouver was incorporated — a raging fire, which the tiny fire brigade had no real equipment to fight, obliterated the entire town. You can't fight a wind-whipped blaze covering entire city blocks with axes, buckets, shovels and ladders.

As you'll learn in this book by Alex Matches, who himself had thirty-three years of experience as a firefighter for Vancouver, the city's ability to fight major fires took a leap in August of 1886, two months after the Great Fire, when a brand-new fire engine arrived from the John D. Ronald Company of Brussels, Ontario. Even so, until the city could find horses to pull that engine, our hard-working firefighters often had to drag it to fires themselves over very rough roads — and sometimes to places where there were no roads.

Alex told the story of the equipment used by our early fire brigades in a 1974 book titled *It Began with a Ronald: A Pictorial History of Vancouver's Firefighting Equipment*. He appeared on my CBC radio show back then, talking about that book. More than thirty years after it first appeared, it is still selling!

Now Alex has expanded the department's history, bringing it up to the present day and enlarging the scope of the story by telling in more detail of the major fires that have burned their way into the city's memory. Some of those conflagrations actually reshaped the city: a 1915 fire on the original Cambie Street Bridge led to a push for a replacement; a fire in 1937 in the offices of the *Vancouver Sun* forced them to move to new quarters (a building we still know as 'The Old Sun Tower'); and the huge Pier D fire of July 27, 1938, which became the biggest blaze for the department since its founding fifty-two years earlier. Old-timers still recall the March 6, 1945, explosion in the harbor of the SS *Greenhill Park* and the subsequent fire, and the July 3, 1960, BC Forest Products drama, VFD's first five-alarm fire and a blaze that covered an area equal to six city blocks. Then there was the department's first six-alarm fire at the Fraser Arms Hotel in 1988. And so it goes.

Many of us literally owe our lives to the men and women of the Vancouver Fire Department. And many of the Department's personnel have given their lives in that service. Alex Matches pays tribute to them. He has been gathering material for this new history for more than three decades. From the axes, buckets, shovels and ladders of 1886 to the ultra-modern equipment and well-trained personnel of the twenty-first century, the Vancouver Fire Department has made enormous strides.

It's a story told nowhere else, and a story that needs telling. And in this book we have the right man to tell it.

— CHUCK DAVIS
Vancouver writer, historian and broadcaster

Preface

My thirty-three year career on the Vancouver Fire Department began in January 1962. After a two-week training course with the training officer and the crews in various firehalls around the city I was sent to No. 6 Fire Hall, in the West End for my probationary period of one year. At twenty-six years old, I was older than most of the junior firemen on my shift and accepted the fact that I was the rookie, the probationer, the PK (piss-kid). I was expected to learn everything I had to know from these guys as well as clean up after them in the kitchen and bathroom, make the coffee, stoke the furnace and hot water bogey, do the shopping and errands and answer the phone, before the second ring. By the end of my probation and first year on the best job in the world, I had been through lots of learning situations, some funny, some sad, some stressful and dangerous but mostly routine. By the end of my fifth year and shortly after I became the driver of No. 6 truck, I was transferred to No. 20 Fire Hall, in South Vancouver. I quickly settled in at No. 20, a one-piece hall with a pump and a four-man crew, and I began to broaden my experience with the many medical situations that I didn't have at No. 6. (During that time, No. 6 didn't respond to medical calls.)

After three years working with a great crew and attending many interesting alarms in an area of town that I felt comfortable in, I got the word that I was being transferred back downtown, this time to No. 1—headquarters…the Big House…the Big Smoke!! My initial reaction to going downtown was one of disappointment but soon turned into a feeling of elation. The crew consisted of an assistant chief, two captains, one lieutenant and usually about a dozen men who manned the three rigs, a pump, a hose wagon and an aerial ladder truck. I was the hose wagon driver and was kept busy attending more than twice as many medical and fire calls than I ever did at No. 20.

This book began as a labor of love over thirty-five years ago, during my time at No. 1 Fire Hall. One day I discovered a closet in the lecture room that had boxes and shelves full of papers, band music, journals and assorted odds and ends. Among the piles and boxes were many old photographs of the early days of the fire department, all pictures I had never seen before. The members of the VFD weren't made aware of the department's history when they came into the service and I was no exception. I took it upon myself to separate, organize and then identify the various photographs and other printed materials into chronological order. I learned about the people, the firehalls, the fire apparatus and many of the big fires and events that took place over the years. Then I began collecting photographs, scrapbooks, journals and other items of interest, on behalf of the department and after all these years have amassed a large collection of archival material, which forms the basis of this book.

I hope you will enjoy my efforts and find the history of the Vancouver Fire Department and "Vancouver's Bravest" (a name adopted by Local 18 in the 1980s) as interesting as I did.

— ALEX MATCHES
May 2006

Acknowledgements

An undertaking of this size would be impossible without the help of many people.

First, to the active and retired members of the Vancouver Fire Service, my thanks for your help with photos, information and encouragement, and your patience, too (I had hoped to have this project completed five years ago). Special thanks to Captain Reg Watts, Captain Rob Jones-Cook, Firefighter Ryan Cameron and Firefighter Shane MacKichan who always made the time to answer my questions and supply photographs. And to the fire buffs, whose interest and knowledge in all fire department topics is legion, especially to long-time friend, Walt McCall, whose help is always appreciated.

Thanks to the many photographers who recorded our history through the years, from the pioneer picture takers, to freelance, commercial and news photographers—far too many to list—we owe you our gratitude for your efforts.

My appreciation also for the courtesies extended by the staffs of the Vancouver City Archives (Archivist Carol Haber), the Maritime Museum (Executive Director Jim Delgado), the Sun/Province Library (Sandra Boutilier), the Vancouver Public Library (Andrew Martin) and I must thank historian, author, broadcaster, journalist Chuck Davis. His knowledge of Greater Vancouver facts and trivia is astounding and second to none. Thanks Chuck!

Lastly, my heartfelt thanks to my wife, Arlene, (who is my computer-literate "right arm") and my son Sandy (also computer-wise), for their valuable contributions.

Hope you enjoy our efforts…

DEDICATION

Respectfully dedicated to Vancouver's firefighters,

1886 to 2006: Vancouver's Bravest.

Their names are listed in Appendix II.

A portion of the proceeds from the sale of this book will be donated to the Vancouver Fire Fighters' Charitable Society.

From Humble Beginnings

(1867–1899)

When the talkative riverboat captain John "Gassy Jack" Deighton stepped ashore on a nameless beach in September 1867, with an Indian wife, a dog and a barrel of whisky he became the first saloon keeper in the little townsite, a motley collection of buildings that became known as Gastown. Even after it became officially known as the Town of Granville in 1870, it was without any form of fire protection, except for the axes, buckets, shovels and ladders of the property owners.

The town of Granville 1870. Captain Jack Deighton's hotel and saloon is the L-shaped building on the shoreline. The lot with the two buildings is the site of the home and office of the first public official, Constable Jonathan Miller. The small building is the jail.
Photo: Courtesy City of Vancouver Archives

TOWN of GRANVILLE
BURRARD INLET, B.C.

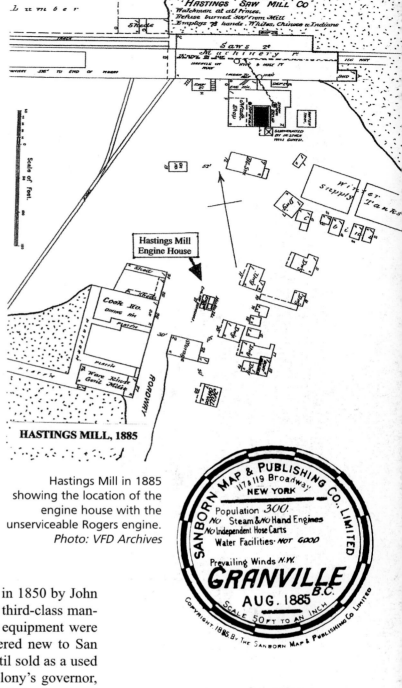

Granville was a fast-growing little town on the south shore of Burrard Inlet and was described by early citizens as "a small oblong of land, less than twenty acres, about four blocks long...boxed in by tall trees and a profusion of undergrowth, including luxuriant skunk cabbages!" It consisted of single-family dwellings, rooming houses and hotels, a butcher shop and packing house, a bakery, a Chinese laundry and other small merchants, a customs house and jail, a post office, a church, a school and several saloons. The majority of the population was men who worked at logging and making lumber out of the big trees at the nearby Hastings Sawmill. Known as Stamp's Mill when it began operating in 1865, it was the first industrial site in the area.

In August 1885 the town had a population of more than 300, but still lacked the means to fight a fire other than with the tools found around people's homes and places of business. At this time, a survey was done by a New York company to assess the town's state of preparedness in the event of a fire. The survey found that the town had "*No* Steam Engines & *No* Hand Engines, *No* Independent Hose Carts, Water Facilities *Not Good*, Prevailing Winds *NW*." Also noted was that the only real fire engine, kept in a shed at the Hastings Sawmill, on the outskirts of town, was "Not Serviceable."

Hastings Mill in 1885 showing the location of the engine house with the unserviceable Rogers engine.
Photo: VFD Archives

The Hastings Mill Engine

This engine had definitely seen better days. It was built in 1850 by John Rogers of Baltimore, Maryland, and was described as a third-class manual pump. Named *Telegraph* (early fire companies and equipment were often given names) the engine was one of three delivered new to San Francisco where it served the Monumental Company until sold as a used pump to Victoria on Vancouver Island in 1858. The colony's governor, James Douglas, responded quickly to the pleas of Victoria's businessmen for some form of fire protection for their substantial investments when he ordered the purchase of British Columbia's first two pieces of fire apparatus from San Francisco. Following years of hard use in Victoria by its first fire company, the Tiger Company, the Philadelphia-style engine was placed in storage in 1862. Described as "useless" in the fire chief's report in August 1865, it was rebuilt in 1866 and kept in reserve. In October 1873, it was sold to Hastings Mill and shipped to Granville.

In the years before the formation of Vancouver's fire brigade, the employees of Hastings Mill took part in local parades dressed as comic firemen in blackface, pulling the old engine. They called their group the Dark Town Fire Brigade (DTFB) and their motto was "We Git Dar." Their worn-out, old engine was broken up in the late 1880s and some had sou-

The Hastings Mill Rogers engine was similar to this Philadelphia-style engine.
Photo: VFD Archives

venir walking sticks made out of the wooden brake arms, which were the pump handles.

Another old, well-used Victoria engine served the small fishing community of Steveston, south of Vancouver. In March 1897 the town council purchased it for an unknown amount. An 1862 Button second-class

The Dark Town Fire Brigade and their engine. This benevolent group raised money for charity with their comical parade antics.
Photo: CVA FD P1

manual pump, it was built in Waterford, New York and was the second engine used by Victoria's Tiger Company. This engine (wrongly believed to have been used by the Vancouver and Portland Fire Departments) was sold in June 1901 to Atlin, BC, near the BC-Yukon border. Atlin's Board of Trade purchased it after two major fires the previous year destroyed two hotels, two general stores, a photography shop, the assay office and the Board's own office — with the loss of forty cases of French champagne valued at $1,200! This historic Button engine is currently on display in Atlin and its likeness is shown on the badge of the Victoria Fire Department.

Badge of the Victoria Fire Department.
Photo: Courtesy Victoria Fire Department

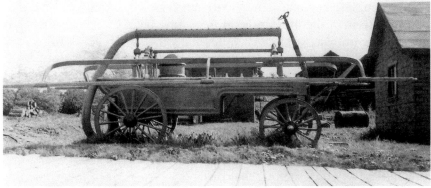

The Button engine in Atlin, BC as shown in the 1960s. *Photo: VFD Archives*

Vancouver is Incorporated

By January of 1886 Granville had a population of nearly 1,000 people and was quickly growing, largely due to an earlier announcement by the Canadian Pacific Railway (CPR) that Vancouver, instead of nearby Port Moody, would become the western terminus of the new transcontinental railway.

Over the previous few months, the residents of Granville had begun preparing their petition to the provincial government in Victoria to enact legislation to charter a city, which was to be named "Vancouver" after Captain George Vancouver, the English explorer. The city was named by William Van Horne, then vice-president of the Canadian Pacific Railway. The petition was signed by 125 men, and presented to parliament in Victoria on February 15, 1886. On April 6, the lieutenant governor assented and the city was incorporated.

Election Day

The first election was held on May 3 and 467 electors, again all men, placed their ballots in the single ballot box at the city's court house and elected Malcolm Alexander MacLean as the first mayor. As there was no voters' list, all who came voted, and it was said that some came early and often!

The mayor and ten new aldermen held the first council meeting on May 10. In his opening remarks to the first city council, Mayor MacLean is reported to have said, "We require immediate protection from fire and delay on this matter endangers a large amount of valuable property. A communication will be laid before you today in regard to a fire engine, which, if you have time, had better be discussed at this meeting." No action was taken.

At the second council meeting, held May 18, Sam Pedgrift, the manager of The People's Shoe Store, went before council and asked that they provide equipment and uniforms for the fire brigade that he was forming. No money was available, but the council passed by-law no. 6 providing for the institution of a fire brigade with the members to elect their own officers and further stated that, "The members of the brigade will be exempt from military and jury duty."

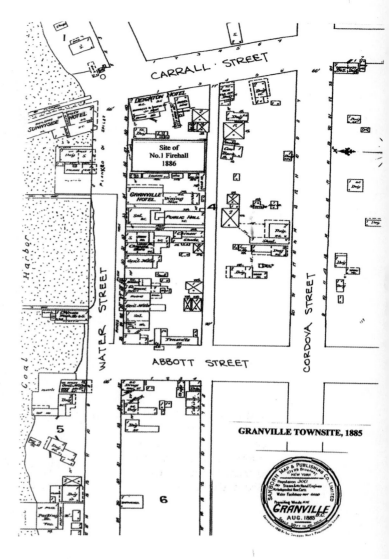

The Granville Townsite in 1885 showing the location of the 1886 No. 1 Firehall.
Photo: VFD Archives

Malcolm Alexander MacLean, the first mayor of Vancouver, 1886–87.
Photo: Courtesy CVA

A Fire Brigade is Formed

On the evening of May 28, a meeting was held at George L. Schetky's men's clothing store to look into organizing a city fire brigade. The pro tem chairman was G.W. West with C.M. Steinbeck as acting treasurer and Fred Germaine as acting secretary. Some of the citizens who attended the founding meeting and made up the original volunteer brigade were:

Sam Pedgrift	Frank Gladwin	Thos. F. McGuigan
Robert Rutherford	J. Hoskins	Wilson McKinnon
A.L. Dunlop	James Moran	John A. Mateer
W.W. Griffith	Peter Larson	Andy Tyson
J.E. West	John McAllister	J. McDonald
Stephan H. Ramage	William J. Blair	Ken Smith
C. Gigier	Thomas W. Lillie	Frank W. Hart
W.N. Dawsey	Hugh Campbell	William Chadwick
Gabriel W. Thomas	William S. Cook	John Garvin
Robert Leatherdale	E. Duhamil	G.E. Upham
Ernest C. Britton	C.M. Hawley	Barney Beckett
W. McGirr	Fred Germaine	John W. Campbell
Frank Curlow	George Schetky	G.W. West
C.M. Steinbeck	John M. Stewart	

John Stewart, the police chief, was an honorary member. At the meeting each of the members was assessed dues of "two bits" per month, elections were held and Sam Pedgrift was elected fire chief of the Volunteer Hose Company No. 1.

Although they were still without any kind of firefighting equipment, they were prepared to fight fires as a team, with some discipline and a chain of command but only equipped with household tools, buckets and whatever they could find to use. In spite of their lack of real equipment and "masheens" ("machines"), on June 2 a hook and ladder company was formed as part of the brigade with William Blair as foreman. Frank Curlow became secretary and George Schetky was elected the new treasurer, with $200 from public subscription.

The new hook and ladder company chose to be identified by red shirts with white trim, as the hose company had earlier chosen blue shirts with white trim.

The Great Fire

The story of Vancouver's Great Fire of June 13, 1886 has been told many times, many ways. The facts are that the weather had been extremely dry for many weeks before that fateful Sunday morning. Several small brush fires were burning away slash near the CPR roundhouse site, under the control of about forty CPR workers and contracted crews, when winds began to come up from the west, fanning the clearing fires. Over the morning, from about ten o'clock, these fires began to spread and the smoke got thicker. The crews on the site were soon having difficulty controlling the growing fires because of increasing winds, as well as the terrific upward movement of heated air caused by the fast-growing fires. Soon, large burning brands (hot, fiery, airborne particles) were falling on the city and the fiery attack on the helpless settlement was coming from all directions.

Townspeople beat out the burning brands as best they could and a group of people, little knowing what the day had in store for them, protected the roof of the newly built Regina Hotel by wetting down blankets.

Just after 2:00 p.m. the call of "Fire!" went out, and people on the western edge of town began hurriedly removing possessions from their homes and buildings. Dr. Robert Matheson began furiously ringing the St. James Church bell to alert the people to the fire. Charlie Tilley, the son of the owner of Tilley's Stationery Store, rescued the city's twenty-line telephone switchboard. Within minutes, gale-force winds forced the people to run for their lives from the wall of flames bearing down on them. It was said that the city didn't just burn, it exploded, and everything in its path was consumed.

Pioneer Alderman W.H. Gallagher said, "The inferno went down the wooden sidewalk faster than a man could run." A man driving a horse and wagon along Carrall Street between Water and Cordova perished in the center of the street and all that was found were a few bones and the steel

The Great Vancouver Fire June 13, 1886.
Photo: Courtesy CVA Historical Journal No. 3, January 1960

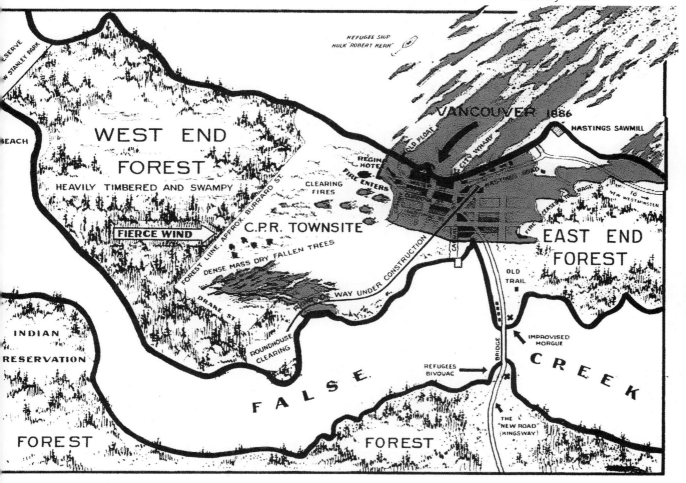

tires of the wagon. Three people who sought refuge in a well were found asphyxiated when the fire sucked the air out, killing them. And, as related by Charles G. Johnson, the poll clerk, "The fire met us like a wall. Our clothes commenced to burn and our handkerchiefs dried up like chips...we were about to give up...I spotted a small gravel patch and I proposed that we lie down...so we lay flat on it, which was not very hot at the time. Bailey could not stand this and said he was going to get through at any cost...and after running around for a few seconds, he dropped and burned up before our eyes."

People were frantically running everywhere, calling for help and for family members. Many survived by running into Burrard Inlet and swimming to safety. Others built makeshift rafts, or hung on to small logs or anything that floated to get away from the heat, and many of these found refuge on the *Robert Kerr*, an old coal hulk at anchor in the inlet. Another group made its way to a ship at anchor but were stopped by the watchman from coming aboard. He was quickly convinced that his life wouldn't be worth much if he didn't help them.

The men working on the land clearing on the roundhouse site were driven into False Creek and were rescued by some Natives camping on the south shore.

From the time the first building caught fire until the fire burned itself out, it had swept through more than 100 acres and completely destroyed the new city, except for the Regina Hotel at the corner of Water and Cambie Streets, the Bridge Hotel at the north end of the Westminster Avenue (Main Street) Bridge, and eight or ten adjacent homes and small buildings. The Hastings Mill buildings at the north foot of Dunlevy Avenue, just east of town, were spared when the fire burned itself out before reaching them.

The survivors of the Great Fire camped on the south side of False Creek.
Photo: CVA GF P6

"It was all over in forty-five minutes; a grand but awful sight," said a young girl who had witnessed the conflagration from her small boat while floating in Burrard Inlet, near Deadman's Island.

The disaster destroyed almost 1,000 buildings and left more than 2,000 people homeless. A silent and sorry procession of survivors headed to the areas immediately north and south of the Westminster Avenue Bridge seeking refuge and comfort. Word had gotten out to nearby New Westminster, and the Hyacks — the New Westminster volunteer firemen — and a labor union, the Knights of Labour, collected food, blankets and clothing.

Late in the afternoon, a man on horseback rode into Vancouver to let the people know that help was on the way from New Westminster. Mayor MacLean and Police Chief Stewart sent messengers throughout the area to let people know that help was coming.

Some of the survivors made their way across the inlet to Moodyville (North Vancouver), where they were cared for at local hotels and the Masonic Hall.

Before sunrise the next morning a small building near the Bridge Hotel was converted into a morgue, and the remains of twenty-one persons were collected; only the three who died in the well were recognizable. Over the following days other remains were found and a sad procession of people came by, hoping not to find missing loved ones among the dead.

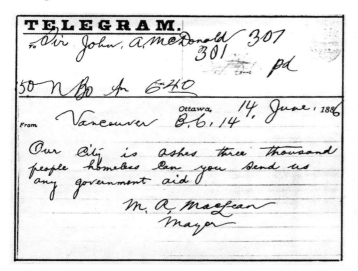
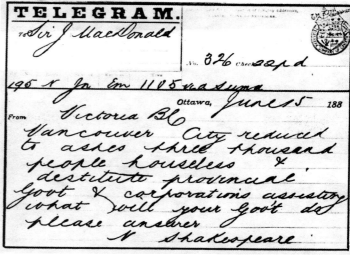

In the days after the fire, the mayor and others sent telegrams to the federal government in Ottawa requesting assistance for the destitute survivors. Immediate aid came from Moodyville and New Westminster and, later, more relief supplies came by steamer from Victoria and Seattle.

The Hastings Mill management and employees also came to the aid of the needy. They fed, nursed and helped in any way to console them. Richard Alexander, the manager (also the local magistrate and unsuccessful mayoralty candidate), invited all who needed lumber to rebuild to "help yourself and take what you need."

Almost nothing remained of the old Town of Granville, alias "Gastown" the first City of Vancouver. In the following days, over the still-smoldering ashes of the fire, the new Vancouver arose.

The question of who struck the first match on that fateful day will never be known, even though many came forward years later claiming to be responsible, including a volunteer fireman who had a contract to clear slash.

Twenty years later when some streetcar tracks were being replaced on Hastings Street near Carrall, some bones and a gold watch were found; the watch identified the owner. It is possible that more people perished in the fire that day but the exact number will never be known, as many people likely just left town and never came back.

Telegrams were sent by Mayor MacLean and N. Shakespeare, the Victoria Member of Parliament, to Prime Minister Sir John A. MacDonald requesting aid for the fire victims, June 14-15, 1886.
Photo: Courtesy CVA

THE DAILY NEWS

Vol. 1   VANCOUVER, B.C., JUNE 17, 1886   No. 12

THE FIRE

Probably never since the days of Pompeii and Herculauneum was a town

WIPED OUT OF EXISTENCE

so completely and suddenly as was Vancouver on Sunday. All the morning the usual pleasant breeze from the ocean was spoiled by smoke from fires in the portion of the townsite owned by the C.P.R. Co. west of the part of the town already built, but no alarm was felt in consequence. The place wherein these fires existed was until two or three months ago covered with forest. A large force of men had been engaged in clearing it. The trees were all felled, and the fallen trees, stumps, etc., were being disposed of by burning here and there in separate heaps. A few weeks ago, during a gale from the west, the city was filled with smoke and cinders from these fires, and fire reached close to several outlying buildings, but after some fighting, danger was averted. This doubtless, tended to lull the people into a sense of security on Sunday. It was about two o'clock in the afternoon that the breeze which had been blowing from the west

BECAME A GALE

and flames surrounded a cabin near a large dwelling to the west of the part of the city solidly built up. A few score men had been on guard with water and buckets, between this dwelling and the cabin. but when the wind became a gale they were forced to

FLEE FOR THEIR LIVES

and in a few minutes the dwelling was a mass of flames and the whole city was filled with flying cinders and dense clouds of smoke. The flames spread from this building to adjoining ones with amazing rapidity.

THE WHOLE CITY WAS IN FLAMES

less than forty minutes after the first house was afire. Of course this being the case, a number inevitably

PERISHED IN THE FLAMES

It is to be feared that the seven whose bodies were recovered constitute only a fraction of the whole number who perished. The total number of victims and their identity will probably

NEVER BE KNOWN

With the exception of Mrs. Nash and Mr. Craswell, the bodies recovered were all burned to crisp and barely recognizable as human remains. Mr. Craswell's body was found in a well wherein he took refuge and died of suffocation. A young man named Johnson, and his mother, were found in the same well. Johnson was dead and Mrs. Johnson has since died.

The body of Mr. Fawcett, the soda water manufacturer, was identified by his wife by means of his watch-chain.

Persons living near the Harbor and in the eastern part of the city hurried toward the wharves at the Hastings Mill, and crowded upon the steamers moored to the wharves. On the steamers and wharves, while the city was a mass of roaring flame, were gathered hundreds of frightened and excited men and sobbing women and children. Anon there emerged from the dense smoke one and another,

GASPING AND BLINDED,

with singed hair and blistered hands and faces, who had struggled almost too long to save property. A considerable number of people were surrounded by the fire and cornered near J. M. Clue & Co's store, and their only means of escape was to make rafts of the planking in a wharf at that place, and push out into the harbor. The wind was blowing fiercely, making the water rough, and the party were in no little

PERIL OF DROWNING

They made their way to a vessel which was at anchor in the harbor, and the watchman on the vessel, with all the proverbial insolence and stupidity of "insect authority," refused to let the party come aboard. He very soon perceived, however that his refusal "did not count," and that his very life would "not count" for much if he attempted to keep the people off the vessel, and surrendered unconditionally. Those who witnessed the conflagration from the water describe the sight as

APPALLING AND WONDERFUL

beyond description.

Many of the large number who lived nearer False Creek than the harbor, and made their way toward the body of water, had a hard struggle to escape with their lives. Mr. Joseph Templeton got through only with the assistance of others. Mr. Martin, of the Burrard hotel, barely escaped with his life, and was prostrated when he reached a place of safety. John Boultbee and C. G. Johnson saved their lives by lying down and

BURROWING THEIR FACES

in the earth. Both are still suffering from their injuries. Everyone suffered not a little from the blinding and suffocating smoke. Families were separated, and agonized women ran wildly about crying for missing children or husbands. Many men were completely crazed, and did not recover their senses for hours. The disaster was one of the most sudden and terrible which ever in the history of the earth has overtaken a community. The number of

BUILDINGS DESTROYED

is estimated at from 600 to 1000. In the west end of the city one building alone remains. In the east end are the Hastings mill, which was saved by the wind veering to the north, and the dwellings of Mr. R. H. Alexander and Ald. Caldwell. On the banks of False Creek two hotels and eight or ten other buildings escaped. This is all that remained of the city of Vancouver on the morning after the fire.

LOSSES

There were probably not more than a score of people in the city on Sunday who did not lose something in the shape either of goods, personal property or buildings. Comparatively few were insured, and these only for a fraction of the value of their property. Hundreds who last week were possessed of considerable property and getting well started in prosperous business, were left by the fire literally penniless. Some were peculiarly unfortunate, and deserve all sympathy and assistance. Excepting live stock and such articles as the people carried about their persons, the value of all the goods and chattels saved from the fire perhaps does not exceed $1000.

THE DAILY NEWS

Harkness & Ross, Publishers

THURSDAY, JUNE 17

THE "NEWS"

The Caldwell block, wherein the NEWS office was situated, was one of the first to be overtaken by the fire, and not even a scrap of paper was saved. Like nearly all others who had started in business in the new city, however, we perceive that the fire, whatever may be its effect upon individuals, is to the city as a whole not a very serious matter; in fact it can scarcely impede the progress of Vancouver at all. A few months, or even a few weeks, will restore the city to as good a basis as it was on before the fire. We have therefore determined to continue the publication of the DAILY NEWS. It will appear in reduced form (we hope however, to present four pages in a few days) until new material can be obtained.

RISING FROM THE ASHES

Following is a partial list of buildings already commenced:

WATER STREET

W. Dufour & Co. store, readymade clothing and gents' furnishings.

White & McKinnon, St. Lawrence hall, cor. Abbott street.

T. D. Cyrs, Granville hotel, old site.

Scoullar & Co., store, hardware, old site.

CARRALL STREET

T. O. Allen, Bodega saloon, old site.

Gravely & Spinks, real estate office, near old site.

Mr. Campbell, store, next to Gravely & Spinks office.

Templeton & Northcott, store, groceries and provisions, north of old site.

Four other buildings in progress on Carrall street.

Tremont house, in tent on old site.

OPPENHEIMER STREET

J. A. Finney, auction mart, cor. Carrall.

H. Mattern, butcher, old site.

Several small buildings are in course of erection further east on Oppenheimer street.

CORDOVA STREET

Martin & Balfour, Burrard hotel, cor. Carrall.

Thos. Dunn & Co. store, hardware.

Grant & Arkell, general store.

British Columbia Stationery and Printing Co., store.

The NEWS office.

Rand Bros., real estate office.

J. A. Brock & Co., photograph gallery.

M. Maconie, restaurant.

McLean & Grant, store, wholesale liquors and cigars.

Robertson & Co., real estate office.

T. Coleman, barber shop, already opened.

Bunning & Kelso, will build a restaurant on the old site of shop.

ABBOTT STREET

Lawson & Pyatt, restaurant.

POWELL STREET

Seven buildings are in course of erection on Powell street.

HASTINGS STREET

D. McPherson, C.P.R. hotel, three stories, old site.

P. Carey, hotel, old site.

A. G. Ferguson, cottage, opposite late Burrard hotel.

Cook & Neelands, commission merchants.

Several small buildings are in course of erection on Hastings street.

ALEXANDER STREET

F. W. Hart will rebuild his furniture store on the old site.

Strouss & Co. will build a two-storey brick building between Alexander and Powell streets.

Many others who have not yet located or are unable to get building material will commence building in a few days. It is understood that Mr. A. G. Ferguson will build a fireproof block of seven stores on the site of his old block. Ald. Hemlow will build a temporary structure on the old site of the Sunnyside and erect a brick building for the hotel in another part of the city.

NOTES

Buildings are already going up in all parts of the city.

The total value of the property destroyed is variously estimated at $1,300,000.

There are many idle men in the city, but workers are hard to get.

Thanks to the promptness of New Westminster, none who remained in the city Sunday night lacked food.

Those who remained in Vancouver Sunday night slept some in the mills, some in the outbuildings near False Creek, some in boat houses, some in the open air; and some slept not at all, but paced about the bare and blackened streets all night.

Loss of life excepted, the disaster, bad as it is, is not irrepairable. Few will be much the worse after a year or two, and with the assistance of friends and the encouragement of fellow losers, even the heaviest losers may still hope to grow and prosper with the town. "Never say die."

Carriages and other conveyances from New Westminster took hundreds of the homeless people to the Royal City after the fire, where they were hospitably received and generously treated. Several hundred went in boats to Moodyville and Hastings. In these places too, with a few exceptions, the people were generous to the sufferers.

The buildings formerly used by Mr. Spratt in manufacturing oil from herring have been appropriated by the relief committee for boarding and lodging the homeless. The upper portion will be fitted up for the accommodation of women and children, and will be in charge of Mrs. Muir. The ground floor will be for men. Mr. C. G. Johnson will superintend the whole building.

The relief fund is growing rapidly. Besides their generous hospitality on Sunday, the ladies of Moodyville have contributed $13 in cash. New Westminster, through Mr. Laidlaw, has contributed $400 by private subscription. H. F. Keefer has contributed one ton of flour; D. Oppenheimer $100; H. Abbott, $500; Mr. McPherson, $50; T. B. Spring, $50; the city of Toronto has raised $1000, and contributions are expected from other eastern places.

The Daily News describes the fire of June 13. *Photo: VFD Archives*

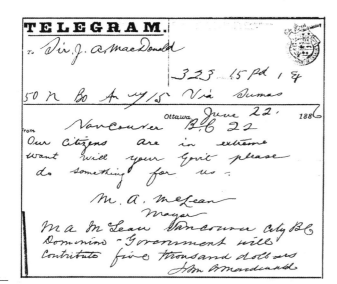

Mayor MacLean again asks for aid and receives $5,000 from the federal government. *Photo: Courtesy CVA*

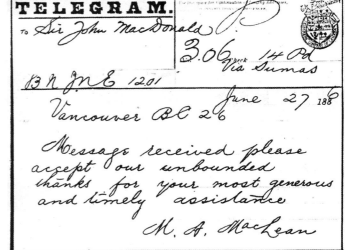

Telegram acknowledging and thanking Sir John A. McDonald for the funds. *Photo: Courtesy CVA*

The New Fire Engine

Eight days after the fire, on June 21, the city council instructed the city clerk (and volunteer fireman), Thomas McGuigan, to order a 3rd-size, 600-gallon (2,800-liter) per minute steam fire engine and 2,500 feet (750 meters) of 2½-inch (65-mm) hose from the John D. Ronald company, (also known as the Brussels Steam Fire Engine Works) of Brussels, Ontario. Also ordered were four Silsby hose reels and a fire bell for the tower of the yet-to-be built firehall. Plans were made for the construction of five underground cisterns, each to hold up to 10,000 gallons of water (45,500 liters) for firefighting purposes. To say that the city was now fire-conscious would be an understatement.

The total cost for the new engine and related equipment was $6,860 for which Mr. Ronald agreed to extend time to pay. There was no debate as to the various engines available and the choice of the Ronald was probably based upon the decision made by the City of New Westminster when they purchased the same model in 1885.

This is how the *Vancouver News*, Monday, August 2, 1886, described the new engine:

> The Ronald steam fire engine, size No. 3, which left Brussels, Ontario, July 11, arrived at Port Moody on the evening of Friday last, the 30 inst. The cost of it coming by express freight over the CPR is $300 a low rate considering that the price of transporting a similar engine, which was conveyed from the same point, in February last year, to New Westminster was $450 — although that road has been completed only a few weeks the competition has already commenced and both passenger and freight rates are being largely reduced. On Saturday, the Fire, Water & Light committee, after deciding that it would be much cheaper to have the engine brought by road, than by water, instructed Berry and Rutherford to send a team of four horses to Port Moody for the purpose of hauling the steamer to this city. At 10:30 yesterday morning, Mr. Ronald, who has been waiting at Port Moody for the arrival of the engine, made a start with his charge and reached New Westminster quite safely and without any untoward incident. A halt was made there for one hour and a half in order to give the horses a rest and to refresh the inner man. At 6:00 o'clock the engine arrived in this city and was located temporarily on the vacant lot of ground opposite the relief committee room on Water Street. Mr. Ronald said that the horses hauled the engine with ease there being no difficulty whatever in getting it over the road. He approved of it being hauled from Port Moody as the journey had lubricated the wheels and set them in good working order. Today it will be thoroughly cleansed and polished. Steam is to be got up tonight and an exhibition given of its protective grapple with the fire fiend. The engine has a handsome appearance, being well and substantially made. Its weight is 5000 lbs; the wheels are of extra large size and set in elliptic steel springs, its draft and movement being light and easy. It is capable of throwing one, two and four large and powerful streams a great height or distance, the apparatus accompanying the engine being ample for the complete protection of the city.

The 2500 feet of rubber hose and the four hose carriages make the equipment as perfect as it can be, and we believe that there is no city of the same size and population as Vancouver which can boast of having such a complete system of protection from fire as our town now possesses. The figures below give the cost of the engine and the accompanying apparatus:

Steamer	$3,800
Four hose carts, ordinary cost $200 each	700
2,000 feet of rubber hose, at $1.18 per foot	2,360
Freight charges (estimated)	300
Incidentals, cost of erecting engine house, etc. – estimate	340
Grand Total	$7,500

Mr. Ronald has agreed to give ten years credit for the sum of $6,860 due to him, with interest at the rate of 7 per cent per annum.

The first engineer for the new fire engine was Edward T. Morris who was hired at $60 a month, effective August 2.

The whole town turned out to see the testing of the new engine. At 6:00 p.m. on August 3 it was pulled out to the end of the Cambie Street wharf and, under the direction of Mr. Ronald, a fire was lit in the boiler. Within three minutes, sufficient pressure had been built up to allow the proud volunteers to give a blast on the whistle. At seven minutes, with forty-five pounds of pressure built up, the first line was charged, the nozzle was opened and the city aldermen and citizens were christened with a stream of salt water! The firemen apologized; they were sorry; it was an accident, they said. The engine performed to everyone's satisfaction and was accepted.

At the meeting of the volunteers following the testing of the new engine it was decided that the brigade would be known as the Invincible Hose Company No. 1, and the Vancouver Hook & Ladder Company No. 1.

Now that the engine had been tested and accepted by council, on August 6 it was officially named *M.A. MacLean* in honor of the city's first mayor. Until the firehall was built, it was to be kept under canvas on an empty lot on Water Street. Then on August 9, the engine and other equipment were placed in a nearby rented building with quarters for engineer Morris. At the weekly meeting of the volunteers on August 10, the Hook and Ladder Company felt that their name was absurd because they didn't have a hook and ladder yet, so they voted to become the Reliance Hose Company No. 2.

Under the direction of Chief Pedgrift, the volunteers began honing their firefighting skills by hauling their new engine and hose reels out

One of the hose reels and the Ronald engine showing the handles for pulling it.
Photo: CVA FD P41

every few days to practice laying hose and watering down the dusty, unpaved streets. Each man probably wished for the day when the brigade could test its mettle against the "fire fiend."

Spratt's Oilery Fire

The volunteers' chance to fight their first real fire with the new engine came around 11:00 p.m., August 11 at Joseph Spratt's Fish Oil Refinery (Spratt's Oilery) described in the newspapers of the day as "a considerable distance from town." By the time the brigade mustered enough of its numbers to haul the two and a half-ton engine about ten city blocks uphill, along a dark, winding and bumpy dirt path through the forest, it's no wonder that there wasn't much to save. However, they did prevent a few nearby houses (which really weren't in too much danger) from catching fire. The newspapers reported that the new fire engine was "a superior article and the firemen, though inexperienced, worked with a will for an hour and a half and saved an adjacent warehouse and office." The estimated fire damage was $50,000 and the cause was unknown.

The refinery had been in operation from 1881 to 1884 processing the fish oil from the abundant herring that were harvested from the inlet, sometimes by dynamite. Located at the foot of a steep bank (at the north end of today's Burrard Street), the facility never reopened.

SPRATT'S OILERY, 1885

Spratt's Oilery as shown from the survey of 1885.
Photo: VFD Archives

The City's First Ball

A historic event took place on August 24 when the city's volunteer fire brigade hosted the first-ever public ball in the city. Held in Gold's Hall, on Water Street near Abbott, the purpose of the ball was to celebrate the arrival of the new engine and to raise funds for uniforms and to cover the temporary expenses in the organization of the brigade. The *Vancouver News* described the event as follows:

> On entering the hall last evening a brilliant sight met the gaze. The platform was prettily decorated: a Canadian ensign, kindly lent for the occasion by Mr. Power of Moodyville, was hung at the back, and in front were two sections of hose with the shiny brass nozzles crossed. A banner inscribed with the word 'WELCOME' surmounted the whole, 'INVINCIBLE HOSE COMPANY NO. 1' and 'RELIANCE HOSE COMPANY NO. 2' being lettered on the side. The walls of the hall were decorated with flags of all nations (lent by the captain of the German bark *Gotha*, now lying in the harbor), and numerous boughs of the evergreen fir.
>
> Among the guests were Mayor and Mrs. MacLean, Mr. and Mrs. Oppenheimer, Mr. and Mrs. Ronald, as well as Chief and Mrs. McColl of New Westminster and Assistant Chief Webb, also of New Westminster. The dance programme consisted of 24 reels, waltzes, quadrilles, schotisches, as well as other dances. After the dance, at midnight, the majority adjourned to the St. Julien Restaurant for a sumptuous banquet of many various meats, salads, sauces and dessert.

Volunteer Fire Brigade Badge.
Photo: VFD Archives

The ball was a complete success and very much enjoyed by the 250 people who attended, and the money that was raised allowed the volunteers to purchase the shirts, helmets and equipment that they needed.

The *Vancouver News* on August 26 reported that the fire bell for the tower of the new firehall had arrived. Ordered through hardware merchant Thomas Dunn & Company of Vancouver, it was cast by the U.S. Bell Company of Troy, New York. Weighing 500 pounds (225 kg) with a 250-pound (115-kg) frame, the bell cost $150 including $45 duty. Fire bells were hung in the towers of No. 1 and No. 2 Firehalls and were used to alert the townspeople and to call the volunteers to a fire. Their disposition is unknown, but it is likely that they were installed in local churches.

In early September more men joined the ranks of the volunteers. They were: John H. Carlisle, Reverend Father Henry Glynne-Fynes Clinton (Father Clinton), Norman Dawsey, Arthur Clegg, C.F. Perry, David Biggar and Dr. Robert St. James Matheson.

On September 15 Alexander W. Cameron replaced E.T. Morris as the engineer of Engine No. 1.

No. 1 Firehall

By September 17 the building of the new firehall was well underway. The site selected was the most historic piece of property in the city: Lot 2 Block 2, on Water Street. The cottage built on this lot in 1871 was the first Customs House as well as the home of the only provincial official on Burrard Inlet, Constable Jonathan Miller. It later became the first courthouse and jail (seldom used, it was said to have had two cells and no locks on

No. 1 Firehall with the new addition.
Photo: VFD Archives

WATER St - LOOKING WEST FROM CARRALL St - MAY. 1888 -

the doors) and on the first election day in the city it was the first polling station, and for a few days afterward served as the first city hall.

The new firehall, which cost $743, was 25 feet deep by 34 feet wide (8 x 11 m) with a 60-foot (20-m) high tower, 12 feet (4 m) square. It opened around October 14 and was seemingly found to be too small. The main floor accommodated the engine and other apparatus and the first floor had sleeping quarters for up to twelve men, but plans were made immediately for an addition with room for more equipment and men. (Over the years, the address of the firehall has been given in various publications as 12 Water Street or 32 Water Street, but old city directories show that the correct address was 14 Water Street.)

Around this time Sam Pedgrift sold his property and left town with the fire brigade's funds from a recently held minstrel show. He and the money were never seen again, and it was said that he headed south of the border with his wife and two daughters. William J. Blair succeeded Pedgrift as chief, and then Wilson McKinnon held the position for a short time. On December 28 an election was held and John Howe Carlisle, age twenty-nine, beat John A. Mateer by one vote, and remained the city's fire chief for the next forty-two years.

Carlisle had arrived in Vancouver in April 1886 and established a cartage business. In January 1887, along with his position as fire chief, he was also appointed the license, fire and health inspector and was paid $75 per month. He kept his hauling business until appointed the city's first, full-time paid fire chief in September 1889.

Between the fall of 1886 and the spring of 1887 the construction of firefighting cisterns was completed. The five underground tanks were strategically located around the city at points within reach of the site of any potential fire scene. They were located at the intersections of Water and Carrall, Cordova and Columbia, Hastings and Abbott, Cordova and Water, and Granville and Dunsmuir Streets.

In late May 1887 the volunteer brigade was tested by another bush fire, very much like the one that destroyed the city the year before. The fire was burning west of town along Hastings Street. The brigade set up the engine at the cistern at Hastings and Abbott and laid over 2,000 feet (600 m) of hose. When the fire, which was burning south toward English Bay, had been knocked down enough, the engine and hose lines were moved to the tank at Granville and Dunsmuir Streets. Around midnight, after fighting the fire all day, a wind change helped the men put it out and they were able to return to quarters about 3:00 a.m. Twenty-four hours later they had a small rekindle in the same area but by 9:00 a.m., with the help of a large shovel corps, the fire was completely extinguished.

The First Train Arrives

The city was in a festive mood on May 23, decorated with flags, bunting and evergreen boughs and awaiting the arrival of the first transcontinental train from Montreal. The men of the VFB along with their guests, the Hyacks of New Westminster, dressed in their finest uniforms, paraded with Engine No. 1 and their hose reels to the train station with the city band and a large group of citizens. At 12:45 p.m., amidst the cheers of the

John Howe Carlisle

Born: November 4, 1857.
Hillsboro, New Brunswick
Joined VFD: September 1886
Elected Volunteer Chief: December 1886
Appointed Paid Chief: September 1, 1889
Retired: January 31, 1929
Died: November 28, 1941

No. 1 and No. 2 Hose Companies in parade uniforms. Foreman Robert Rutherford and Foreman John Mateer (reclining) are holding speaking trumpets. *Photo: CVA FD P7*

gathered crowd, CPR locomotive 374, with bell ringing and whistle blowing, pulled the first transcontinental passenger train into Vancouver.

In June the first hook and ladder truck, which had been ordered by the city council in November 1886 at a cost of $1,150, arrived and was placed in service in the now-enlarged No. 1 Firehall. Supplied by the Waterous Engine Works Company of Brantford, Ontario, it was hand-drawn and carried a complement of wall and extension ladders as well as lanterns, axes and other small equipment. William McGirr organized the first ladder company and Pete Larson was elected foreman. Around this same time, a small library and reading room opened in the new firehall, Vancouver's first public library in a city-owned building. It was started with a gift of books donated by merchant David Oppenheimer, who later served as the city's second mayor, from 1888 to 1891.

With the new Hotel Vancouver now open uptown at Georgia and Granville Streets and with the growth of the city heading into that area from the original Gastown site, plans were now being made to build a

The Vancouver Volunteer Fire Brigade and their new engine with the New Westminster volunteers (kneeling front) before their march to the CPR station to welcome the first train into the city. *Photo: CVA FD P21*

second firehall. A second Ronald engine, identical to the original, was ordered, as well as two more hose reels. Chief Carlisle went to city council to urge them to buy a team of horses for Engine No. 1 so that his men could perform more efficiently at fires. His request was turned down but taken "under advisement."

The following new members joined the volunteers in 1887: A. McDonald, William Hamilton, John Taylor, John McKenzie and William Saunders; in early 1888, Angus D. McKenzie, A.M. Montgomery, Bert Hall, W. L. Hayward and George Brown.

The strength of the brigade was now as follows: Fire Chief J.H. Carlisle; Assistant Chief John Garvin; Chief Engineer Alexander W. Cameron; Robert Rutherford, Foreman, Hose Company No. 1; John Mateer, Foreman, Hose Company No. 2; Pete Larson, Foreman, Hook and Ladder Company No. 1; Fred Upham, President; Thomas (Barney) Beckett, Secretary; Angus McKenzie, Treasurer, and fifty volunteers.

The new hook and ladder at No. 1 Firehall with Chief J. Carlisle. *Photo: CVA FD P41*

The No. 1 Hook and Ladder crew in dress uniforms. Foreman Pete Larson is on the right, middle row. *Photo: CVA FD P27*

When council agreed to purchase a team of horses for Engine No. 1, Thomas Simpson was hired as the first teamster.

No. 2 Firehall

When No. 2 Firehall opened at 724 Seymour Street on August 3, its new Ronald engine, along with one of two new hose reels (the other new hose reel was placed in service at the City Hall) would still be hand-drawn, much to the crew's dismay. Once accepted and placed in service, the new engine was named *Joseph Humphries* for the city alderman and chairman of the Fire, Water and Light Committee.

The first engineer of Engine No. 2 was William Hamilton, who was officially named the brigade's assistant fire engineer, under Alex Cameron.

By the end of August the Vancouver Water Works had laid the water main from the Capilano River watershed on the North Shore under the First Narrows. On March 25, 1889 the first flow of water came into the city mains. By mid-April hydrants were tested, and by June 25 the cistern water system was replaced by water mains and fire hydrants.

The first team of horses learning how to pull the engine.
Photo: CVA FD P42

The Part-Paid Fire Department

A major milestone in the brigade's history occurred September 1, 1889 when the city council approved plans for a part-paid fire department, con-

The 1888 Ronald Engine No. 2, the *Joseph Humphries*, at Firehall No. 2 on Seymour Street. Engineer William Hamilton is standing in front with Fireman W. Saunders standing on the engine.
Photo: Bentley Collection: Dominion Photo Company

No. 2 Firehall on Seymour Street, c. 1890. *Photo: VFD Archives*

sisting of Chief Carlisle, two engineers, two stokers, two drivers, six full-time men and ten "call men." The chief was paid $75 per month, the engineers $60 a month, the other men $15 a month and the call men were paid about $1 for fires attended. Most of the single men lived in the firehalls. The new paid department now had two drivers, Thomas Simpson for Engine No. 1 and Arthur Clegg for the new Hose Wagon No. 1. The hose wagon had just been received from the Silsby Manufacturing Company of Seneca Falls, New York. Purchased at a cost of $920, it could carry 1,000 feet (300 m) of 2½-inch (65-mm) hose, more efficiently than eight or ten men pulling a loaded hose reel. Engine No. 2 would still be hand-drawn.

No. 2 Engine drafting from the cistern at Granville and Dunsmuir, 1888. *Photo: Courtesy CVA*

Hose Reel Champions

Always encouraged to take part in athletics, the early volunteers entered many hose reel competitions with fire brigades from nearby towns and cities and became very proficient at it. On September 19, at the competition in Tacoma, Washington, Vancouver's Alert Hose Team (the team's competitive name) won the North American Hose Team Championship by defeating twelve American teams, including Rochester, New York, which had held the honor for many years. The team won four races, set a new world record for speed, and received $4,000. Each member was awarded ribbons and a medal at the meet, and on their return to Vancouver each received a gold medal at a civic reception and dance. The sixteen-member team continued to compete for many years.

Chief Carlisle's buggy and horse, Tanner, with the new Silsby hose wagon. September, 1889 at No. 1 Hall. *Photo: CVA FD P40*

No. 1 Hose Wagon inside the firehall, showing the quick-hitching harness. *Photo: CVA FD P58*

Hose Reel Championship medal.

Back row, left to right: Steve Ramage, A. Montgomery, J. McKenzie, D. Biggar, W. Saunders, W.L. Hayward, and A. McDonald, trainer.
Front row, left to right: R. Rutherford, F. Gladwin, B. Hall, George Brown, John Garvin
Photos: CVA FD P10.

The New Fire Alarm Boxes

The installation of the Gamewell Fire Alarm System with fifteen fire alarm street boxes was begun in 1889, and on February 25, 1890 the first test was made from Box 16 at the corner of Hastings and Granville Street. The boxes were located as follows:

Box 3 Beach Avenue and Granville Street
Box 4 CPR Shops, Drake Street
Box 5 Drake and Granville Streets
Box 6 Smithe and Granville Streets
Box 7 Helmcken and Richards Streets
Box 8 North end, Cambie Street Bridge
Box 9 Cambie and Georgia Streets
Box 10 Homer and Robson Streets
Box 12 Dunsmuir and Granville Streets
Box 13 Dunsmuir and Homer Streets
Box 14 Seymour and Pender Streets
Box 15 Homer and Pender Streets
Box 16 Hastings and Granville Streets
Box 17 Hastings and Richards Streets
Box 18 Seymour and Cordova Streets

The missing numbers 1, 2 and 11 were used to designate alarms (first and second) and the fire alarm "strike-out" (meaning the fire is under control or out) signal (1-1). These signals were received in the firehalls via bells and punched reel-to-reel tapes (see chapters 3 & 7). The fire department telephone number at this time was 89.

The compiling of city fire alarms and fire loss records began in 1890. That year the department recorded 110 alarms with a total loss of $17,300, with insurance paid on $13,180.

Fire alarm street box and firehall gong.
Photos: VFD Archives

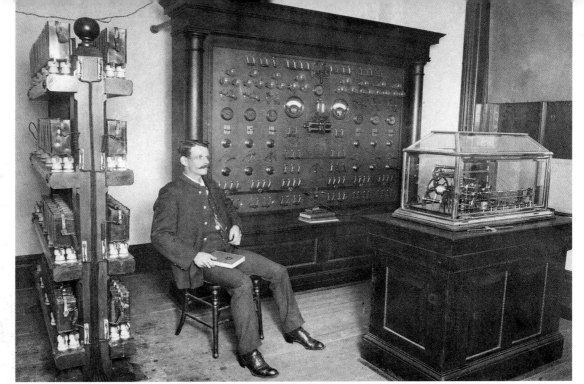

Electrician Chris Barker with Vancouver's fire alarm system, showing the main panel and glass encased repeater.
Photo: VFD Archives

Chief Carlisle went before council once again to urge them to purchase horses for Engine No. 2, "as the firemen become exhausted dragging the equipment to the fire," he said. In March 1892 he was successful in his bid and John Smalley was hired as Engine No. 2's first driver. From that time onward, when a piece of apparatus was purchased it came with the necessary horsepower.

By May of 1892, No. 2 put its new hose wagon in service. Because of the concern about the cost of Hose Wagon No. 1, it was suggested that perhaps a local builder could build a wagon at a better price. Vancouver carriage builder F.K. Winch built the new hose wagon, meeting the criteria, for $525.

When the city was created in 1886 the boundaries were roughly Alma Road on the west, 16th Avenue on the south, and Nanaimo Street on the east, with the bay and harbor on the north. In September 1892 the provincial government created the new municipality of South Vancouver, and at the same time Burnaby, to the east, was also incorporated. In the northeast corner of the city was Hastings Townsite, which extended between Nanaimo Street and Boundary Road and south to 29th Avenue from the waterfront. It became part of the city on January 1, 1911.

Hose Wagon No. 2 and Engine No. 2 with their new teams of horses, May 1892. John Smalley is the engine driver.
Photo: CVA 677-457

No. 3 Firehall

No. 3 Firehall opened on 9th Avenue, west of Westminster Avenue in the heart of Mount Pleasant, on November 1, 1892, with one of the hand-drawn hose reels. This hall would be so equipped for the next year until its new hose wagon was placed in service. (In 1910, 9th Avenue was renamed Broadway and Westminster Avenue became Main Street.)

VFD buildings have traditionally been referred to as "Firehalls" but No. 3 was the only one that had signage calling it a FIRE STATION.

The first chemical engine on the department arrived and was put in service at No. 2 Firehall on December 16. With a capacity of 100 gallons (450 liters), the Morrison Duplex added to No. 2's capabilities in the downtown area. (The chemical engine was a large version of the portable soda/acid extinguisher).

At 1:30 a.m. on June 5, 1893 tragedy struck the fire department when John Smalley, the twenty-five-year-old driver of Engine No. 2 was killed when he fell from the seat of his rig while responding to an alarm. Buried in Mountain View Cemetery Firemen's Plot, he was the first of thirty-two Vancouver firefighters to die on duty.

A Fully Paid Fire Department

On July 1, 1893 the Vancouver Fire Department became a fully paid department with a complement of thirty men working out of three firehalls. October 31 saw the arrival of the new Hose Wagon No. 3. Built by the St. Charles Omnibus Company of Belleville, Ontario, it cost $500 and was able to carry 1,200 feet (370 m) of 2½-inch (65-mm) hose. The first horses used to pull this rig were borrowed from the water works department because the city hadn't purchased the required horses yet.

On February 18, 1894 Vancouver firemen, on duty twenty-four hours a day, seven days a week, except for occasional time off at irregular periods and for meals, petitioned the Fire, Water and Light Committee for one consecutive twenty-four-hour period off each week. Their request was denied.

An international depression slowed the growth and expansion of the city over the next few years, until about 1898. The electric streetcar system, which had been operating since June 1890, and the electric lighting company, were both offered to the city, but

Vancouver-built Hose Wagon No. 2 shown at a fire, Burrard & Dunsmuir c. 1900. *Photo: VFD Archives*

No. 3 Fire Station (as shown in 1911). *Photo: VFD Archives*

Hose Wagon No. 3
was put in service October 31, 1893.
*Photo: Bentley Collection,
Dominion Photo Company*

The Morrison Chemical Engine, from a sales
catalogue engraving.
Photo: VFD Archives

Logo of the Firemen's
Benefit Association.
Photo: VFD Archives

on voting day the public turned down the referendum to purchase them. The streetcar system was largely responsible for growth in areas like Mount Pleasant, Fairview, Strathcona and the West End, and this growth would eventually require the fire department to expand into these areas, too. After the BC Electric Railway Company purchased the street railway system, it began double tracking, and rapid growth of the city followed.

The VFD answered an average of seventy alarms a year during these depression years, as follows: 1894 – 58 alarms; 1895 – 97 alarms; 1896 – 64 alarms; and 1897 – 62 alarms. Then the Klondike Gold Rush of 1898 brought prosperity to the city and was the beginning of a new boom that would last well into the twentieth century. The calls in 1898 reached a record 131 alarms, the highest annual number to date.

In the late evening of September 10, 1898 the major part of the City of New Westminster was destroyed in a disastrous fire that leveled almost sixty city blocks of business and residential buildings. The fire started in a riverfront hay storage warehouse and spread to two sternwheeler riverboats, the *Edgar* and the *Gladys*, which drifted downriver, setting fire to every wharf they touched. The raging fire then jumped Front Street and was quickly spread uptown by fierce winds.

By dawn the next day, hundreds were left homeless and damage was estimated at $2.5 million. Only two brick buildings were left standing and it is recorded that the VFD crew and equipment, under the direction of Chief Carlisle, were credited with saving one of these buildings.

The Firemen's Benefit Association

The Firemen's Benefit Association (FBA) of Vancouver was formed on November 14, 1898 with a donation made by the Harry Lindley Opera Company, which gave a benefit performance with proceeds "to aid injured or disabled firemen." The first Firemen's Ball, held February 10, 1899 at the City Hall, raised an additional $474.30 for the FBA and became an annual event for many years.

In April, William H. McPhee, one of the city's early firefighters, left the VFD to become chief of the New Westminster Fire Department.

Hugh Campbell, one of the most popular men among the early volunteers and a member of the championship hose reel team, quit the VFD on July 3, 1899 and went to work for the CPR. Hugh died in 1956 at the age of 89.

In September 1899 the new Waterous 1st-size* 1,000-gallon (4,500-liter) per minute engine replaced No. 2's 1888 Ronald engine that went to No. 1 Hall. Earlier, on March 1, 1899, the original 1886 Ronald engine was sent from No. 1 to No. 3 Hall, on a "semi-reserve" basis because of a cracked cylinder. (*Steam fire engines were rated as first size (or class) at 1,000 gallons per minute capacity; second size, 800 gpm; third size, 600 gpm, etc. Larger sizes were rated as Extra First size and Double Extra First size. American sizes and manufacturer's capacities and ratings varied until standards were developed about 1906 by the Fire Underwriters and the National Fire Protection Association [NFPA].)

The last apparatus purchases of the year, and the century, included another chemical engine, a 120-gallon (545-liter) Champion, and the department's first aerial ladder, a 75-foot (24-m) Hayes truck, each with a two-horse team. With all of these new rigs going into No. 2 Firehall, it became apparent that the old hall was too small and had to be replaced.

No. 2 Champion Chemical Engine, 1899. *Photo: VFD Archives*

The 1899 Waterous Engine No. 3 at a working fire.
Photo: VFD Archives

The VFD's first Aerial, an 1899 Hayes shown drilling on Seymour Street across from No. 2 Hall. *Photo: CVA FD P35*

Vancouver's veteran volunteer firemen, 1886.
Front row: left to right: John W. Campbell, Wilson McKinnon, A.W. Cameron, Hugh E. Campbell, J.H. Carlisle, John A. Mateer, Andy Tyson, J. McDonald, Thomas W. Lillie.
Second row: C. Gigier, W. Dawsey, John McAllister, Gabe Thomas, Barney Beckett, W. McGirr, John Garvin, Stephen Ramage, G. F. Upham, Ernest Britton, C.M. Hawley.
Back Row: Robert Leatherdale, E. Duhamil, James Moran, Frank Gladwin, Peter Larson, J. Hoskins, William S. Cook. (Photo taken 1898.)
Photo: VFD Archives

The Golden Years to Post World War I

(1900–1919)

The years 1900 to 1910 in Vancouver's history are referred to as The Golden Years. The land boom was said to be unequalled and everyone speculated in one way or another, hoping to strike it rich, and many did. The city had telephones (1885), electric lights (1887) and streetcars (1890) and was far ahead of most other North American cities. The growth of the fire department more than tripled during this period, with the addition of seven new firehalls, more equipment and, of course, many more men.

In the late winter of 1899–1900 Dawson City, Yukon lost a steam pump while fighting a fire. Details of the loss are sketchy, but one story said that while drafting through the river ice, the heat of the engine melted the ice enough for the engine to fall through and sink, badly damaging it. An order for a new engine was placed with the Waterous agent but because he was unable to deliver it for several months, he arranged for Vancouver to send their new 1899 engine to Dawson. He loaned the city a 1st-size Silsby engine until a new replacement was available.

On September 7, 1900 the 1,000-gpm Waterous (Engine No. 3) was sent to Dawson City on the CPR's Salvage Steamer *Tees*, via Skagway, Alaska, the White Pass and Yukon Railway and down the Yukon River on a sternwheeler. On March 26, 1901 a new Waterous, identical to the first Engine No. 3, arrived in Vancouver and was placed in service at No. 2 Firehall.

1901 Waterous Engine No. 3. *Photo: VFD Archives*

The men of the Vancouver Fire Department, 1901. Top row, left to right: Tom Tidy, N.L. Ross, Capt. James Lester, Chief J.H. Carlisle, Capt. C.W. Thompson, J. Davidson, J. McMorran. Third row: W. Brownlee, Andy Gill, John Campbell, J. McInnis, Ed McKeating, Chris Barker. Second row: W.G. McDonald, Harry Duncan, William Flood, F. Barker. Front Row: Dick MacAuley, Ed Mitchell, W. Jordon, Fred Murray, John Courtney, Capt. J. Moran, W.D. Frost, Norm McPherson, A.W. (Dad) Cameron, D. Scott, Arthur Clegg.
Photo: VFD Archives

The Trorey Cup. *Photo: VFD Archives*

The Trorey Cup

Participation in athletics was always a priority on the department, and all of the three firehalls had a handball court with fierce competition on each of them. In 1900 George E. Trorey, a prominent city jeweler, presented the fire department with a silver cup for the annual handball competition. The winners, Chief Charlie Thompson and William Flood, were the first to have their names engraved on the cup and were presented with a handsome set of cufflinks as a souvenir of the event. The challenge for the cup lasted many years and it now sits, relatively unknown, in the department's trophy case. Competition continues to this day in many other team and individual sports, including soccer, hockey, curling, racquetball and golf.

On March 29, 1902 James A. Lester left to become the chief of the Dawson City, Yukon, Fire Department. He joined the VFD in April 1894.

On April 3, 1903 a tender was awarded to demolish No. 2 Firehall to make way for the construction of a new modern hall. In early May the men and equipment were moved out to temporary quarters at 845 Granville Street.

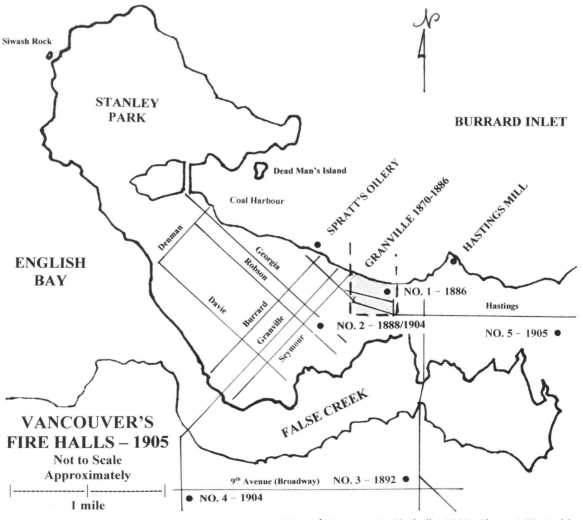

Map of Vancouver's Firehalls, 1905. *Photo: VFD Archives*

No. 4 Firehall

On January 1, 1904 a modest, one-bay, two-story firehall was opened in the Fairview district at 9th Avenue and Centre Street (southeast corner), now known as Broadway and Granville Street, with a two-horse combination hose/chemical wagon. John W. Campbell, one of the original volunteers, was the captain of the hall and remained there until his untimely death from a heart attack in March 1910 at age forty-seven. At the time of his death, he was the most senior member of the department and the holder of badge No. 1. This firehall was considered to be far out of town, and at different times the crew kept a deer fawn and a bear cub as mascots.

Firehall No. 4, 1904. Note the deer on the hose wagon. *Photo: VFD Archives*

The bear mascot at No. 4 Hall. The men are:
Rear, left to right: Archie McDiarmid, Noel Layfield, Joe Hillier
Front, left to right: W.H. McKechnie , Capt. John Campbell
Photo: VFD Archives

The New No. 2 Firehall Opens

At 9:00 a.m. on January 18 No. 2's engine and truck crews moved into the new hall. The hose wagon and the remainder of the men and equipment moved back in on the afternoon of January 29. On February 1, with much fanfare, the new, five-bay, three-story fire department headquarters building was officially opened. On February 13 the hall journal noted that, "Barker cut juice in on the new switchboard and repeater (*sic*)," in the alarm office.

A postcard of Headquarters, Firehall No. 2. *Photo: VFD Archives*

Almost one year later, on January 1, 1905, the new state-of-the-art fire alarm office opened on the third floor. Equipped with the latest Gamewell Company components it became a must-see for visiting firemen. Chief Electrician Chris Barker, who was promoted to that position on July 1, 1899, was now designated Superintendent of the Fire Alarm Office.

No. 5 Firehall

By February 1, 1905, No. 5 Firehall in the East End, at Vernon Drive and Keefer (now Frances) Street, opened with a two-horse hose/chemical combination wagon. A year to the day later, a new Waterous 2nd-size 800-gpm (3,600-liter) steam engine (Engine No. 5) arrived to complete the fire fighting complement in the district.

The department strength was forty-six men in mid 1905, consisting of the Fire Chief Carlisle and Assistant Chief Thompson, five captains, three steam engineers, two chemical engineers, the fire alarm superintendent, one electrician (alarm operator), eleven drivers, sixteen hose men and

Firehall No. 5 with its hose wagon, 1905.
Photo: VFD Archives

ladder men, one stoker, one chief's driver, two relief men, and a blacksmith.

No. 3 Hall received a new engine to replace their old one, the original 1886 Ronald, which was now nearly twenty years old. The new engine was a 1905 Waterous 3rd-size, 600-gpm (2,700-liter) engine (Engine No. 4), which was placed in service August 8, and because of its small size, only required a two-horse team.

Engine No. 4 at the Alberta Lumber Company's $250,000 mill fire on False Creek, April 17, 1914.
Photo: VFD Archives

No. 2 daily journal on August 29 notes that "hose horse Pat shot, having a broken leg and condemned by Dr. Hart," the department's veterinarian. On November 4, 1910 Assistant Chief Thompson's journal noted that Dr. J.B. Hart had died.

March 16, 1906 saw the opening of the new No. 1 Hall at Gore and Cordova Streets, which was identical in design to the new No. 2, except it had only two stories.

On the death of Alexander Cameron, at age 61, Robert George Forsythe became the department's chief engineer, on April 23, 1906. He joined the VFD July 1, 1903.

Firehall No. 1, shortly before its opening, 1906.
Photo: VFD Archives

Motorization Begins

The year 1907 was an exciting one for the VFD with the acquisition of more pieces of fire apparatus that included a Seagrave hook and ladder truck and a third size "Canadian" 600-gpm (2,700-liter) engine, built by the John D. Ronald Company. Chief Carlisle also replaced his buggy and horse, Billy, with a two-cylinder, 22-hp Model B Buick touring car, which marked the beginning of the end of the horse-drawn era. Billy was put out to pasture and was occasionally used to pull a hay wagon for his keep, but was never able to be ridden. Well-cared for, the old horse was put down at age twenty-nine after he fell and injured himself.

The "Canadian" engine, built by the John D. Ronald Company, 1907.
Photo: VFD Archives

Chief Carlisle and his horse, Billy.
Photo: VFD Archives

Much impressed by what the motorcar could do, the chief began investigating the availability of motor fire apparatus. In spite of much criticism, he purchased the first three motor apparatus built by the Seagrave Company of Columbus, Ohio.

The Seagrave Company began building ladders in Detroit, Michigan in 1881, then moved into hand-drawn equipment, then horse-drawn rigs, and finally they built motor apparatus of their own design. Chief Carlisle convinced the city fathers that ordering motor fire apparatus the costs would be cheaper than the care and feeding of horses. His first order consisted of two AC-53 (Type C, air-cooled, four-cylinder, 53-hp) hose wag-

The new Seagrave apparatus and Chief Carlisle's 1907 Buick at No. 2 Hall. *Photo: VFD Archives*

ons and an AC-53 chemical wagon, costing $4,950 and $6,970, respectively. Upon their arrival in Vancouver in December 1907, drivers were hired and the men began learning how to operate them.

The first Seagrave hose wagon (Registration No. 2382) went into service at No. 5 Firehall in early 1908, with Fireman Harry Stephens as the first driver. Stephens later went to Kamloops, BC to teach drivers when that city purchased two Seagrave rigs. He never returned to Vancouver and ended up staying there until retirement.

Fireman Frank Lucas standing on running board with pipe in hand has just delivered the new Seagrave hose wagon to No. 5 Firehall. *Photo: CVA FD P12*

No. 6 Firehall

When people in the then upper-class West End heard that a firehall was going to be built in their midst, many began to voice their objections to having such a noisy installation, requiring yet another stable, in their neighborhood. When they were told that there would be no horses, only the very latest in automotive equipment, they thought that was just fine.

On March 1, 1908, with much excitement, the hall opened at Nicola and Nelson Streets, with Captain Ed Mitchell in charge. Because it never had horses in it, No. 6 is said to be the first firehall in North America to be built specifically for motor apparatus.

Firehall No. 6, March, 1908. *Photo: VFD Archives*

No. 6 Hall's new Seagrave rigs. *Photo: VFD Archives*

Its first rigs were the second Seagrave hose wagon (Reg. No. 2383) and the Seagrave chemical engine (Reg. No. 2384) and the new drivers for them were William Tiffin and Gordon McGuire. The new hall soon became another must-see for visiting firemen, and later in the year when a fire chief's convention was being held in Victoria, many of the delegates and their wives made a special trip over to Vancouver via steamship to inspect this new installation.

No. 7 and No. 8 Firehalls

Earlier, on January 1, 1908, No. 7 and No. 8 Firehalls quietly opened; the former at West 5th Avenue at Yew Street in the Kitsilano district, on the city's west side and No. 8 at Victoria Drive and Pandora Street on the east side, near the city limits.

Firehall No. 7, c. 1910. *Photo: VFD Archives*

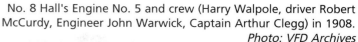
No. 8 Hall's Engine No. 5 and crew (Harry Walpole, driver Robert McCurdy, Engineer John Warwick, Captain Arthur Clegg) in 1908.
Photo: VFD Archives

Firehall No. 8, c. 1910. *Photo: VFD Archives*

The Amoskeag self-propelled steam pump on Cordova Street in front of Firehall No. 1.
Photo: VFD Archives

The Amoskeag Self-Propeller

Around this time, the chief had an opportunity to purchase an Amoskeag 1,200-gpm (5,400-liter) self-propelled steam engine. Built by the Amoskeag Manufacturing Company and Manchester Locomotive Works of Manchester, New Hampshire, it was the last of twenty-four built between 1867 and 1906. The only one of its kind in Canada (and west of Chicago), Carlisle bought it based upon the success of similar units used by New York City (five), Hartford, Connecticut (four), Boston and Chicago with three each. It went into service at No. 1 Hall on April 24, 1908.

On May 14 the chief reported to the Fire and Police Committee that the engine had exceeded its capacity of 1,200 gpm by delivering almost 1,450 gpm at tests. One small problem he reported was that the traction spikes on the wheels tended to pick up pieces of the wooden block roadways and fling them about, so he had the spikes removed. The chief continued, "but the principal difficulty, however, was that it produced a stream of flying sparks as it passed down the street, which could cause fires, but would be remedied by installing a spark arrester on the stack."

Engine No. 6 was the last steamer delivered. Here it is shown at an unknown fire in No. 3's district.
Photo: VFD Archives

Engineer John Warwick was promoted First Engineer of the Amoskeag, March 1, 1908. He joined the VFD May 10, 1906. Contrary to what some believed, this very reliable (not experimental) piece of equipment served the city well for many years. It was kept in reserve until the early 1940s, and then it was sold to the Crow's Nest Coal Company in Fernie, BC. Used for years in that town, this unique piece was scrapped in 1953 after an offer to return it to the City of Vancouver was turned down.

The last horse-drawn steam engine (Engine No. 6) was received and placed in service at No. 3 on November 30, 1908. It was a 2nd-size Waterous 800-gpm (3,600-liter) pump and the only one delivered with rubber tires. After the horses were retired in 1917, this engine was equipped for towing behind a motor apparatus, as required.

No. 9 and No. 10 Firehalls

On July 1, 1909 No. 10 Firehall opened at West 13 Avenue and Heather Street in central Fairview with the horse-drawn 1907 Seagrave Hook & Ladder, Model 9A, and the old (1892) No. 2 Hose Wagon. Then on October 11 the new No. 9 opened at Salsbury and Charles Street, in Grandview, with an old, used hose wagon and the 1892 Morrison chemical engine.

The 1909 annual report stated that the fire department had thirty-four horses and that motor apparatus, which the chief continued to purchase from the Seagrave Company, would replace them. Once again he was criticized, this time for favoring the Seagrave Company. He denied this, saying that his aim was to get the finest equipment available. During 1909, the VFD received two more Seagrave hose wagons, a chemical engine and the first motorized tractor aerial built by the Seagrave Company. It was a model AC-90 Type E (air-cooled four-cylinder, 90 hp), with a 75-foot (23-m) wooden aerial. Weighing 8.5 tons, it cost $14,000 and went into service at No. 1 Hall on Sept. 18, 1909.

Firehall No. 9 opened in Grandview with an old hose wagon and the 1899 chemical engine, shown here in 1911.
Photo: VFD Archives

Firehall No. 10 opened July 1, 1909 in the Fairview district near the Vancouver General Hospital. Hall is shown here in 1913.
Photo: VFD Archives

No. 1 Hall's 1909 Seagrave chemical.
Photo: CVA 354-17

A record number of alarms were recorded for the year at 359, and three major fires accounted for two-thirds of the total fire loss for the year. The multiple alarm fires at a furniture store, a grain elevator, and a lumber mill amounted to losses totaling more than $215,000.

The chief requested that council give him a fire warden instead of the police sergeant acting in that position because he was only able to spend a small part of his time on fire prevention. He also requested in the annual report a fireboat, another auto aerial, two more auto hose wagons and two auto chemical engines, two firehalls to replace the inadequate Nos. 3 and 4 halls, twenty more street boxes, a new car for himself and a car to replace Assistant Chief Thompson's horse and buggy. The report noted that the city currently had 126 fire alarm boxes and 826 fire hydrants. Lastly, the chief reported twenty-nine appointments to the department for 1909 with nineteen resignations.

The new Seagrave tractor aerial, the first one built, is shown at No. 2 Hall. *Photo: VFD Archives*

Carlisle ordered Seagraves, in spite of the criticism of the alderman who continued to claim that the apparatus were useless. Three chemical engines were ordered for Firehall Nos. 2, 3 and 5, and a hose wagon for No. 3. The sales success of the Seagrave Company necessitated the opening of a Canadian subsidiary plant, the W.E. Seagrave Company of Walkerville (Windsor), Ontario, from which these latest rigs were ordered.

The year 1910 was a bad one for the horses. In March the assistant chief reported that the team on No. 9 Chemical Wagon should be disposed of as unfit, with one of the others, Doc, on No. 9 Hose Wagon laid up with a bad leg. On June 8 one of No. 3 Engine's horses died of pneumonia, and the engine had to be pulled by the remaining two horses until August 5 when a new replacement team was purchased. At the end of June, with one of No. 5's horses sick, the engine was ordered to answer only any large or multiple alarm fires, with nearby halls covering the smaller fires. In August, several other horses were transferred around the department and two, Fritz from No. 3, and Blackie from No. 5, were retired to the city works department.

Occasionally, much to their new owner's dismay, these retired fire horses would sometimes respond to fire calls along with their old friends.

On April 25 the fire alarm system was placed under the control of the city electrical department, and at the end of June Fire Alarm Superintendent Chris Barker unexpectedly resigned and was replaced by electrician Lloyd Dunham, who became the chief electrician. Dunham had come on the VFD on April 1 as a lineman/electrician.

Chief Electrician Lloyd Dunham, No. 2 Hall. *Photo: VFD Archives*

The Cottrell Warehouse Fire

On October 30, 1910 the Cottrell Warehouse at 349 Railway Street, near

The Cottrell Warehouse fire attracted many spectators. *Photo: CVA 354-18*

The 1909 Seagrave aerial working at its first major fire. *Photo: CVA 354-20*

the waterfront, was destroyed in a three-alarm fire. The alarm came in from Box 37 at 11:45 a.m. and the fire had "a good hold when the department arrived with dence (*sic*) volumes of smoke and flame issueing (*sic*) from all the windows and outlets." Cause was reported as spontaneous combustion; the new five-story structure was gutted and the contents were destroyed. Damage was set at $130,000 in this largest fire of the year.

No. 11 Firehall

January 16, 1911 saw the opening of Firehall No. 11 on East 12th Avenue and St. Catherines Street with a two-horse hose wagon. Richard (Dick) Frost was the first captain.
Photo: VFD Archives

The new No. 11 Firehall opened at East 12th Avenue and St. Catherines Street and served the large commercial and residential area between No. 3 and No. 9 Firehalls.

The new No. 4 Firehall opened February 2, 1911, at 1475 West 10th Avenue.
Photo: VFD Archives

Another honor was bestowed upon the Vancouver Fire Department in 1911 when a "Commission of London men" surveyed various fire departments around the world and rated the best three brigades "as regards to equipment and efficiency" to be those of London, England; Leipzig, Germany; and Vancouver.

Personnel totaled 145, including the fire chief, the assistant chief, twelve captains, ten lieutenants, six steam engineers, seven chemical engineers, one fire warden, one machinist, an auto expert, a secretary, blacksmith, carpenter and veterinary surgeon; a superintendent of the fire alarm system who had two telephone operators and four linemen; 29 drivers, 65 hose men, ladder men and stokers, working out of eleven firehalls. The department once again had more than tripled in size over the previous five years and the fame of the VFD and its chief had spread to all parts of the world.

On April 12, 1912 Fireman John M. McKenzie was killed when he fell from a ladder at a drill at No. 1 Firehall. He had joined the department May 1, 1909.

Fireman John M. McKenzie.
Photo: VFD Archives

The New No. 3 Firehall Opens

The much-needed replacement for the old No. 3 Hall opened on East 12th Avenue and Quebec Street on June 24. Equipped with the 1905 600-gallon (2,700-liter) Waterous engine, a 1910 Seagrave AC-53 hose wagon (air-cooled, 53 hp), a 1910 Seagrave AC-53 chemical engine and the 60-foot (18-m) Hayes aerial (shortened from 75 feet [22 m] in July 1911), the new firehall had a crew of fifteen men.

The Vancouver Firemen's Football Club 1910-11.
Back row, left to right: L. Ledwell, C.W. Thompson, W. McKechnie, J. Fitzpatrick, H. Painter, T. Moffatt, G. Lefler, N. Fox, L.O. Hillier, S. J. Gothard.
Middle row: A. Stewart, A. McDiarmid, C. Durant, Mayor L.D. Taylor, J. Loftus, Chief J. H. Carlisle, C. Jewett, J. Grant, J. McLellan.
Front row: T. Hoggarth, F. King, P. May.
Photo: VFD Archives

V. F. D. No 3

Firehall No. 3 opened in Mt. Pleasant.
Photo: VFD Archives

In 1912 the VFD had more than thirty horses on its roster, most of which have passed into history with little known about them. There was Fritz, Pat, Dan, Brandy, Billy, Tanner, Sport, Tom and Jerry to name but a few. Three of the most interesting were at No. 3 Firehall where Spark, Flash and Star pulled the big 1,000-gallon Waterous Engine No. 3. When the bell sounded, the horses responded from their stalls to the front of the rig, positioning themselves under the harness, ready to be hitched up. With the kindling lit under the boiler they left the hall in a cloud of smoke and dust. This team had a reputation of being able to out run the motor rigs and get about four or five blocks from the hall before they could be overtaken.

Other fire companies could undoubtedly make a similar claim, except that one of No. 3's horses, Flash, the middle one, was blind! When the bell hit, Spark went up the right side of the engine, Star went up the left side and stood off to the left, leaving room for Flash to find his way, then Star would give him a gentle push into position. Once hitched up, they took off with Flash being steered along by his teammates.

No. 3's horses on the street in front of the hall. The three center horses are Spark, Flash and Star.
Photo: VFD Archives

The Webb Couple Gear Electric eighty-five-foot aerial was placed in service at No. 2 on July 6 and remained there until placed in reserve in 1932. This unusual apparatus had an electric motor in each wheel that was powered by a generator using a six-cylinder 90-hp gasoline engine. The aerial it replaced was the Hayes horse-drawn ladder, which went to No. 3.

The 1912 Webb Aerial at Firehall No. 2.
Photos: VFD Archives

Another view of the 1912 Webb Aerial at Firehall No. 2.
Photos: VFD Archives

Following the opening of the bridge, the Duke of Connaught, a son of Queen Victoria, inspected members of the Vancouver Fire Department. *Photo: VFD Archives*

On September 20 the governor-general of Canada, the Duke of Connaught (1911–1916), along with the Duchess and their daughter, Princess Patricia, officially opened and named the Connaught Bridge, more commonly known as the Cambie Street Bridge.

The American-LaFrance Apparatus Arrive

December 20, 1912 saw the arrival in Vancouver of Canada's first order of American-LaFrance fire apparatus, built in Elmira, New York. Chief Carlisle was notified that his equipment had arrived and he went down to the freight yards to watch the unloading. He discovered four fire rigs on the train instead of the three that he had ordered. He thought the factory had made a mistake then he learned that the fourth unit belonged to the neighboring municipality of Point Grey. The hose wagon/chemical combination, Registration No. 206, was that department's first piece of custom fire-fighting equipment.

The *Vancouver Daily Province* described the new American-LaFrance fire trucks:

> Auto Equipment Tested Yesterday Made Sixty Miles an Hour
> With A.W. Plimpton, of Elmira, New York, at the wheel, the hose wagon did sixty miles an hour without particular difficulty on a level street, and mounted a grade of 20 per cent at a speed of fifteen miles an hour. The test was satisfactory to the firefighters who participated. The new hose carts are equipped with a new style tire, the "airless," composed of an ordinary casing with sections of soft India rubber in place of an air filled tube to give resiliency. By a peculiar goose-neck construction of the front axle, the machine is able to turn in a space smaller than that required by an ordinary touring car. Two of the new hose wagons and a chemical will be placed in service in Vancouver immediately and another combination chemical will be installed in Point Grey.

Although incorporated as a municipality in 1908, Point Grey had relied on a volunteer brigade, with aid from the city if required, but, effective October 1, 1912, hired its first paid fire chief, R.C. Tait, and three paid men. When the new hose wagon was accepted, three more paid men were hired at $85 per month and construction of a new firehall was underway. By the end of the year, the new chief was suspended (cause unknown) and Arthur Turner, from Calgary, was appointed and remained the fire chief for many years.

The VFD had 583 alarms in 1912 and two multiple-alarm fires accounted for more than half of the year's total loss. The first on April 3 in the 300-block of West Hastings involved several stores with a loss of more than $154,000 and the other, at Commercial Motors and the Champion and White building products site in the 900-block Main Street, suffered $200,000 in losses including several horses.

Canada's first delivery of American-LaFrance fire apparatuses is previewed at Firehall No. 2. Point Grey's hose wagon/chemical combination is shown far left. *Photo: CVA 99-5*

For one year only, in 1913, white cap covers were issued for summer use. The crew at No. 10 is:
Rear, left to right: Charles McLennan, William Plumsteel, James Johnston, Ed Erratt.
Front row: John (Baldy) Balderston, Capt. Tom Botterell, Harry Moth, Harry Painter.
Photo: VFD Archives

In 1913 members of the department were issued with white cap covers for summer use. Pictured is the crew at No. 10 Hall showing off their new look.

No. 12, No. 13 and No. 14 Firehalls

The city was expanding out into the suburbs and more new firehalls were built to keep up with this growth. During 1913 three halls were opened. In July No. 14 Firehall was opened at Slocan and Cambridge Streets in the northeast corner of the city, which had grown considerably since the completion of the nearby exhibition grounds. In 1910, the Vancouver Exhibition Association opened and at the same time the BC Electric Railway Company (BCER) extended its streetcar lines to serve the exhibition, which also encouraged people to build and settle there. Known as Hastings Townsite, the area became part of the city in 1911. When No. 14 opened, it had the 1905 combination hose wagon and the 1892 Morrison chemical engine in service.

Firehall No. 12, 1985. *Photo: VFD Archives*

On October 13, 1913 two more firehalls opened. The new No. 12 at 8th Avenue and Balaclava, near the then city limits at Alma Road, on the West Side, opened with a hose wagon, the 1893 St. Charles and the 1899 Champion chemical engine. Then on the same day, maybe to prove that firemen weren't superstitious, the new No. 13 opened on the city's southern boundary, on East 24th Avenue and Burns Street (later changed to Prince Albert Street). It was equipped with the 1904 combination hose/chemical wagon that originally served at No. 4 Firehall.

Over the years people wondered why No. 13 Hall faced 24th Avenue and not the much busier 25th Avenue (King Edward Avenue) directly behind it. The decision makers at the time felt it was more appropriate to have the hall facing the city instead of facing the municipality of South Vancouver across the street. Little did they know at the time that in less than twenty years the cities would amalgamate.

Architecturally, Firehalls No. 11 through 15 were built on the same design with variations in rooflines, balconies, and windows but all had basically the same floor plan.

Firehall No. 13, 1913. *Photo: CVA SGN 992*   Firehall No. 14, 1974. *Photo: VFD Archives*

In January 1914 two more American-LaFrance rigs were added to the VFD roster. The first of these was a 1913 Type 12 hose wagon, which went into service at No. 5 Firehall, replacing the first 1907 Seagrave hose wagon, one of the originals, which now went to No. 8. The second rig was the 1913 American-LaFrance Type 14 city service ladder truck that went to No. 9 Hall where it remained until 1955. Then in April, No. 4 Hall received a new Seagrave Type WC-80 (water-cooled, 80 hp) city service ladder truck.

No. 5 Hall's new American-LaFrance hose wagon. The crew members are, left to right: Archie McDiarmid, Oswald Mottishaw, Pete Milne, Hector McKenzie, Capt. Jack DeGraves. This hose wagon would be lost in the Pier D fire, July 1938. *Photo: VFD Archives*

Canada Went to War

On August 4 Britain declared war on Germany and as part of the British Empire, Canada was also at war. Over the next several months, sixty members of the department answered the call, twenty-two of whom would be wounded and eleven would be killed. The Honor Roll lists the following members who died serving their country:

Alexander M. Alexander, Hoseman, No. 1

Roy H. Broderick, Truckman, No. 9

Frank Hooper, Hoseman, No. 1

John Horn, Hoseman, No. 1

Percy B. May, Auto Driver, No. 5

Alexander McDonald, Truckman, No. 1

William D. Reid, Hoseman, No. 6

John Robb, Truckman, No. 2

Thomas H. Snowden, Hoseman, No. 11

John Wedderburn, Chemical Engineer, No. 2

Alexander F. Yule, Truckman, No. 9

Fireman Alex McDonald died in Montreal on November 11, 1918, the day the war ended, from the effects of mustard gas poisoning received at the front.

Once again the department was shocked by the on-duty death of a member, this time with the death of Fireman Albert D. Stewart, age twenty-four, who accidentally fell down one of the pole-holes at No. 3 Hall, on September 11, 1914. He had joined the VFD one month earlier, on August 11. Following a departmental funeral he was buried in the Firemen's Plot at Mountain View Cemetery.

There were several multiple-alarm fires during the year. One notable fire was in the Denman Arena built by Lester Patrick (of NHL fame), downtown at Denman and Georgia Streets, with damage exceeding $50,000. When the rink was built in 1911, it was the largest indoor ice rink in the world.

The Mainland Building Fire

Another fire of note was in the Mainland Building at 852 Cambie Street, March 20, 1914, in the warehouse district, which has seen many large fires over the years. The alarm came in just after 6:00 p.m. and within minutes the six-story building was heavily involved. The loss was estimated at more than $300,000 to the tenants, which included wholesale grocers, storage and transfer companies and manufacturer's agents, such as Wrigley's of Canada and Lowney's Chocolates Company. In total 125 men fought the fire with eighteen pieces of equipment from ten firehalls. (At that time the department had 201 men and fifteen firehalls).

The World (a Vancouver paper of the era, no longer in existence) reported the next day that, "Had there been a fire call sent in last evening from another part of the city it would have been impossible for the firemen to have controlled either, and for a city the size of Vancouver to have no fireboat is a farce in the full sense of the word." Chief Carlisle had been requesting a fireboat annually for the previous five years.

The miracle of the incident was the fall of Fireman Neil MacDonald, of No. 1 Company. He was working on an aerial at the front of the building when he somehow became tangled up in the hose line he was directing down onto the fire, slipped and fell about forty feet (13 m). On the way down, he struck the ladder, then fell onto the hood of the truck and bounced to the ground, landing on numerous hose lines. Bleeding from his injuries, he was treated and loaded into the chief's car and taken to Vancouver General Hospital (VGH). The news of the day reported, with less sensitivity than would be seen today, that, "his face was badly battered, the nose entirely smashed, his front teeth broken, and a deep gash at the back of his head." He was listed in good condition by VGH, and after four months recuperating he returned to duty on July 13.

On August 18 Chief Engineer Robert Forsythe was thrown from No. 2's big Waterous engine while responding to an alarm. As the engine was turning at Pender and Beatty Streets, it slid on the pavement and toppled over. He suffered painful cuts and bruises but was able to return to duty after two weeks. The team and driver escaped injury.

Fireman Neil MacDonald, No. 1 Firehall.
Photo: VFD Archives

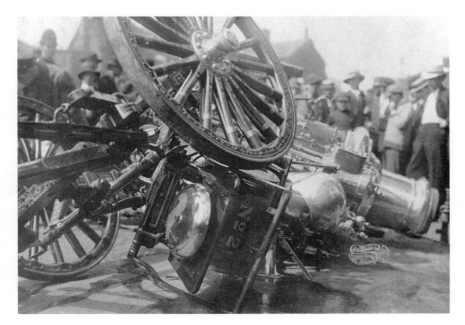

No. 2 Engine overturns en route to a fire.
Photo: VFD Archives

No. 15 Firehall

On September 15, No. 15 Firehall opened on East 22nd Avenue and Nootka Street in the south Renfrew area, with one of the early Seagrave hose wagons. Very soon afterward a horse-drawn hose wagon replaced the motor rig because the poor road conditions in the district often found the Seagrave firmly stuck in the mud. (In 2006 the city agreed to keep and restore this historic firehall, the last remaining to have used horses.)

No. 15 Firehall as shown in 1950.
Photo: VFD Archives

In October 1914 the city reduced fire department wages; 15 percent reduction for wages under $100 and 20 percent for wages over $100.

The First Motorized Engine Arrives

On January 20, 1915 an American-LaFrance 1,250-gpm (5,600-liter) motorized pump, the first on the VFD, was put in service at No. 2 Firehall. With a six-cylinder 225-hp engine, this huge machine, weighing five tons (22,000 kg) had cost $12,950. It would be many years before pumps with this capacity would be common on many fire departments. Four days earlier the Point Grey Fire Department put its first motorized pump in service. Also an American-LaFrance, it had a capacity of 625 gpm (2,800 liter), was powered by a six-cylinder, 105-hp engine and cost $7,500.

The VFD's first motorized engine was this 1914 American LaFrance, which saw service at No. 2 Hall until 1951.
Photo: VFD Archives

Two Bridge Fires

Two mysterious and suspicious fires occurred during the night of April 26, 1915, on the Connaught (Cambie Street) Bridge and the Granville Street Bridge. At 4:30 a.m. a police constable turned in the alarm for the Connaught Bridge where a fire was burning about 100 feet (30 m) north of the swing span, under the tar-coated decking planks. By the time the

Fire damage to the Connaught (Cambie Street) Bridge put it out of service for almost a year. *Photo: CVA BR P16.2*

The bridge collapse injured Capt. Plumsteel, No. 3 Hall and a BC Electric lineman. *Photo: Courtesy Ray Acheson*

first companies arrived, the bridge was burning furiously with the fire spreading northward. Because of the length of the 4,000-foot (1,200 m) bridge, long hose lays were necessary to reach the fire, and the first-in crews had to remove the heavy planking and cut holes in the decking to get at the fire. A second-in engine company responded under the bridge and began drafting water from False Creek and attacked the fire from underneath.

Little headway was made because of high winds, and after a two-hour battle a ninety-foot section of the bridge fell into the creek taking two men with it. One, a BC Electric Railway lineman who was disconnecting streetcar trolley lines, suffered painful cuts and bruises in the fall. The other, a fireman, Captain William Plumsteel of No. 3 Company, had only minor injuries and remained on duty. Both were plucked from the water by a launch standing by in False Creek. The damage to the bridge was about $80,000 and the bridge was closed for almost a year, reopening March 15, 1916.

The second fire, on the Granville Street Bridge, came in at 6:35 a.m. The fire was found burning under the decking and sidewalk also near the swing span. This much smaller fire was quickly extinguished.

There was some conjecture that "alien enemies" had attempted to burn both structures and on May 15 the acting mayor received an anonymous note that said, "We have just started, we and our allies will burn things to der ground. Deuschland auber alles. For der Fatherland." The police were given the letter and began an investigation but did note that a German who could write such perfect English would at least be able to spell "Deutschland uber alles" correctly.

Talk following the fire, among other things, included the necessity of a fireboat. Louis D. Taylor, the flamboyant mayor and good friend of the fire department, (he served five terms between 1910 and 1934, liked good cigars, and red bow ties), took an "I-told-you-so" attitude, recalling the discussion of the fire department estimates of the previous day where the fire chief once again requested a fireboat. Some aldermen were of the opinion that a fireboat would have made no difference and one suggested that, "a foot-walk built under the bridge floor with a line of hydrants could be used for fire-fighting purposes."

No. 6 Rigs Collide, Fireman Milne Killed

At 11:40 p.m. June 1 the downtown halls were tapped out (bells sounded for an alarm) to respond to Box 56, at Mainland and Helmcken Streets, in the warehouse district. Somehow, at No. 6 Firehall the box number was misread as Box 66, at the CPR Station, Train Platform. Within minutes the chemical engine (the original 1907 Seagrave) left the hall and headed up Nelson Street to Burrard where it turned left and headed down towards the CPR Station. Barely seconds later the driver of the 1912 LaFrance hose wagon headed north on Nicola Street to Pender, then right towards Burrard Street, the short delay caused by the need to prime the engine from the main gas tank. The hose wagon's route was the preferred way to go, as it was a flatter route than the up and down route taken by the chemical.

At the intersection of Pender and Burrard Streets the two vehicles collided, each crew stating later that they were unable to hear the siren of the other rig. The heavier chemical engine struck the hose wagon on the right rear corner sending it crashing into a tree on the boulevard. On

No. 6 Hose Wagon struck a tree after colliding with the chemical engine.
Photo: VFD Archives

impact, the chemical engine spun around on the streetcar tracks in the intersection and flipped over, pinning Fireman Charlie Milne, twenty-five, under equipment and wreckage. The uninjured firemen treated the injured and called for a police ambulance. Milne was taken to nearby St. Paul's Hospital where he died of massive head injuries. The others suffered fractured ribs and kneecaps, cuts and bruises, and one man was knocked from the tailboard of the chemical engine into the path of the hose wagon, which ran over him, crushing his thigh. Fireman Milne joined the VFD in October 1913, and was the uncle of the department's carpenter, Captain Bruce Milne, who retired in 1985.

The two rigs are shown in this inside picture at Firehall No. 6.
Photo: VFD Archives

The fire call that No. 6 should have responded to went to a three-alarm warehouse fire that caused more than $150,000 damage. When No. 2 Company arrived, no fire was showing and the person who had pulled the box had left the scene. Within a short time the crew was alerted by the sound of breaking glass and found a fire burning on the fourth floor of the six-story Percival Building. The fire very quickly extended through the roof, but two hours later it was brought under control. The occupants of the building, at 1150 Hamilton Street, were various manufacturers' agents and import companies, including the Canadian General Fire Extinguisher Company agency. There had been a major fire in the building in March 1914, with over $50,000 in damage and losses.

The local newspapers reported "Resignations Hurt Fire Department" on August 7, 1915. During the year, thirty-six men enlisted to go to war; thirty-three new men were hired, a dozen had resigned and manning was down thirty men from the full complement in 1914.

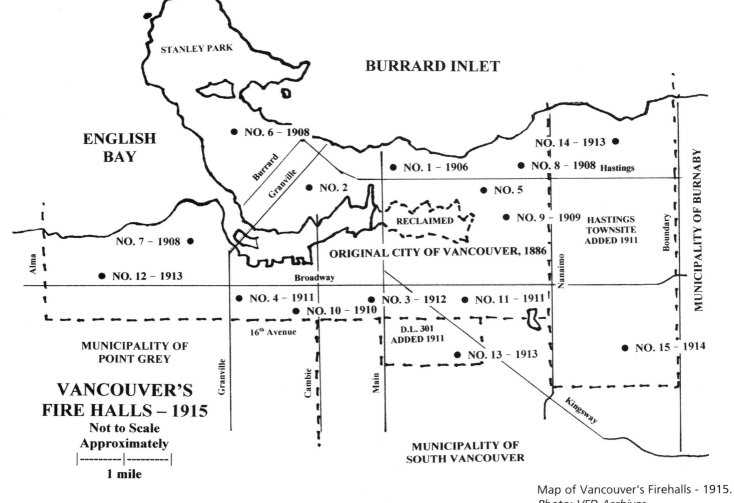

STANLEY PARK

BURRARD INLET

ENGLISH
BAY

● NO. 6 – 1908

● NO. 14 – 1913 ●

● NO. 1 – 1906

● NO. 8 – 1908 Hastings

Burrard

Granville

● NO. 2

● NO. 5

RECLAIMED

● NO. 9 – 1909 HASTINGS
TOWNSITE
ADDED 1911

MUNICIPALITY OF BURNABY

NO. 7 – 1908 ●

ORIGINAL CITY OF VANCOUVER, 1886

Alma

Boundary

Nanaimo

● NO. 12 – 1913

Broadway

● NO. 4 – 1911

● NO. 3 – 1912

● NO. 11 – 1911

● NO. 10 – 1910

16th Avenue

D.L. 301
ADDED 1911

MUNICIPALITY OF
POINT GREY

● NO. 15 – 1914

VANCOUVER'S
FIRE HALLS – 1915

Granville

Cambie

Main

● NO. 13 – 1913

Kingsway

Not to Scale
Approximately
|----------|----------|
1 mile

MUNICIPALITY OF
SOUTH VANCOUVER

Map of Vancouver's Firehalls - 1915.
Photo: VFD Archives

The Firemen Join a Union

The firemen's union, originally formed in 1911, was short-lived when the city learned of its existence and offered the men a $10 increase to disband it, which they did. In 1914 wages were reduced by 15 to 20 percent. By early 1916 the union was reorganized (with part of the 1914 wage cuts restored, a first-class fireman made $80 per month), the firemen affiliated with the AF of L, received Charter No. 15363, and joined the local trades and labor council. Once again the city tried to encourage the firemen to abandon the union idea; the city said there was no reason to have such an organization, as any problems between the two parties could be handled without one.

The union's request for one day off every four was rejected and in June 1916, with a wage increase pending, the city "discovered" that to rank a fire captain with a police inspector would cost "a big sum" as the city put it. At the same time the firemen requested a semimonthly pay system, which was granted, and that the city supply them with canvas night trousers (protective gear). When told that supplying the trousers would reduce their wage increase, the firemen withdrew their request. The heavy-duty "bunker pants" were issued for the first time by the city in the late 1970s.

On July 14, the firemen requested the city to look at salaries and suggested parity with the police wages, which were $1,140 annually to the firemen's $960. To further their argument, they pointed out that policemen work a nine-hour day, have one day off in fourteen, fourteen days annual vacation (twenty-one days for officers, in summer months) and enjoyed "advancement for heroic conduct."

Firemen, on the other hand, worked twenty-four hours a day with three hours off per day for meals and had one day off in seven; had fourteen days annual vacation for all ranks, with their vacations scheduled January through December, and they had no similar advancement system.

The union then petitioned the city in December to submit a plebiscite to the voters, asking for a two-platoon system. Their request was refused by the city. The following year a plebiscite presented by the new city council was defeated and it would be two more years before the two-platoon system began.

In the fall of 1916 a used, four-cylinder, 80-hp, chain-drive Stearns auto was purchased for $300 from a local automotive shop and converted into a squad wagon. Using an old Seagrave hose wagon body, this rig was used to transport up to twelve men to fires requiring more manpower, anywhere in the city, particularly during meal hours. (Meal hours were 6:00–9:00 a.m., 11:00–2:00 p.m. and 5:00–8:00 p.m.) The squad was used until mid 1917, then used temporarily at No. 12 Hall as a hose wagon when the last team of horses was retired from that hall. Later it was relegated to service vehicle status into the early 1920s and then disposed of.

The shortest recorded career on the VFD was one by H.E. Horne, who started work March 16, 1917 at 8:30 a.m. (late, as the shift started at 8:00) and ended when he left the firehall at 10:00 a.m. and never came back!

No. 13 and No. 15 Firehalls Close

On June 8, 1917, Firehalls No. 13 and 15 were closed because of restraint, reorganization and motorization. They remained closed until June 1, 1927. During these closures, the hall's quarters were rented out to civilian families, who acted as caretakers, and any extra space was used for fire department storage.

Firemen Go On Strike

After many months of trying to negotiate the one-day-off-in-four, the two-platoon system and pay parity with the police (even though one-half of the 1914 wage cut had been restored), the men met at No. 2 Firehall at 10:00 a.m. Saturday, July 7. After much deliberation, they decided that at 2:00 p.m. they would strike, unless their request for the one-day-off-in-four was granted in the meantime. Many were prepared to get their demands or look for other work.

At 11:30 a.m. they informed Mayor McBeath of the decision of the Firemen's Union and began removing their belongings from the firehalls. About twenty-five officers remained on duty to deal with any outbreak of fire.

Mayor McBeath asked the men to put the strike on hold for forty-eight hours, as most of the aldermen were out of town and a meeting could not be convened. The mayor could not agree to the proposal of the one-day-off-in-four but stated that he would recommend that the council should immediately enter into negotiations "with a view to an early and amicable settlement." The firemen could not agree, in spite of the mayor's insistence that they were prejudicing their case in the eyes of the citizens.

The mayor approached the officer commanding the local military detachment regarding organizing a temporary fire brigade; he told the mayor he doubted that he could be of any assistance and suggested that the attorney-general in Victoria be contacted. Once again, the mayor was unable to get help and was further advised by the manager of the Returned Soldier's Club that "returned soldiers would not, on any account, act as strike breakers."

Believing that the strike could last the weekend, fire insurance underwriters put the following in the afternoon edition of *The World,* July 7:

> Underwriters Issue Warning to Public—
> The Underwriters, recognizing the big conflagration hazard existing in the city, due to the dry and hot weather conditions, wish to warn the public to use great care during the time there is no fire protection in the city, as a very small fire might easily lead to a general conflagration which would wipe out the largest part of the city.

After a strike of three hours, the mayor agreed to the terms of the firemen, "to recommend to council that the one-day-off-in-four be granted; should a tie vote exist, be the tie-breaker for the firemen, and promise no discrimination against the men when a settlement is reached." To say the mayor was reluctant to these terms would be an understatement; he told *The World,* "they had me by the throat, with my back against the wall." By 5:00 p.m. all the arrangements had been made and the men went back to work in the firehalls until 6:00 p.m. Monday, by which time the council would have made a decision.

The vote by council on Monday, July 9 that would have ended any further strike came after nearly three hours of weary discussions. After a motion, an amendment and an amendment to an amendment, each point was voted upon and defeated by a vote of five to four.

Then a strange thing happened: one of the aldermen left the chambers because he didn't want to be a party to an "illegal procedure," namely, retaking a vote count on the firemen's issues after other business had been transacted, and other votes taken. The votes were recounted at four to four and Mayor McBeath cast his vote for the firemen, as promised, settling the issue.

Council begrudgingly ratified the firemen's demands. The one-day-off-in-four, by Chief Carlisle's estimate, would have an annual cost to the city of $13,000. In spite of the action being described as "unholy," "unethical" and "unlawful," the city council also restored the remainder of the 1914 wage cuts. Some said that failure to do so would have resulted in the continuance of the strike.

Many people continued discussions on the issue, with the firemen being censured for what they had done; some still believed that they should have been dismissed or imprisoned for civil disobedience and endangering life and property.

The Board of Trade said (after the fact) that they would have offered to mediate the dispute and one of the aldermen said that, "They've put up with these conditions for the past 26 years, surely they could put up with them for another six months, anyway."

During the three-hour strike, despite the very hot weather, not one fire alarm came in. The headline of the BC Federation of Labour paper during the week of July 9 read: "Vancouver Firemen Get Tired of Piffling Conduct of City Council and Resort to Direct Action to Enforce Their Demands — A Three-Hour Strike Brings Council to Terms and Raises the Most Pitiful Squawk Ever Squawked — Firemen Get One Day Off Duty in Every Four."

On July 12 Chief Carlisle was at the meeting of the Fire and Police Committee and presented a number of recommendations for the reorganization of the fire department. These, in part, would permit the firemen's request for the one-day-off-in-four to be adopted without increasing the expenses of the department. It recommended that all horse-drawn apparatus be placed in reserve, that the horses be disposed of and a new motorized engine be purchased. After some discussion, with comparatively few alterations the report was adopted.

During the question period one of the aldermen asked why this plan hadn't been put forward earlier, and another said that the fireman's strike likely wouldn't have happened had this been done. The first alderman was told that a previous suggestion had been "side-tracked" two years earlier by council to make way for a plan that was to be developed by some of the aldermen and was never completed. The proceedings ran smoothly to this point, then Assistant Chief Thompson, evidently smarting from some of the criticisms recently directed at his department, spoke with some warmth.

"The trouble in the past," he began, "has been caused by some of the aldermen interfering with recommendations contained in the reports from the head of the department. There has been altogether too much interference on the part of aldermen with the administration of Chief Carlisle, and some of them are trying to do it even now. The council has dallied with this question and that was the cause of the trouble. A fireman walking down the street today is looked upon as a snake in the grass by the citizens, as a result of the impression given by the action of the council and some of the newspapers. An impartial board of investigation," the deputy went on, "should be appointed to find out whether it is the firemen or the city council that is to blame for the trouble which led to the strike."

The deputy sat down to the sound of a pin dropping then a crescendo of babble began building. Quiet was once again restored (is that a faint smile on the weathered face of the old chief?) with the chairman reminding everyone that there was no motion before the committee, "So all this conversation is out of order," he said.

Chief Carlisle proposed using motorized apparatus to tow steamers. This is the only known photo of this arrangement. There are no records to indicate if it was successful.
Photo: VFD Archives

The chief's report, besides advising of the purchase of a new pump, recommended placing all of the steam engines in reserve, which would release fifteen men for other duties, and by doing away with the need of sixteen horses, which could be sold for about $2,600, there would be additional savings of about $1,400 per month for feed. The cost of running motorized apparatus compared to horse-drawn was about one-tenth. As well, motor apparatus could cover a much larger area in less time, which would enable the fire department to close some firehalls, with

No. 2 Engine racing along Georgia Street symbolically illustrates the end of the horse-drawn era of the VFD.
Photo: VFD Archives

more savings. The new pump would respond to all areas and the reserve steamers would be rigged for towing by motorized hose wagons, as needed. Also, some of the motor rigs would be turned into combination hose/chemical wagons, and he proposed closing the machine shop at No. 2 Firehall in favor of the city shops.

And finally, he suggested that the Workmen's Compensation Board (WCB) pay sick firemen's wages rather than using the fire department's sick fund.

The start of the one-day-off-in-four system and full motorization began July 19, 1917.

Full Motorization: The End of an Era

Now that motorization had been completed, Chief Carlisle felt it was appropriate to record this history-making achievement, so on December 19, 1917 he had every in-service piece of fire apparatus assemble beside the CPR freight shed on Pender Street for a photograph. All the crews and their rigs, twenty-three in all including the two chief's automobiles, posed for the "panoramic" camera. Well, almost every piece of in-service apparatus. In a couple of photos taken before the panoramic, the Amoskeag self-propelled steamer can be seen among the other rigs. It's not known whether the old steamer took a run or was just excluded from the chief's milestone event.

All the department apparatus were assembled for a photograph to commemorate full motorization.
Photo: VFD Archives

Fireman George J. Richardson.
Photo: VFD Archives

The first IAFF logo. *Photo: VFD Archives*

International Association Is Formed

On February 25, 1918 meetings began in the American Federation of Labor (AF of L) building in Washington, DC, for the purpose of forming an international union of firefighters. Thirty-six men attended these meetings from sixteen states, the District of Columbia, and the province of British Columbia. Fireman George J. Richardson, at age twenty-four the youngest of the delegates, attended as one of the table officers of the City Firemen's Union, Local No. 15363, AF of L. The union had elected him as their delegate to attend the Washington meetings, and gave him the mandate to vote for the new union as the International Fire Fighter's Union.

At the conclusion of these meetings on February 28, the International Association of Fire Fighters (IAFF) was formed, that name being favored by the majority, with headquarters to be located in Washington, DC. George Richardson was elected the International's Vice-President for Canada and the Vancouver Fire Department became Canada's first and only member of the IAFF and received Charter No. 18.

Richardson got permission from Chief Carlisle to take his two-week annual vacation to attend the meetings, and the City Firemen's Union paid his wages for the extra ten days that he required.

On April 2, with President Charles Watson in the chair, the members voted to change the name of their union to the International Association of Fire Fighters, Vancouver Local #18 and all ranks up to captain were invited to join, with some abstaining.

No. 10 Firehall Closes

April 4, 1918 saw the closing of No. 10 Hall because of motorization and it was felt that, as it was only about ten blocks away from both No. 3 and No. 4, the covering of its district would not be compromised. There was, however, some concern for the nearby Vancouver General Hospital, one block away.

The new city council, under recently elected Mayor R.H. (Harry) Gale — whose platform was "cooperation, get-together and community spirit as a means of advancing the city to one of the greatest on the coast" — granted a $10 across-the-board increase to the members of the fire department. But between March and July, 140 men resigned to take jobs with shorter hours of work and better conditions, leaving the department dangerously under strength.

Ongoing special meetings of the city council and the union executive were held to find a solution to the problem of the under-manning. The union members recommended the adoption of the two-platoon system (day shifts/night shifts) but the city could not agree. Finally, after much deliberation, with the union showing the advantages and efficiency of the system, it was accepted with plans to start on October 1, 1918, making the VFD the first in Canada to adopt it. This change in the hours of work resulted in the hiring of thirty-two more men, increasing the strength of the department from 155 to 187 men.

The two platoons, designated A and B, worked ten-hour day shifts, fourteen-hour night shifts, and twenty-four-hour Saturday shifts, then one day off, a system which, spread out over several weeks, averaged eighty-four hours per week. On October 2, the union began holding A and B shift meetings so all members were able to attend all monthly meetings.

The Steveston Fire

Steveston, BC, the small, fishing community on Lulu Island at the mouth of the Fraser River, had a major conflagration on May 4, 1918 requiring outside help because it had no fire brigade. Help came from Vancouver and the Point Grey Fire Department, but not much could be saved. The fire, which started in a small shack behind one of the many salmon canneries in the area, destroyed three canneries, a bakery, the drug store and the post office. The homes of over 500 people, mostly Chinese cannery workers, were also lost and the total fire damage amounted to almost $500,000.

Point Grey's pump at the Steveston fire. The empty hose-bed shows they have laid all their hose.
Photo: VFD Archives

The Crash at "Death Corner"

The worst accident in the history of the VFD occurred when No. 11 Hose Wagon struck a streetcar at the corner of East 12th Avenue and Commercial Drive, killing four of its five-man crew.

The intersection was called "Death Corner" following an accident the previous January 20 when a car was struck by a BC Electric Railway interurban tram, killing three people. Following this accident, Police Chief William McRae urged city council to install a signal device at the intersection, but council shelved the proposal.

The crew of the ill-fated No. 11 Hose Wagon. **Rear, left to right:** Torquil Campbell, Otis Fulton, Captain Frost, Lieutenant McKenzie, Donald Morrison. **Front left:** Leonard Farr.
Photo: VFD Archives

At 4:00 p.m. May 10, 1918 a telephone alarm was received at No. 11 Hall for a bush fire at Lakewood Drive and Broadway. The crew of Captain Richard Frost, Driver Otis Fulton, Lieutenant Colin McKenzie, Firemen Donald Morrison and Torquil Campbell, headed east along 12th Avenue from the firehall. As the rig neared the intersection, streetcar No. 173, traveling down the grade from the right, was also nearing the intersection. (As the rule of the road hadn't yet changed in Vancouver, both vehicles were traveling along the left side of the roadway.) Witnesses reported that both vehicles were traveling between thirty and forty miles per hour (50–65 km/h) and it was felt that the streetcar motorman either didn't see or hear the fire rig, or he was racing to beat it through the intersection.

The accident scene on Commercial Drive.
Photo: VFD Archives

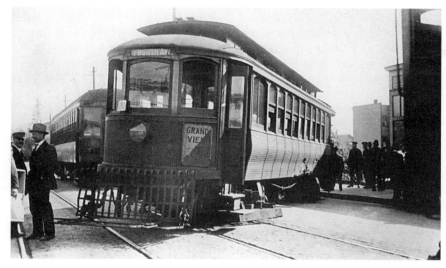

The streetcar No. 173 was derailed on impact.
Photo: VFD Archives

Fireman Campbell, feeling that a collision was imminent, yelled at his brother firemen to jump, which he then did. The others rode the hose wagon into the intersection where it struck the streetcar at the left rear wheels, throwing men and equipment in all directions.

Captain Frost and Otis Fulton flew from their seats, struck the side of the streetcar and were instantly killed. Lieutenant McKenzie and Don Morrison were thrown into the air on impact from their tailboard positions. Both men hit the trolley wire and fell to the tracks, after the car had gone by, with such force that both succumbed to their injuries. All died of massive head injuries.

The impact of the collision made the streetcar careen down the street causing it to derail and with its momentum it traveled more than 250 feet (75 m) before it stopped. None of the passengers on the streetcar was injured.

Within a short time of the accident, the police patrol arrived on the scene to investigate and the motorman was taken to the police station to give a statement.

The remains of the hose wagon, a shop-built 1912 Oldsmobile, were removed to the firehall. The only member of the crew not on duty that day was Fireman Leonard A. Farr, who stayed at the hall all evening as other firemen, neighbors and local kids came by to offer condolences. Sadly, on December 26, Farr was a victim of the 1918 influenza epidemic.

A poignant aside to the accident was that Captain Frost, a forty-four-year-old bachelor, had decided to resign from the department after almost twenty years to become a rancher and farmer. Due to leave in about two weeks, earlier in the day he had made some final arrangements for the delivery of some stock to his property. He was promoted to Captain on January 1, 1911, and was the hall's first captain when it opened January 16, 1911.

Driver Otis Fulton, thirty-five, was a widower with two children at home; Donald Morrison was thirty-three and married; Colin McKenzie was thirty-four and married; and Torquil Campbell, age thirty, was married and had a young family. Torquil Campbell lived to enjoy retirement and died in 1976 at age eighty-eight.

The remains of the hose wagon.
Photo: VFD Archives

All firehall flags were ordered flown at half-mast until following the funeral, which was held Monday, May 13 at No. 2 Firehall on Seymour Street. Nearly 7,000 people filed by the four caskets lying in state with the solitary figure of Chief Carlisle standing beside them. The caskets were topped with each man's helmet and seated behind the single row were

The funeral at No. 2 Firehall.
Photo: VFD Archives

platoons of firemen, relatives, friends and civic officials.

At the completion of the service by three ministers, the coffins were placed upon the 1913 American-LaFrance city service truck, two on each side, for the procession to the Mountain View Cemetery Firemen's Plot. The funeral cortege with the police pipe band and the Salvation Army brass band wound its way up Seymour Street to Georgia, down Granville, along Hastings Street to Main Street, then south to Mountain View Cemetery.

All along the way people on the street paid their respects and many shopkeepers closed their doors and drew the blinds until the long procession had passed by.

The funeral cortege.
Photo: VFD Archives

The Coughlan Shipyard Fire

On May 15, with the tragedy at "Death Corner" still uppermost in everyone's mind, Fireman William J. Cameron of No. 12 Hall was struck and killed by falling timbers at the J. Coughlan and Sons Shipyard fire.

The alarm came in at 2:30 a.m. for a fire in the boiler shop caused by an exploding acetylene tank, and before long the fire was a "general alarm" which was considered a four-alarm fire (although never used, a five-alarm fire was the largest at that time). In the three hours it took to knock the fire down, Cameron was killed; the 8,800-ton steamer *War Chariot* was destroyed, *War Charger* was badly damaged. Totally destroyed were the No. 4 building ways, the fitting-out wharf, the boiler shop and the drawing office, stopping all work on a huge shipbuilding program for the British government. The estimated loss was between $1.5 and $2 million. Two ships, *War Camp* and *Alaska* were saved when they were set adrift in False Creek from where they were later secured and docked.

Had there been a wind that night, it was felt that the Connaught Bridge, west of the fire scene, would have been lost, as well as hundreds of nearby homes. Mill owners and other property owners located around the creek were called and advised to get as many of their employees as possible to help deal with flying fire brands. As a result, many small, potentially dangerous fires were extinguished.

The question of the need of fireboats was once again brought to the fore by many of the creek-side property owners. The city was criticized for recently closing down nearby No. 10 Firehall in April, and the critics said that more men would likely have been on the scene of the fire a little sooner had that hall been in operation. However, it was learned that just as the shipyard fire was being reported, a downtown fire call came in for No. 2 Hall delaying that company's initial response with Firehalls No. 3 and 4.

Coughlan's Shipyard fire. *Photo: CVA 354-52*

The shipyard, the largest industry in the city at that time, was building the largest ocean-going vessels in Canada and employed some 2,300 men with a monthly payroll of more than $200,000. The yard had contracts for ten similar ships but had to close for a short time in order to clean up after the fire.

A civic funeral was held for Fireman Cameron, twenty-four years old and single, and he was buried in the Firemen's Plot at Mountain View Cemetery. He was born near Montreal and came to Vancouver in 1913, joining the VFD on September 10, 1917.

The Fallen Firemen's Monument

In the summer of 1918, the Firemen's Benefit Association erected and dedicated a monument on the Firemen's Plot in Mountain View Cemetery to the memory of nine fallen brothers. Made of granite, the monument is topped with a broken column twelve feet (3.5 m) tall with the firemen's names in lead lettering and the evocative caption "OUR FIREMEN" on the base. By 1928 the sixteen graves on that site were filled. In the years that followed, two other burial sites were purchased and developed in other sections of the cemetery.

The Fallen Firemen's Monument in Mountain View Cemetery.
Photos: VFD Archives

The 1918 Spanish flu pandemic killed 50,000 Canadians and claimed an unknown number of city firemen, but it is known that Frederick Sentell, age thirty-seven, died in 1918; John (Baldy) Balderston, age thirty, died on December 29, 1918, and Ralph M. Gibbons, age thirty-one, who died on January 17, 1919, were all confirmed victims, along with Leonard Farr of No. 11 Hall. "Baldy" was the father of retired (1974) District Chief Bill Balderston.

The Imperial Oil Fire

Retired Assistant Chief Alf Smethurst (VFD 1918–1960) liked to remember his first large fire, which was at the Imperial Oil Company's office and storage yard on Smithe Street. The three-alarm fire, in the summer of 1919, started behind Canada's first gas station, Automobile Filler No. 1

and caused many thousands of dollars damage. The gas station opened in 1906/07 and dispensed gasoline from a converted hot water tank through a short length of hose.

Imperial Oil fire. *Photo: CVA 354-50*

The first fatality of the 1919 hunting season was the fire department's blacksmith, Robert J. Moore. He and his son and a friend were in their hunting shack when a tree blown down by a heavy wind crashed through the roof, killing Moore instantly. The father of four had joined the VFD June 7, 1901.

Fire Department blacksmith, R. J. Moore. *Photo: VFD Archives*

The Roarin' Twenties, Amalgamation and Jubilee Year

(1920–1936)

The morning after the Balmoral
Apartment fire, June 20, 1920.
Photo: CVA 99-3238

The Balmoral Apartment Fire

On the warm evening of June 20, 1920, the druggist on the corner of Davie and Thurlow Streets thought he saw smoke coming from one of the basement windows of the Balmoral Apartments, at 1148 Thurlow Street. Within moments, a neighbor next door to the apartment heard a muffled explosion and a cry of "Fire!" He immediately called the fire department.

Fire Alarm Operator Tom Burke, who tapped out Firehalls No. 2 and No. 6 to respond, took the alarm at 10:44 p.m. Both companies were somewhat hampered by the gathering crowd, but they were able to have several hose lines on the fire within five minutes of their arrival. Soon, the fire had gone through the roof of the six-story building. Assistant

Chief Thompson said that he could see the fire as he left No. 2 and thought that there must have been some delay in turning in the alarm, but the fire had spread very quickly.

At 10:54 p.m. the chief put in a second alarm then called for a third alarm at 10:59. Ladders were raised to effect rescues and for operating hose lines, and other lines were advanced from the ground floor to the roof, floor by floor, until the fire was knocked down. Finally struck out at 2:10 a.m., the fire had claimed five lives.

The first loss of life occurred when a man jumped to his death from an upper floor before the arrival the fire department. The next man was encouraged to jump from the fourth floor into a life net being held for him by a group of misguided bystanders who had taken the nine-and-a-half-foot (3-m) Browder life net from No. 2 Truck. He jumped, hitting the edge of the net and fell to the sidewalk, striking his head. He was then carried across the street to St. Paul's Hospital where he later died of his injuries. Chief Thompson ordered the men not to use the net again as they were untrained and not using it properly.

The next two victims, a man and a woman, were found on the sixth floor. They had died of smoke inhalation, and it was felt they would have survived had they not refused to go to the roof and await rescue with one of the survivors who had encouraged them to do so.

The last victim was the building janitor, Mr. S.A. Spencer, who, upon discovering the fire in the basement, ran through the building alerting people to the fire because the building did not have a fire alarm system. He managed to get through the building before he perished on the sixth floor, a hero who had saved many lives.

It was never determined exactly how the fire started, but the rapid spread was caused by combustibles in the furnace room, the lack of a fire door at the bottom of the elevator shaft, open stairways, no fire gongs, and flammable wall coverings on the first two floors. Total damage was $93,600.

Star Fireman/Athlete Injured

The headline in the paper read, "Driver Turned For Street Car; Gene O'Connor Is Dying." On February 1, 1921 just after 10:00 p.m., Hose Wagon No. 1 responded to an alarm at Box 24, Abbott and Hastings. The driver, while trying to avoid hitting a streetcar at Carrall and Cordova Streets, swung his rig wide, skidded onto the sidewalk smashing two large plate glass store windows, and then struck a pedestrian who suffered a fractured leg, scalp cuts and internal injuries.

Fireman Eugene Major Connor (not O'Connor), popular local sportsman and wrestler, was thrown from the rig and struck his head on a lamppost, fracturing his skull. He was rushed to the hospital by police patrol wagon and was not expected to live. The fire call turned out to be of a minor nature.

After missing 188 shifts at work, which were covered by his brother firemen, Connor returned to work. His recovery was no doubt a testament to his excellent physical condition as a wrestler. He was also a well-

VANCOUVER FIRE FIGHTERS' CRACK ATHLETES

TRAINING DOES IT
And our Fire Laddies are well trained. Sturdy, muscular, peppy, they're the right kind to fight a fire, and win the fight.

Fireman Eugene Conner (doing handstand) is shown among the department's many athletes. **Left to right:** Ralph Ravey, Lorne Foley, Tom Burke, L.O. Hillier, Gill Martin, Alf Smethurst, R.E. McLaren, J. Stewart, Archie McDiarmid, J. Wright, Tom Anderson, Jack Anderson, Fred Taylor, Gene Conner. **Inset** - Deputy Chief Charlie Thompson. *Photo: VFD Archives*

known daredevil and often entertained by walking on his hands on the cornices of downtown buildings. Gene was born in Sulphur Bluff, Texas, joined the VFD on September 30, 1918, and retired in 1955 as a captain. He died September 19, 1983 at the age of eighty-eight years.

Connor was one of many athletic fireman of that era. At the 1920 Olympic Games in Antwerp, Belgium, Fireman Archie McDiarmid, an athlete on the ten-member Canadian team, was the designated flag bearer for the opening ceremonies parade. For some unknown reason, a Canadian flag wasn't available, so Archie led the team into the stadium carrying a bare flagpole. He came fourth in the fifty-pound hammer throw.

Archie McDiarmid with some of his trophies and awards.
Photo: VFD Archives

Captain Thomas Tidy.
Photo: VFD Archives

The Fire Warden's Branch Is Formed

In 1921 the Fire Warden's Branch (fire prevention inspectors) was organized to replace the police sergeant who had been appointed fire warden years earlier. The first chief fire warden was veteran fire captain Thomas Tidy, who had joined the VFD in 1893, and his second-in-command was Captain Jack DeGraves. The remainder of the branch consisted of four lieutenants, nine inspectors and two clerks.

January 1, 1922 the "Rule of the Road" came into effect in Vancouver requiring all drivers to now drive on the right-hand side of the road. The VFD only had two fire apparatus on its entire roster with other than right-hand steering wheels; the new 1922 White chemical wagon at No. 2 Hall and No. 2 Truck, the old Webb aerial, which had its steering wheel right in the middle! No major driving problems were recorded.

On June 14, 1922, Chief Carlisle became the first recipient of the Appreciation Medal of the Native Sons of British Columbia, in a colorful ceremony that took place on the steps of the courthouse. With the mayor and other civic dignitaries, a parade of firemen, the police pipe band, school children and citizens in attendance, the chief accepted the medal and the many tributes praising his good citizenship and fine service to the community.

Captain J.H. DeGraves.
Photo: VFD Archives

In the evening a banquet was held in the Hotel Vancouver, and the Native Sons announced that it was their intention to hold a similar citizenship ceremony every June 13, on the anniversary of the Great Fire of 1886.

Point Grey No. 2 Firehall was opened September 1, 1922 on the corner of West 12th and Trimble Street. The PGFD began serving that neighborhood in 1915 from a temporary hall, which wasn't much more than a garage, at 4411 West 11th Avenue. After about a year at that location they moved into rented quarters at 10th Avenue and Sasamat Street. Then on June 6, 1922 the owner of the property gave them notice to vacate and within three months the new quarters were built and ready to move into on 12th Avenue.

PGFD Firehall No. 4, at 70th Avenue and Hudson Street, opened October 22, 1922, but only after the final payment of $2,250 had been paid to the contractors. (It appears that the municipality had some trouble paying its bills.)

Chief Carlisle accepts the Appreciation Medal from the Native Sons of B.C. *Photo: Stuart Thomson: VFD Archives*

Point Grey No. 2 Firehall. *Photo: CVA 354-62*

On April 8, 1923 Chief Carlisle was once again honored, this time by receiving the King's Meritorious Service Medal, awarded by King George V. This was the first time that a Canadian fire chief had received this decoration that was presented by the king's representative in British Columbia, Lieutenant-Governor Walter C. Nichol, in a special ceremony held in Stanley Park.

As president of the Dominion Fire Chiefs' Association (DFCA), Chief Carlisle hosted their 15th Annual Convention in Vancouver, July 31 through August 3.

In 1923, the VFD had eleven firehalls, (although city council had closed four firehalls, Nos. 5, 10, 13 and 15, under their retrenchment policy) and twenty-seven pieces of apparatus, including eight hose wagons, two chemical engines, four combination hose/chemical wagons, two aerial ladder trucks, two city service trucks and three pumps, with a combined pumping capacity of 3,400 gallons per minute. The old Amoskeag self-propelled engine was in reserve and the chief, assistant chief and the three district chiefs all had cars.

The manpower consisted of 191 men and with the five senior chiefs; it included twenty-six captains, four engineers, six fire wardens, two secretaries, a mechanic and three assistants and 144 firefighters. The rank of lieutenant had been temporarily discontinued during reorganization.

Fire Alarm Office at No. 2. *Photo: CVA 354-25*

Vancouver's Fire Alarm System

The Fire Alarm Office (FAO), located on the top floor of No. 2 Hall, on Seymour Street, received more than 80 percent of its alarms via telephone through the emergency number, Seymour 89. The system had 318 boxes on thirty-seven box circuits and all alarms came through the fire alarm system and were relayed by the central station operator to the firehall due to respond.

The four operators on duty operated two large switchboards, one of which was always recharging. When the operators were alerted by the electric master clock that the board in operation had to begin its recharge cycle, then the changeover to the charged board took place. Power to recharge the batteries on the DC system was supplied by the power plant at the nearby Hotel Vancouver, and should that fail, there was a gas engine-powered generator in reserve. Maintenance to the system was ongoing and was the responsibility of a foreman and four inspectors who ensured that the system was always operating properly.

A twenty-four-hour record of alarms received, department activities, apparatus movement, and alarm system maintenance was kept on a daily log sheet.

An example of a box alarm received at the FAO is as follows: When the street box was activated, a signal was sent from the box, through the circuit, to a panel in the alarm office which sounded a small gong and lit up a small red light. Simultaneous to the action of the gong and blinking light, a paper tape on a reel-to-reel was stamped with the date and time (hour, minute and second) that the alarm was received. On another register, a relay operated a master punch that perforated the tape with an orderly row of holes that indicated the box number. For example, if the box number was 126, then the tape would be punched, thus:

The first single hole indicated number "1", the second two holes indicated number "2", and the six holes indicated number "6". The operators would identify box number 126 on the master list as, HASTINGS MILL, NORTH FOOT OF DUNLEVY STREET. Since No. 1 Hall was the closest to that location, the operator tapped out (alerted) No. 1 Company, as well as any supporting companies assigned to that box.

In the firehall watch-rooms there were two sets of registers or "jokers" that were similar to the ones in the FAO. The first was the primary register, with ¾-inch paper tape and the other was a secondary register with a ½-inch tape, both running reel-to-reel. The tap-out signal sent by the FAO activated the 18-inch station gong, BONG-BONG, BONG-BONG, which alerted the crews to respond to their rigs. As the tap-out was sounding, the box number was being punched out twice on the primary tape, thus:

The captains receiving an alarm in the watch-room at No. 1 Hall.
Photo/clipping: Vancouver Sun

and then followed by two rounds on the secondary tape:

The captain(s) responded to the watch-room, counted the punches, making sure that they were all 1-2-6, checked the number on their district's master list, then "coded out" by throwing a switch that advised the FAO that they had received the alarm and were responding to the box location.

For a telephone alarm, the FAO, after receiving the information from the caller, tapped out the firehall that was to respond, then transmitted a THREE-THREE signal, twice over each of the primary and secondary registers:

On receiving the THREE-THREE, the captain picked up the telephone and the alarm operator gave him the location and information regarding the alarm, which he repeated back to the operator and, once verified as correct, the captain coded out and the company responded to the fire.

Multiple, or greater alarms, were transmitted as follows: a second alarm was punched TWO-TWO-TWO; three-alarm was punched THREE-THREE-THREE; four-alarm was FOUR-FOUR-FOUR, etc. When a multiple alarm was transmitted, all halls were tapped out and the size of the alarm was punched out on the tapes, followed by two rounds of the nearest box number.

When a fire became a greater alarm, on the orders of a chief, all halls were tapped out and any apparatus movement — either to the fire or to fill-in at other quarters — was done by each company, referring to the "control card" being used. These control cards were numbered and found listed on every hall's master list of alarm boxes. The cards told what each company's movement would be; to either respond to the fire or to change quarters and "fill-in" at another firehall. Travel to fill-in was non-emergency.

When the fire was considered under control then it was "struck out" by the chief officer in charge at the scene. The FAO then transmitted the strike out with a ONE-ONE signal, once on each tape. If there were two or more incidents at the same time, they would be struck out in the order that they came in — second fire or incident, ONE-TWO; third fire or incident, ONE-THREE, etc. In the days before the apparatus had two-way radios, the chief transmitted greater alarm signals and strikeouts from the street box using the small telegraph key in the box.

Point Grey headquarters opening. Original Point Grey HQ on the left. *Photo: VFD Archives*

New Point Grey Headquarters Opens

On April 30, 1924 the Point Grey Fire Department opened its new head-quarters hall at 38th Avenue and Cartier Street. Built at a cost of around $35,000, it was officially dedicated June 1, 1924, which was also the date that the PGFD began the two-platoon system.

Around this same time, the South Vancouver Fire Department opened its new No. 2 Firehall at Wales Road and Kingsway.

On September 10, 1926 another department milestone took place when the last piece of fire apparatus painted white was placed in service at No. 6 Firehall. It was a 1926 American-LaFrance, 840-gpm (3,800-liter) Type 12 pump, built in Toronto and cost $15,600.

It is said the reason that all subsequent orders of fire apparatus would be red instead of white is that the white rigs were very difficult to see in the fog.

Opening of South Vancouver No. 2 Hall.
Photo: VPL 6404

The Firemen's Band is Formed

The Firemen's Band was formed in February 1927 and the first bandmaster was Jean Coupland, a musician who wasn't a fireman. The city gave the band a start-up grant that was used to buy instruments and music. The band's first public appearance was in Vancouver's Diamond Jubilee Parade, July 1, 1927.

On June 1, 1927, after being closed for ten years, both No. 13 and No. 15 Firehalls were reopened with motorized apparatus, and a week later No. 15 received a new 1927 Studebaker hose/chemical combination, the only one of its kind on the fire department. It was built in the department machine shop.

Only known photo of the VFD's
1927 Studebaker combination.
Photo: VFD Archives

The first official photo of the Firemen's Band taken at the Exhibition Grounds June 1927. Pictured among the bandsmen are: a union president (Lucas), two centenarians (Betts, Smethurst) and four who died in the line of duty (Barnett, Ellis, Jenkins, Wilkins).
Photo: VFD Archives

VANCOUVER FIRE DEPARTMENT BAND.
JUNE 28TH 1927

J. LEATHERDALE. A. LeCLAIR. A.E. WILKINS. L. LEAVY. F.G. LOCKE. J. EVANS. C. HODSON. W. BETTS. R. SMETHURST. R.G. SHAW. G. HARROD. T. ANDERSON.
A. KING. J. LYON. T. WARNE. C.R. HILLIER. R.H. WHIPPLE. H. MCELWAINE. J. TRAVIS. F. JENKINS. W. MIDDLETON. E. WILLIAMSON.
M. PILLAT. M. McLEOD. E.W. BARNETT. G. FORD. M.K. McLENNAN. H. ELLIS. G. ROSS.

The Royal Alexandra Apartment Fire

The worst loss-of-life building fire in Vancouver occurred on Friday, July 8, 1927, in the six-story Royal Alexandra Apartment building at 1086 Bute Street. Eight lives were lost in the fire.

The first alarm came in at 1:43 p.m. and with No. 6 Hall located just four blocks away it was the first company to arrive, followed closely by No. 2 Hall. Immediately upon the arrival of Chief Thompson, a second alarm was called as the fire had already gone through the roof, and the twenty-four or so men on scene obviously needed help. Then moments later at 1:51 p.m. a third alarm was called, bringing in response from more than half of the city's fire-halls.

Except for the time of day, the scenario was almost identical to the earlier Balmoral Apartment fire in 1920: a six-story building, rapid spread of fire, daring rescues and death.

The Royal Alexandra Apartment building fire, July 1927.
Photo: VFD Archives

It took time for the needed aerial ladders to arrive on scene from Nos. 1 and 2 Halls so the first-in firemen laid the necessary hose lines and effected rescues where possible, with the help of several policemen and citizens.

When No. 2 Truck, the 1912 Webb, eighty-five-foot (26-m) aerial arrived, it was set up in front of the building and was used to pluck people from the upper floor windows, many of whom were suffering from smoke inhalation and burns. One woman had fashioned a rope out of blankets and slid down from the sixth floor to the fourth, where she dangled until a ladder could be put up to rescue her. The department's only other aerial was the seventy-five-foot (23-m) 1909 Seagrave from No. 1

The Webb Aerial at front of building.
Photo VFD Archives

Hall, Gore and Cordova, which responded on the second alarm. Using ground ladders from the two aerials and No. 4's Seagrave city service truck, many parts of the building were accessed for rescue and fire fighting. (It was said that this was the last fire at which the Browder life-nets were "officially" used for rescue purposes. They were still carried for many years on trucks, but were only used by firemen for drills and displays until all were disposed of during the late 1970s.)

More than a dozen civilians and twenty firemen suffered painful burns and cuts. A group of nurses from St. Paul's Hospital, which was a block away, hurried to the scene and treated the injured, with the help of doctors, a local druggist and bystanders.

Within an hour the fire was under control, with lines stretched throughout the interior of the building towards the top floors. At 4:20 p.m. the fire was struck out.

The eight victims were a married couple visiting from Winnipeg; a young mother and son, age nine, on vacation from Edmonton; a young brother and sister, aged ten and twelve, respectively; a sixty-six-year-old woman visitor; and the building's housekeeper.

The cause of the fire was a highly flammable varnish remover that was being used by a painter in Room 404. He said he had just finished stripping the floor of the suite and was cleaning up, when a pail out in the hallway, containing his materials, suddenly exploded into a mass of flames. When he couldn't put out the fire, he immediately phoned the fire department, yelled to the building engineer for help and tried to get a house line operating. Because of the flames and heat, he had to aban-

Rear of building showning fire damage to top floors.
Photo: VFD Archives

don his plans and he then activated the fire alarm and got out of the building. The fire completely destroyed the top three floors of the building, with damage estimated at more than $200,000. The rapid spread was attributed to a shaft venting system that opened to the roof for the apartment's built-in beds.

An analyst, giving expert testimony at the inquest, stated that flames would leap over seventy-five feet in a fraction of a second on a floor coated with the volatile paint remover that had been used in the apartment. The painter's only explanation of how the fire started was possibly someone passing down the hallway had discarded a match into his pail of materials. The painter was charged with manslaughter, but the outcome of the case is not known.

No. 2 Hall's tractor aerial at a house fire on Cambie Street. *Photo: Jack B. Thompson*

The question of old, out-dated fire apparatus was also the subject of the inquest and steps were taken to begin replacing more of them immediately. On August 18, 1927 No. 2 Hall put its new truck in service. It was a LaFrance Type 17 tractor-drawn aerial with an eighty-five-foot (25-m) ladder and was a much-needed addition to the fast-growing downtown and West End districts.

Three Vancouver apparatus being prepared in the paint shop at the LaFrance Company of Canada in Toronto.
Photo: VFD Archives

No. 6 Hall's new 1928 LaFrance Type 31, front-wheel drive aerial.
Photo: VFD Archives

In the spring of 1928 more new apparatus arrived. Among the first were two LaFrance Type 145 Metropolitan 840-gpm (3,800-liter) pumps and a LaFrance Type 75 combination hose/chemical. By July a new LaFrance Type 31, front wheel drive eighty-five-foot aerial was in service at No. 6, and a new LaFrance Type T-70-6 city service ladder truck was in service at No. 12 Hall.

The Pierre Paris Tannery and
BC Box Company fire.
Photo: CVA 99-1720

Paris Tannery and BC Box Plant Destroyed

Described as the fiercest fire in many years, flames destroyed the BC Box Company and the adjacent Pierre Paris Tannery in a very hot mid afternoon blaze on August 28, 1928. Two major structures were destroyed within an hour and the resulting damage totaled more than $200,000 and caused several firemen to suffer minor injuries and heat stroke. The fire required more than half of the fire department's resources to respond. The cause was not determined.

The Fireboat *J.H. Carlisle*

On September 1, after almost twenty years of recommending to the city council the need of a fireboat, No. 16 Fireboat Station, on the north shore of False Creek opened with the new *J.H. Carlisle* in service.

The boat was fifty-five feet long by fifteen feet wide (17 x 4.5 m), had a draft of about five feet (1.5 m) with a capacity of 4,500 gpm (20,450 liters). The boat was funded partly by the property owners around the perimeter of the creek. It was launched August 14 at the Burrard Dry Dock Company yards in North Vancouver and christened by Miss Iris Gibbens, the daughter of Acting Mayor P.C. Gibbens. (Miss Gibbens, who later married a city policeman, became Mrs. McIlroy and was the mother of retired Captain Neil McIlroy, who joined the VFD in 1964.) The first pilots of the new fireboat were Captain R. Frank Ross and Captain John A. McInnis, with relief pilots George McInnis and Hector Wright. All were federally licensed mariners.

Iris Gibbens launching fireboat *J.H. Carlisle*.
She was told she had to break the champagne bottle on the first swing for good luck. She did, and sprayed the official party.
Photo: CVA 99-1971

The Tunic Request

The union went to the chief on December 14 and requested that he consider a uniform change from the old-fashioned high-collared tunics to the more modern, double-breasted lapel style. After all, they'd had this same style since 1889 — almost forty years. After due consideration, the chief's response was that he felt if the men had the newer style that they would be inclined to wear their uniforms out in public, on their leisure time, as suits. He denied their request.

In 1936 the double-breasted tunics were issued, and after more than seventy years they're still being worn on parade and dress occasions.

Fire Chief Frank Raymer and his Point Grey firefighters before amalgamation. *Photo: VFD Archives*

Amalgamation Day

On January 1, 1929 the municipalities of Point Grey and South Vancouver joined the City of Vancouver, giving the city a total of forty-four square miles. As a result, the VFD grew by six firehalls and about eighty men. South Vancouver's headquarters Hall No. 3 on Wilson Road (later changed to 41st Avenue) and Draper Street became VFD No. 17 and SVFD No. 2 Hall at Wales Road and Kingsway became VFD No. 20. PGFD headquarters hall became No. 18, their No. 2 became No. 19; No.

3 became No. 21 and PGFD No. 4 in Marpole (70th Avenue and Hudson Street) became VFD No. 22.

The two fire chiefs, William E. Clark of South Vancouver, and Frank Raymer of Point Grey were appointed the district chiefs in their respective areas. No. 17 Hall became the DC's hall for #3 District and No. 18 was the DC's Hall for #4 District. (Chief Clark was the father of VFD Captain E.C. (Bim) Clark, who retired in 1984.)

South Vancouver and Point Grey fire captains reverted to firemen ranks as most only averaged about ten years service, which was considerably less than city fire captains. Their future promotions would be based upon their positions on the new integrated seniority role.

The firemen's unions of South Vancouver, Local 259 and Point Grey, Local 260 had earlier joined the Vancouver union, Local 18, IAFF, on June 1, 1928, and their local numbers were abandoned.

The fire alarm box system for Point Grey (forty-one boxes) and South Vancouver (fifty-nine boxes) were connected to Vancouver's system giving the city a total of 475 boxes.

The following fire apparatus were put on the new VFD roster:

From South Vancouver: two 1914 Seagrave hose/chemical combinations; a 1927 Studebaker city service ladder truck; and a 1928 Ford Model 'A' roadster, the chief's car.

From Point Grey: a 1912 American-LaFrance hose/chemical combination; a 1914 American-LaFrance 625-gpm pump; a 1914 American-LaFrance city service ladder truck; a 1922 Packard 625-gpm (2800-liter) pump; two 1926 American-LaFrances, a 625-gpm pump and a city service ladder truck; a 1926 Dodge Brothers-Graham bush fire/service truck; and a 1924 Studebaker roadster chief's car.

South Vancouver fire fighters with Fire Chief William Clark posed in front of SVFD No. 3, which, upon amalgamation, became VFD No. 17. The sixth man from the left in the back row is Ralph Jacks who became Vancouver's ninth fire chief.
Photo: CVA 354-84

Charlton William Thompson

C. W. Thompson

Born: March 25, 1875,
Simcoe, Ontario
Joined VFD: June 17, 1895
Appointed Chief: Feb.1, 1929
Retired: December 31, 1934
Died: July 6, 1957

Carlisle Retires: Fire Chief Thompson Named

Chief J. H. Carlisle's
retirement picture after
forty-two years' service.
Photo: VFD Archives

At the end of January 1929 Chief J.H. Carlisle's retirement became official after more than forty-two years of faithful service to the city. He had seen the city grow from virtually nothing after the Great Fire, to a large and prosperous city known throughout the world, and he had nurtured and guided the fire department from very modest beginnings to one of the finest to be found anywhere. At the age of seventy-one years he probably felt that he still had a few good years left, but it was time to turn the leadership over to someone else, and this is how he informed the city council on January 17 that he was leaving, "As it is the desire of the chairman of my committee, I hereby tender my resignation as chief of the Vancouver Fire Department, and I strongly recommend that Assistant Chief C.W. Thompson be appointed my successor, that Chiefs Clark and Raymer of South Vancouver and Point Grey be retained in the rank of District Chiefs in their districts, and that all captain ranks be revised and that appointments be according to years of service and ability."

Chief Carlisle's long-time assistant Charlton W. (Charlie) Thompson, age fifty-three, became the department's second fire chief

John Carlisle enjoyed many years of retirement and passed away in Burnaby on November 28, 1941 in his eighty-fourth year. He is buried in the Masonic Cemetery in Burnaby.

On May 9, 1929, all firefighters who were driving fire apparatus were told that they must get driver's licenses. The cost would be $1.

On May 15, the citizens voted yes on a referendum for the city to build a new combination fire alarm office and police dispatch building at a cost of $350,000 with the criteria that it be: a) fire-proof; b) not within 150 feet (45 m) of any building; and c) in the geographical center of the city. The site chosen was on Cambie Street at West 20th Avenue.

Two more replacement pieces of apparatus arrived and were placed in service in the fall of 1929. They were the LaFrance Type 112 combination hose/chemical, which went to No. 11 Hall and the 840-gpm (3,800-liter) Bickle "Canadian" pump, which went to No. 6 Hall.

1929 LaFrance Combination. *Photo: Frank Degruchy*

1929 Bickle Pump at No. 6 Hall. *Photo: Jack B. Thompson*

On March 17, 1930 a midday fire occurred at the A.P. Slade warehouse at 157 Water Street. The three-alarm fire was made more difficult to fight by the black, oily smoke produced by sacks of burning peanuts. Considered by many to be one of the city's best spectator fires, the cause of the $25,000 fire was never determined.

The Slade fire looking west to Cordova Street. *Photo: CVA 99-2450*

A.P. Slade Warehouse fire looking east. *Photo: CVA 99-2449*

The inhalator service, which was started by the first aid crews, had been serving the city since 1927 and continued to grow with the training of all members of the fire department. On May 10, 1930 after winning the city first aid competition, the team proudly posed with the Vancouver

Shield, and the *Daily Province* Shield as provincial champions.

By May of 1931, there were 269 qualified members with everything from first-year certificates to an instructor's rating, the latter being held by First Aid Captain Jack Anderson

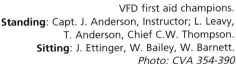

VFD first aid champions.
Standing: Capt. J. Anderson, Instructor; L. Leavy, T. Anderson, Chief C.W. Thompson.
Sitting: J. Ettinger, W. Bailey, W. Barnett.
Photo: CVA 354-390

Funeral of Fireman Herb Ellis.
Photo: CVA 354-90

Fireman Herb Ellis, thirty-eight, a ten-year veteran, was fatally injured at the Beach Avenue drill grounds on April 17, 1930 when he fell about thirty feet (10 m) from No. 2's aerial ladder. He was climbing down when he missed a handhold and fell, striking his head. A war veteran, he joined the department on May 26, 1920 and was one of the original members of the fire department band, in which he played the saxophone. A civic funeral was held at Christ Church Cathedral, April 21. He left a wife and four children, one of whom, son Bernie, served on the VFD to his retirement as a captain in 1981.

Canadian National Dock Fire

The Canadian National (CN) Steamships dock fire of August 10, 1930 was the largest of the year with a dollar loss in excess of $1.25 million. The newly constructed, 1,000-foot (308-m) long structure was within two days of being turned over to the federal government by the contractors when fire destroyed it in half an hour. Nearby No. 1 Company was on scene within minutes but all that could be done was to cover the exposures and three large fuel tanks at the south (land) end of the pier. A general alarm (four-alarm) was called.

The crews had difficulty reaching the fire because of the thick smoke as the fire quickly burned its way to shore. With the new creosote-coated pilings it would have been suicide to attempt to fight it from the dock because of the speed with which the fire traveled to shore after starting at the harbor end. In spite of care taken, more than 1,700 feet (565 m) of hose and several nozzles were lost. Fortunately, the many CN coastal ships were out of port, otherwise they would likely have been damaged or lost.

The decision to rebuild the dock was made and the question of the lack of a harbor fireboat was once again raised. Mayor Malkin stated that he didn't think the city was responsible for protecting property from which taxes were not received.

Canadian National Dock general alarm fire. *Photo: VFD Archives*

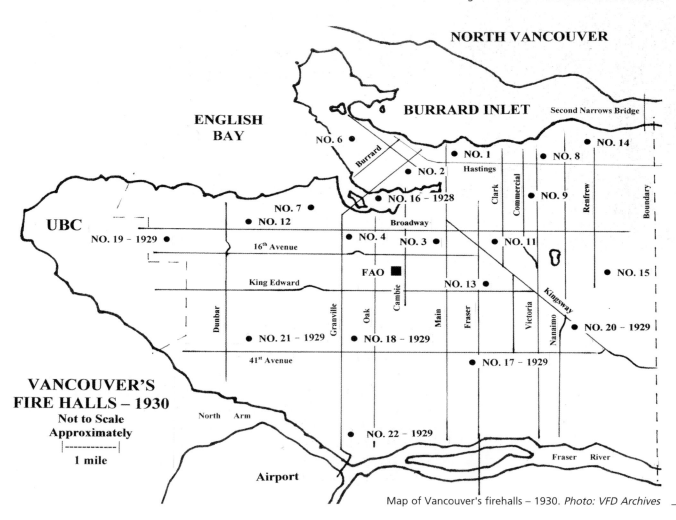

Map of Vancouver's firehalls – 1930. *Photo: VFD Archives*

The First Baptist Church Fire

Most firefighters will agree that most church fires are bad, and difficult to fight because of the large space with a lot of wood and combustibles. The First Baptist Church at Nelson and Burrard Streets in downtown Vancouver was no exception. The alarm on February 10, 1931 came at 7:00 a.m. and went to a second alarm before it was brought under control an hour later. The bell tower and the Sunday school were saved, but the main part of the structure was totally destroyed. The fire appeared to have started behind the organ, cause unknown. Damage was almost $100,000 but the church was rebuilt and still serves that area of downtown today.

The VFD Band advertising an upcoming baseball game on the B.C. Electric's Observation Car. *Photo: Harry Bullen: VFD Archives*

From May 27 through 29, Chief Thompson hosted the First Annual Convention of the British Columbia Fire Chief's Association in Vancouver.

On July 4, at Athletic Park, the firemen's baseball team, one of the city's best and winners of many championships, beat their opponents, Arrow Transfer, 5 to 3, in the first night baseball game played under lights in Canada. Athletic Park was the first Canadian stadium to have floodlights for night games.

Fire Alarm Office at 3637 Cambie Street. *Photo: VFD Archives*

New Fire Alarm Office Opens

The Fire Alarm Office was moved from No. 2 Hall into new quarters at 3637 Cambie Street, on July 11. The emergency phone number was changed at this time from "Seymour 89" to "Fairmont 1234". Shortly after this move the operators, who were under the city electrical department's control, became members of the firemen's union.

During 1931, in an attempt to reduce the incidence of false box alarms, the Fire Warden's Branch sent a member around to the elementary schools with a street box, to show the children how it operated and to tell them what a danger it was to pull boxes maliciously. It was found that this ounce of prevention was worth a pound of cure, and false box alarms declined. Although the program was successful it was discontinued.

In 1932, over bitter opposition, the city council reduced all civic wages by 10 percent, and in 1933 a further 10 percent cut was made, with a thirty-five-day unpaid lay-off for firemen only.

On February 28, 1933 the city announced major budget cuts to all departments with a reduction to the fire department of $202,000. Chief Thompson stated that with this reduction, he would have to let 138 of his 368 men go, close up to nine firehalls, including No. 16 Hall, the fireboat station and take *J.H. Carlisle* out of service, "if the city can induce the False Creek property owners to release the city from its agreement." He further stated that, "these reductions would leave the fire department with 83 men on duty per shift which is totally inadequate to protect the city from fire."

The eventual outcome of the budget cut was that two Firehalls, No. 7 and No. 8, were permanently closed on May 1. There were no lay-offs but the firemen's wages were reduced by 20 percent plus the thirty-five-day unpaid lay-off days, which meant a 14.5 percent greater loss than that of other civic workers.

On September 12 the Dominion Photo Company destroyed 1½ tons (2,250 kg) of nitrate photographic negatives on the beach at Spanish Banks in accordance with the city by-law, which prohibited the storage of more than thirty pounds (14 kg) of such film. This was part of a Fire Warden's Branch program to get rid of this highly flammable film. A photograph was taken of the event — using nitrate film!

On October 16, 1933 Fireman Andrew Grant collapsed while on duty at No. 6 Firehall and was rushed to St. Paul's Hospital in an unconscious

NOTICE

Public Notice is hereby given that, on and after July 11, 1931, the telephone numbers for the FIRE ALARM OFFICE will be changed to the following:

In case of FIRE ONLY, phone Fairmont 1234.

Private Exchange connecting Firehalls, Fairmont 1271.

C. H. FLETCHER,
City Electrician.

Electrical Dept., City Hall, July 6, 1931.

Inside the new Fire Alarm
Office on Cambie Street.
Photo: CVA 354-99

Archibald McDiarmid

Born: December 8, 1883,
Renfrew, Scotland
Joined VFD: September 16, 1907
Appointed Chief: January 1, 1935
Retired: December 8, 1941
Died: August 11, 1957

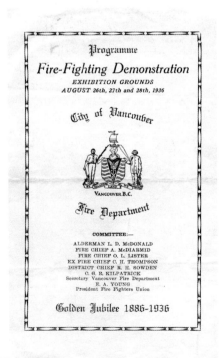

state, where he died. Fireman Grant, in his mid-twenties, left a wife and young daughter. It is believed that he died from the effects of a fire he had attended earlier. A well-known bandsman, he was given a civic funeral at the Salvation Army Citadel under the auspices of the Firemen's Benefit Association. His grandson, Grant Muir, is currently a member of the VFD.

Con Jones Park Fire

A spectacular fire destroyed the Con Jones Park grandstands July 29, 1934. The first alarm came in at 4:30 a.m. and within a very short time, the old, wooden stands that circled the grounds were gone. The caretaker and his family escaped with only their nightclothes, losing everything, including their car. It was believed that a cigarette carelessly discarded after the previous evening's ball game caused the three-alarm fire, which had a loss estimated at more than $32,000. The stadium was rebuilt and opened as Callister Park and became the site of many of the firemen's soccer team games as well as horse shows, rodeos and demolition derbies. It remained an important sports venue until it was torn down in 1971.

The Year 1935

On December 28 the city ratified District Chief Archie McDiarmid's appointment as fire chief on the retirement of Chief Thompson, December 31. He was selected over three more senior chiefs: First Assistant Chief Loftus, Second Assistant Chief DeGraves, and District Chief Plumsteel. With the appointment, AC Loftus retired, DeGraves became First AC and Plumsteel became Second Assistant Chief. Three captains were appointed acting district chiefs, among them Ed Erratt, who would later become the department's fifth fire chief.

The large fires of 1935 included a warehouse fire on Water Street, with damage of more than $500,000, an apartment fire that, fortunately, had no loss of life and two steamship fires, on board the SS *Cape York* and SS *Frederika Lensen*, with damage totaling $75,000. The bunker fire on the *Frederika Lensen* took two days for the fireboat *Pluvius* to extinguish.

Vancouver and the VFD's 50th Anniversary

The year 1936 marked Vancouver's Golden Jubilee, and on July 3, before a crowd of almost 2,000 people, the cornerstone for the new city hall was laid. The Firemen's Band, under the direction of Bandmaster Will Edmunds, entertained the gathering.

July 17 was Vancouver's Jubilee Parade day with the fire department playing a large part. Led by the VFD band, the parade entry consisted of retired Fire Chief Carlisle driving an old horse-drawn Democrat buggy followed by a team of firemen pulling one of the original hose reels, a steam pump pulled by a team of three horses, the self-propelled 1908 Amoskeag steam pump, and a variety of current aerials, chemical wagons and pumps. A group made up of the last remaining 1886 volunteers also took part.

Retired Fire Chief Carlisle in the 1936 Jubilee parade.
Photo: VFD Archives

The Arena Fire

Vancouverites have always had a penchant for remembering dates and times by referring them to the big fires, such as the CN dock fire that occurred at the end of a Depression-designated "Prosperity Week," or the "Great Fire" for the old-timers, or the Balmoral or Royal Alexandra apartment fires. They were remembered much the same way as people today remember 9/11, the Kennedy assassination, the Moon landing, or when their favorite team won the championship. The Denman Arena fire of August 20, 1936 became known as the "The Jubilee Fire."

The arena at Denman and Georgia Streets was opened on December 21, 1911, and was within a few months of celebrating its twenty-fifth year. Built by the Patrick brothers of NHL fame, the building was used for everything from hockey — as the home of the Stanley Cup Vancouver Millionaires of 1915 — to lacrosse, boxing, wrestling, religious gatherings, political meetings, ice shows, and was at one time the scene of a

THE VANCOUVER DAILY PROVINCE

FORTY-THIRD YEAR—NO. 124   OFFICIAL FORECAST: FAIR AND MODERATELY WARM.   VANCOUVER, B. C., THURSDAY, AUGUST 20, 1936 —22 PAGES   PRICE THREE CENTS On Trains, Boats and in the Country, Five Cents

PROBE $500,000 ARENA FIRE
FIERCE BLAZE SWEEPS WEST END BLOCK

TERRIFYING in their intensity, flames leaped hundreds of feet in the air, carrying blazing bits of debris which endangered homes within a radius of four blocks.

HUNGRY flames licked up streams of water as firemen heroically attempted to stem the blaze. The Arena doomed, firemen concentrated efforts in saving the Auditorium.

beauty pageant judged by the famous silent movie star, Rudolph Valentino.

On the evening before the fire it was the scene of an exhibition boxing match between heavyweights Max Baer and James J. Walsh and was attended by a crowd exceeding 4,000 people.

At 1:36 a.m. the first alarm came in for a fire behind the arena, in one of the boat works on Coal Harbour. No. 6 Company responded with No. 2 and DC Beaton, who, on arrival, called for the help of Nos. 1 and 3 Companies. Within minutes, the fire had spread and soon involved several nearby marine businesses, a houseboat and several small boats. A second alarm was called at 1:54 a.m. using Box 1558, then a third-alarm call came at 2:02 a.m. It finally became a four-alarm (general alarm) fire at 2:10 a.m.

During the height of the blaze Nos. 1, 2, 3, 4, 6, 9, 12 and 18 Companies attended, for a total of nine pumps, two hose wagons, three aerials, four trucks and almost sixty-five men, including the senior chiefs. As well, the old National Harbours Board (NHB) steam fireboat, *Pluvius,* was on the scene, but was unable to do much. In spite of the intensity of the fire, the adjacent auditorium was saved, although it suffered some major damage. Also of great concern, because of the many flying brands, was the Horse Show Building situated across Georgia Street from the arena, as well as the many homes in the area. Had there been a wind that

98

night, the city might very well have had its second Great Fire.

A large crowd was attracted to the fire and the police had a big job keeping the area clear of people, many of whom were in pajamas and nightclothes and in a partying mood. Many people also remember this fire as one of the last attended by the old 1908 Amoskeag self-propelled steam pump, which was kept in reserve at No. 3 Hall.

By 4:30 a.m. the fire was knocked down, but many more hours were spent overhauling to ensure that it was completely extinguished. Three firefighters went to the hospital with pulled muscles, burns and cuts. In the following days, the men received much praise for their valiant efforts in newspaper editorials. But Provincial Fire Marshal J.A. (Jerry) Thomas described the arena as "the worst fire trap in the City of Vancouver, ever since it was built."

Captain H.G. Bowering, of the Fire Warden's Branch, investigated the fire and no definite cause was ever determined. The estimated fire loss was more than $600,000.

Fireboat *Pluvius. Photo: Courtesy Vancouver Maritime Museum*

Local No. 18, IAFF, Loses Charter

The International Association of Fire Fighters revoked the charter of Vancouver, Local 18, on October 16, 1936. On April 1, 1935, the union, in an attempt to restore previous pay cuts by the city, had taken a strike vote as a statement of their feelings and by doing so had violated the constitution of the International.

City Fire Fighters' **Union, Local No. 1**

CITY FIRE FIGHTERS UNION NO. 1.

of VANCOUVER, B.C.

MEETING NOTICE.

The next regular meeting of the above Union No. 1. will be held on

THURSDAY NEXT

NOVEMBEP 26th.   NOVEMBER 26th.

10. A.M.   8. P.M.

at

195 - E - PENDER ST.

A full attendance is very important in order to complete our reorganization. Our new Constitution and Bylaws will be up for discussion and adoption, and if you have anything you wish to have inserted in them this is your golden opportunity.

Hall Delegates are urged to turn in their November dues as soon as possible to Bro. Howell Acting Mem/Sec. No. 2. Fire Hall.

Members note,-

This will be the first real meeting of your new Union No. 1. so please turn out in force and give it the start it deserves.

Officers and Committee reports will be interesting, so let all members set aside next Thursday as their Union Day.

fraternally yours,

P.H. Enright

Sec-Treas.

The new association then became known as the City Fire Fighters' Union, Local No. 1, with Ernest A. Young as president.

By this time one of the city's oldest traditions, the annual Firemen's Ball and Supper, was held November 4 in the Hotel Vancouver under the auspices of the Firemen's Benefit Association. This was the thirty-seventh ball held by the FBA.

At the end of 1936 firemen's wages were restored the 10 percent they had been cut in the early thirties, with 5 percent restored on both May 1 and December 1. The $506 cut returned the annual wage back to $1,890. Prior to this change, first-class Vancouver firefighters ranked sixteenth in Canada on wage scales. The city also reinstated the seventy-two-hour workweek, which, in 1932 forced the firefighters to take an extra thirty-five days off without pay.

The Vancouver Firemen's Athletic Club (VFAC) sponsored many department sports, and had a championship hockey team in 1932.
Photo: VFD Archives

New Chiefs, the War and Major Fires (1937–1945)

The *Vancouver Sun* Fire

The headline of the *Vancouver Sun* read, "Big Blaze Destroys Main Building of City Newspaper" and it referred to the fire that had destroyed their business, editorial and classified offices, as well as the photographic department, March 22, 1937.

The first alarm came in at 2:00 a.m. and the downtown companies Nos. 1 and 2 responded to the offices at 125 West Pender Street, with Fire Chief McDiarmid and Assistant Chief DeGraves in charge. Within minutes a second alarm was called.

The fire, which had started on the second floor, had a good hold and quickly spread to the third floor. The fire seemed to be contained on these two floors when a backdraft (the ignition of smoke and gases, causing an explosion) occurred, blowing windows out across Pender Street and causing part of the roof to fall into the lane, narrowly missing a crew of firefighters. At 2:45 a.m. the fire went through the roof and a third alarm was called. There wasn't much to be saved except the buildings on either side, one of which contained the composing/press room of the paper. At one point some pressmen laid two hose lines across the roof of their building in an attempt to stem the spread of the fire in the building next door, but had to abandon their plans.

The single casualty was the janitor, who had tried to stop the fire with portable extinguishers. He suffered minor burns and smoke inhalation, but he managed to crawl to safety after unbolting and unlocking the front door and stumbling out into the fresh air.

Damage was estimated to be over $200,000 to the main building; however, the pressroom was saved, with some slight water damage, and the newspaper came out on time that day. A cause was never determined.

The following day, this editorial appeared in the *Vancouver Sun*:

> We see the massive red fire trucks swinging down the street with sirens shrieking to high heaven and stalwart firemen hanging nonchalantly from the rigs and we swell our chests a little and say, "Pretty fine fire force we have!"
>
> Or we pass a black and gutted building with windows crashed out and great raw gashes in the woodwork and we say, "These firemen sure smash things up."
>
> But that is all speculation, all theoretical appraisal.
>
> Just as you never know a man or a woman until you marry them, similarly you never know the fire department until they get into your house.
>
> Well, we had the firemen in our house very early yesterday morning.
>
> And we have nothing but admiration and praise for them.
>
> They treated every person on the premises, every article of furniture, every inch of woodwork with the utmost care and consideration.
>
> And when they prevented the Vancouver Sun's fire from spreading beyond the four walls of the one office building they did a job of fire fighting that should go down in the history of Vancouver.
>
> It takes what is known as "intestinal fortitude" to be perched up precariously on an unbraced ladder three storeys above the street and direct a heavy stream of water into a blazing and tottering inferno.
>
> It takes skill and intelligence of a high order to confine a destructive fire to a single building when that building shoves up tightly against two other buildings.
>
> We think the Vancouver Fire Department did a job of work yesterday morning that no other department in the whole world could have done better.
>
> To every fireman, from the Chief down to the rawest rookie, we offer our congratulations and thanks.

News clipping of *Vancouver Sun* fire.
Photo: Vancouver Province

On September 18, 1937 the first piece of fire apparatus purchased in over seven years was placed in service at No. 4 Hall. It was a 1937 LaFrance, Type 412-RB, 1000-gpm (4500-liter) pump, the only one of its

The new 1937 LaFrance pump with crew at No. 4 Hall.
Photo: Harry Bullen: VFD Archives

type built in Canada in 1937. Two other pumps of this type were built in Canada, one each in 1936 and 1938.

Also during 1937, Vancouver firefighters began wearing a new composition helmet in place of the old, traditional leather helmet, which had been worn since before the turn of the century.

In February 1938 Elmer R. Sly was installed as the president of the City Fire Fighter's Union, Local No. 1, a position he would hold until 1941.

The City Planning Commission honored retired Fire Chief J.H. Carlisle on March 10, 1938 when they named Carlisle Street after him. Situated near Renfrew and Hastings Streets, it is one of very few streets in the city that doesn't have any addresses on it, only the backs of the buildings that face the adjacent streets. When the sign was replaced in the early 1980s, it had 2800-block on it when it should have read "2900" and was later corrected.

Carlisle Street sign. *Photo: VFD Archives*

Canadian Pacific Pier D Fire

The largest and most famous of Vancouver's many large waterfront fires is the Pier D fire of July 27, 1938. The first alarm came in at 1:46 p.m. for a fire at the northeast corner at the harbor end of the pier. When the first-in companies arrived, it appeared obvious that the pier was doomed; still, hose lines were laid and an offensive attack was taken, but not for long.

CPR engines working the area removed many box-cars away so firemen could have better access. One engine was seen removing twelve boxcars along the

Post card showing Pier D, on right, prior to fire.
Photo: Author's collection

tracks that run inside the pier; by the time it got clear, the last two boxcars were on fire.

At 1:50, CPSS *Princess Charlotte*, preparing for an Alaskan cruise, was backed away from the pier and escaped certain destruction.

A second alarm was put in at 1:53 followed by a third at 1:56 p.m. Four minutes later, more than half of the pier was involved and a fourth alarm was called for at 2:02 p.m. At this time, four firemen on hose lines who were fighting the fire from the wharf deck were forced by the intense heat to jump into the harbor to save their lives. A tugboat crew pulled them from the water. One of the firemen who had to jump was Ralph Jacks, who would become the department's ninth fire chief. The others were Captain H.G. Bowering of No. 2, and Firemen Charlie Fitzpatrick and Theo Warne. The crew of the tugboat each received Royal Humane Society awards for saving the firemen.

At 2:14 p.m. the fireboat *J.H. Carlisle* was ordered to respond to the fire from False Creek, arriving on scene at 2:57. By 2:20 p.m. the last of the many employees attempting to save papers and records were forced from the building, as most of the pier was now ablaze. Firemen and apparatus were being forced to move to safety because the fire had now ignited the viaduct leading to the pier, Pier H and Freight Shed 3, situated parallel to the shoreline, east of Pier D.

When No. 12 Hose Wagon, which had responded on the third alarm, arrived on scene, the crew was instructed by the chief's driver, Reg Hill, on orders of Fire Chief McDiarmid, to lay hose lines up to the front door of the pier. The lines were laid and were soon operating but it was a fruitless effort as the fire was now destroying the front of the building. The heat and smoke forced the firefighters to abandon their lines and run to safety. Then, as if in an act of defiance, the fire allowed the smoke to clear for a moment around the big, neon PIER D sign and façade, and the front of the building came crashing down, burying No. 12's hose wagon, the old 1913 American-LaFrance, under tons of debris.

The driver of the rig, Fireman George Black, was one day short of ten years' service on the VFD that day. Nobody kidded the big, ex-coal miner when he lost his rig, at least not to his face, as firemen usually do to each other. The entry on the Fire Alarm Operator's daily work sheet simply shows this about George's rig, "—-#12 Hose Wagon burned—-."

George retired in 1966 as an assistant chief and enjoyed retirement until his passing in 2003 at age ninety-seven. He was the last remaining member of the department to have worked with every fire chief, as a subordinate or a superior, from J.H. Carlisle, the first chief, to Glen Maddess, the thirteenth chief and, of course he knew the current and fourteenth chief, Ray Holdgate.

Pier D at 1:50 p.m. four minutes after the alarm was turned in. *Photo: Courtesy Maritime Museum*

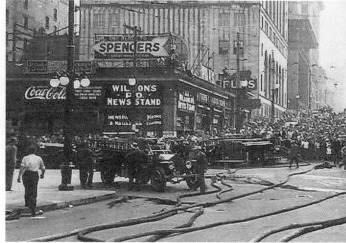

Laying lines, Pier D fire. *Photo: VFD Archives*

Just before the pier façade collapsed. *Photo: VFD Archives*

Showing total destruction of Pier D.
Photos: VFD Archives

George Black with
recovered nozzle.
Photo: Author's collection

The fire was struck out at 4:52 p.m. but for the next four days crews were on the scene overhauling the site to ensure there was no rekindle.

In the days following, the question of the need of a fireboat in Burrard Inlet arose again. Many felt the damage or the loss could have been greatly reduced had a fireboat been handy to the scene, but this is unlikely as the fire had a good hold on the pier.

The old National Harbours Board ninety-four-foot (28-m), 1,540-gpm (7,000-liter) steam fireboat *Pluvius* had been condemned and placed out of service at the end of 1937, so it wasn't available. Steel-hulled, she was built as *Orion* in 1904 in Oslo, Norway, and served as a whaling vessel until converted to a fireboat in the 1930s. The fireboat was renamed *Pluvius* a few years before she was placed out of service and scrapped.

One of the nozzles lost in the fire that day was recovered during dredging operations in 1978. Resting on a mass of melted brass it is still in the open position showing that whoever was using it likely had to drop it and run.

The pier was never rebuilt.

The Fire Defense Report

Following the Pier D fire, a special committee was appointed (from August 15, 1938 through August 10, 1939) by the city council to report on the fire defenses of the City of Vancouver. Little did the committee members know that the Second World War was about to break out and change things, but it is interesting to see what comments and recommendations they made.

High praise was given the crews of the fireboat, inhalator and fire alarm office, as well as firehall crews for the condition and cleanliness of the equipment and apparatus observed during random firehall visits. Aside from a few ancillary items, like larger water mains and a drill school, it was recommended that a new headquarters be built, relocated around the Georgia and Beatty area. It was felt that a new hall should be built near Burrard and Nelson, as well as a new hall on the east side in the area served by old No. 5 and old No. 8. Further, a new one-piece hall was needed in the southeastern section of the city and No. 7 should be reopened, also as a one-piece hall.

They recommended that four new pumps be added to the current roster of twenty-one pieces; that equipment be standardized; and that seven or eight vintage pieces be replaced, with three pieces going to a reserve pool from those replaced. Three new city service trucks should be purchased, with one old one to be kept in reserve, and an aerial be put in service at No. 4. And lastly, that two fireboats be purchased for Burrard Inlet because of the inability of land-based fire companies to deal with major fires along the waterfront.

Tugboat fighting Pier D fire. *Photos: VFD Archives*

A water tower was recommended as well as many turret nozzles, deluge sets, basement nozzles, foam generators, and large diameter, 3-inch (75-mm) hose added to downtown pumps; the addition of sixty men to restore the department's strength to the 1933 level and up to 100 men to more efficiently carry out fire department duties. The report stated, "each piece of apparatus should be manned as follows: Pumps, downtown, 7 men; outside districts, 5 men; aerials, 7 men; other trucks, 6 men; hose wagons or hose/chemical combinations, downtown, 6 men; hose wagons, etc, in outside districts, 4 men and the fireboat, 6 men."

The Fire Warden's Branch should also have "a substantial increase" in manpower to make it possible to comply with the requirements of the Provincial Fire Marshal's Act, and the officer in charge of the branch should have the rank of a district chief.

Most items in the report were never done; some firehalls were built and apparatus were replaced over the next twelve to fifteen years, but the manning would go down.

On September 18 the fire department's emergency telephone number was changed from "Fairmont 1234" (FA-1234) to "Fire 1234" (FI-1234), in preparation for the new dial telephone service, which was to begin in December 1939. All public notices on "break glass" fire alarm systems were reprinted to show this change.

P.D. 18 30M. 2-37.

CITY OF VANCOUVER
PUBLIC O NOTICE

In case of FIRE BREAK THE GLASS in this switch, thereby alarming all occupants in the Building, then telephone the Fire Department FIRE 1234, or if a telephone is not accessible, notify the Brigade by STREET FIRE ALARM BOX

WHICH IS LOCATED AT THE N.E.

CORNER OF   JERVIS & BURNABY   ST.

To operate the Street Box, first break the glass, turn the key, open the door, and pull down the lever, then release at once. If the box is of a type having no glass, pull down door by means of the handle, then pull down lever and release immediately. Remain at the box until the arrival of the Fire Apparatus.

DO NOT HESITATE. Delay may mean loss of life. Do not depend on others to give the warning.

A. McDIARMID,
Chief of Fire Department

Building fire alarm switch notice showing the new "FIre 1234" phone number.
Photo: VFD Archives

On November 10, 1938 the VFD put its last wooden, hand-operated aerial in service at No. 1. Called the *Queen Mary* because of its size, it was also to be the last tractor-aerial (requiring a tillerman) on the department until 1966, when No. 6 received the next one. The eighty-five-foot LaFrance Type 512 tractor-aerial was built in Toronto and cost $23,100. The old-time captains didn't like anything to hinder their exit from the rigs, so they arranged to have the cab doors removed allowing them to jump off the rig unimpeded.

The new LaFrance aerial, commonly known as *Queen Mary*.
Photo: VFD Archives

On November 16 the fire department purchased forty-four more Burrell canister air masks. These "self-contained" masks replaced the two, hand-pumped air masks that had been in use for many years. The men wearing these old masks got air through a long length of hose connected from the pump that was located outside the building and operated by a man who pumped air through the hose to the mask, like a deep-sea diver.

Another waterfront fire that could have developed into a disaster was the Vancouver Ice & Cold Storage fire at Gore Avenue and Railway Street, on December 9, 1938.

The alarm to the four-story-high wooden building came at 1:30 p.m. for a fire in the company's old plant, which contained mostly vegetables, cheese and game. The concern, however, was for the stiff westerly wind that was blowing at the time. Through quick action by the crews, the fire was knocked down by 1:55 p.m., much to everyone's relief. The fire loss amounted to $3,500.

The only casualty at the fire was a fireman who drove the pick end of his fire ax into his leg.

Vancouver Ice & Cold Storage fire.
Photo: VFD Archives

VFD's Last Hose Wagon

The last rig received before the war and the last-ever hose wagon used on the department went into service at No. 1 on March 28, 1939. Built in Toronto by LaFrance, it was a Type 512-BO (with booster-hose system), which carried a bank of four seventy-five-pound tanks of carbon dioxide. This hose wagon was purchased to replace the old hose wagon lost at the CPR Pier D fire. It also had its doors removed on the orders of the captain. Thirty years later the mechanics of the day had to reinstall the doors prior to disposal to the new owners, and couldn't understand why they wouldn't fit. Then they discovered that the aerial and hose wagon doors had somehow become switched and they had unknowingly spent hours rebuilding and readjusting each door to make it fit.

Delivery photo of the LaFrance hose wagon bought to replace one lost in Pier D fire. *Photo: VFD Archives*

Same hose wagon with doors removed per the captain's instructions. *Photo: VFD Archives*

During 1939 the Two-Platoon Act and Hours of Work Act were amended to provide for a sixty-hour workweek for all British Columbia firefighters, and Vancouver civic employees received another 2.5 percent restoration of cut salaries with a remaining 5 percent to be negotiated.

Canada Goes to War Again

On September 10, 1939 Canada entered the Second World War. Within a short time, the Fire Auxiliary (FA), made up of dozens of civilians, was formed and weekly drills were held at various firehalls around the city, with the different fire companies. The auxiliary was outfitted with coveralls, regulation turnout coats, which were knee-length at the time, standard service belts, with bucking strap, spanner, wedge and hydrant key and steel military helmets. They were equipped with a fleet of half-ton panel trucks that had the rear doors removed and were rebuilt to carry hose and small equipment. Trailer hitches were installed for pulling two-wheel portable pumps.

A Bickle-Seagrave portable trailer pump. *Photo: VFD Archives*

Fire Auxiliary members in uniform. *Photo: VFD Archives*

A Fire Auxiliary International panel truck with a Bickle-Seagrave portable trailer pump. *Photo: VFD Archives*

Union Boat Works fire with two fire captains on a hose line.
Photo: VFD Archives

Canadian Firefighters Overseas insignia.
Photo: Calgary Fire History

Excelsior Paper fire, 1531 Main Street.
Photo: Vancouver Sun

At first these outsiders weren't accepted by the firefighters, until they proved they could do the job. The FA was disbanded in September 1945.

Three neighboring buildings of the arena destroyed in the "Jubilee Fire" of 1936 were themselves destroyed in a three-alarm blaze that started at 9:45 p.m. on the evening of May 28, 1940. The fire at the Union Boat Works, which suffered half of the total damage, destroyed a completed seine boat valued at nearly $30,000. The Benson Shipyard lost a partly built government survey boat, valued at $20,000, plus tools, equipment and gear. (Fortunately, just two weeks previous, both Union and Benson's launched two new seine-fishing boats worth a total of more than $120,000.)

Lastly, Sumner Brass Foundry Limited lost its shop, patterns, and many finished boat propellers and fittings. The big Burrard Shipyard and Drydock, immediately to the west caught fire, but it was extinguished before any major damage could be done. The fire, fought by more than 200 firefighters was struck out at 11:05 p.m.

The Oath of Allegiance

A memo from the fire chief to all members on June 26, 1940, ordered: "All fire department members who have not taken the Oath of Allegiance must report to the City Clerk not later than July 5, 1940, between 9 A.M. and 5 P.M."

And on the same memo: "During the summer months, firemen will be allowed to wear their dress shirts (light weight blue, cotton, with long sleeves) after work around quarters has been completed, (instead of the dark blue, heavy wool work shirts, with the long sleeves) but with all buttons done up."

The associate minister of defense in Ottawa announced that the Canadian government was seriously considering a plan submitted by Chief O.L. Lister of the University Fire Department, to send a corps of Canadian firefighters to Britain to assist them in combating fires caused by the war. A corps was later formed and sent to Britain as the Canadian Fire Fighters Overseas Corps.

There were two major fires in 1941. The first resulted in the destruction of the administration building of the Boeing Aircraft Company at the airport on Sea Island on June 1, with damage exceeding $80,000.

Then on June 15 the Excelsior Paper Stock Company plant was destroyed. The first alarm came in at 1:45 p.m. The damage was so complete the fire wardens investigating the fire had nothing to go on. The fireboat *J.H. Carlisle* prevented any spread of fire on the False Creek side of this $40,000 fire. During the time of this fire, three others were in progress: a roof fire in the West End, a car fire downtown and another car fire on Lion's Gate Bridge. These three fires delayed the response of assisting fire companies at the larger Excelsior fire, a situation that doesn't happen very often.

Just before the year ended, on December 8, Fire Chief Archie McDiarmid retired and was succeeded by his first assistant, John H. DeGraves, aged fifty-six, who joined the department on May 1, 1906.

Rescue & Safety Branch Formed

Firefighters had been attending to the first aid needs of citizens since 1927 and on January 10, 1942 the Rescue & Safety Branch was formed and went into service at No. 3 Firehall. The unit responded to ninety-six first aid/inhalator alarms in its first year of operation.

On February 9, 1942 the Fire Warden's Branch reported in the press that in 1941 the fire department answered 742 chimney fires in the city at a cost to the taxpayer of some $37,000. It warned that in future, all chimney fires would be investigated, and if the homeowner hadn't had his chimney cleaned within the last year, charges could be laid.

Because of the war, all fire department members were encouraged to save their old metal uniform buttons and turn them in to be re-used because they were now hard to get.

John Henry DeGraves

Born: January 28, 1885, Albury, NSW, Australia
Joined VFD: May 1, 1906
Appointed Chief: December 9, 1941
Retired: January 27, 1945
Died: July 14, 1957

The first squad wagon of the newly formed Rescue & Safety Branch was this 1940 International panel truck.
Photo: VFD Archives: Foster family

Sterling Lumber Company, at the north foot of Victoria Drive was destroyed on June 21, 1942 in a spectacular fire with damage estimated at $100,000. Several large exposures were saved including the Pacific Wheat Pool Elevators, McKenzie Barge and Derrick Company, BC Marine Engineers and Shipbuilders, and the Excelsior Lumber and Shingle Mill. The two-alarm blaze tied up traffic in the area for over an hour including the arriving CPR Transcontinental Train No. 1, which was halted until four hose lines were removed from the tracks.

Sterling Lumber Company fire.
Photo: CVA 1184-306

During 1942 Chief DeGraves promoted Second Assistant Chief Ed Erratt to First Assistant and DC Cy Ruddock to Second Assistant Chief.

By August 1942 special war-time codes were devised using the telegraph keys in the FAO, firehall watch rooms, and fire alarm street boxes; and were to be used in the event that telephones were knocked out of service, or if it was necessary to transmit messages of a secret nature. The number codes covered everything from an Air Raid Warning (1-6), Change Quarters (3-1), Ambulance Required (10-3) to Today Is Payday (1-8), and every senior officer had a signature code (Fire Chief, 1-2-1). Another wartime addition was the placing of two anti-aircraft guns on the roof of the fire alarm office.

Fall Kills Fire Warden Lou Taylor

At 5:20 p.m. Tuesday, October 20, 1942, while carrying out an inspection at the Canadian National Dock, Fire Warden Louis B. Taylor, age fifty-four, slipped over a railing and fell forty feet (13 m), struck his head on a guardrail then landed face down in the water. The CN employee he was with pulled him from the water and the inhalator crew worked on him, but he could not be saved. Lou joined the South Vancouver Fire Department in 1922, but injured his back at a fire in 1925 and transferred to the Warden's Branch on amalgamation.

The Fire Warden's Branch banned smoking in all Vancouver theatres on February 3, 1943, and on March 12 required that anyone burning garden debris had to have a burning permit. These permits could be obtained at each of the district chief's halls. Permits were not allowed for beach fires or burning of commercial debris, which was to be burned in the city incinerator situated next to the Cambie Bridge.

At the Capilano Shingle Mill fire on April 15, the ladies of the Red Cross brought out the Air Raid Precautions (ARP) Mobile Canteen Wagon for the first time to serve coffee and light snacks to firefighters at the fire scene. This association between the VFD and the Red Cross ladies continued for more than forty years. Fire Chief DeGraves is shown taking a time-out for a coffee.

The Army & Navy Department Store Fire

At 6:00 a.m. on August 28, 1943 the first alarm was raised for a fire on the top floor of the five-story Army & Navy Store on Hastings Street. Nos. 1 and 2 Companies responded, laid lines and laddered the building to get at the fire. Shortly after, the flames were through the roof of the top floor. Fireman Jack Liddell was working a hose line on No. 2's aerial ladder, more than halfway up, when he and his partner got a sudden blast of smoke and heat. Overcome, they tried to get down the aerial, when Liddell slipped and fell forty feet to the ground, suffering a fractured leg and chest injuries. His partner scrambled down the aerial, helped to treat him and sent him off to the hospital.

A few minutes later, just after the injured man was taken away and the aerial was moved to a safer location, part of the top story wall fell to the sidewalk, narrowly missing three firemen on hose lines. Because of the damage it had suffered when the wall fell, the big neon sign now left dangling in the front of the store had to be pulled down for the safety of the crews. Damage totaled about $250,000 and, with the war on, it would be difficult to replace the damaged goods.

Liddell, who joined the PGFD on June 1, 1924, came on the VFD with amalgamation in 1929. Later, he became the department's training officer.

Chief DeGraves at the canteen wagon at a multiple alarm fire.
Photo: Steffens Colmer #37456

The Army & Navy Department Store fire scene showing No. 2's aerial, the damage to the fifth floor, and the neon sign that had to be pulled down.
Photo: Vancouver Sun/VPL Photo #84833

G.L. Pop, Fur Store Fire

The headline in the *Vancouver Sun* read, "21 City Firemen Gassed Battling Fur Store Blaze." The early morning fire on September 4, 1943 destroyed several thousand dollars' worth of fur coats and game trophies at the G.L. Pop Furriers store, 2152 Main Street, in one of the "smokiest fires" in memory. While the men were battling the blaze, they were unable to cope with the fumes and began staggering and collapsing from the heavy concentration of carbon monoxide and ammonia fumes from the burning furs and refrigeration plant.

Over fifty men were affected, with twenty-one being hospitalized. They were initially treated on the street then sent to nearby No. 3 Hall where the city doctor assessed them further and all serious cases were sent to Vancouver General Hospital. The owner, Mr. Pop, was also treated, along with the firefighters.

Tenants in the apartments upstairs began smelling smoke about 9:00 p.m. but the alarm wasn't turned in until 1:17 a.m., by which time the fire had a good hold. Over the next couple of hours several of the apartments suffered fire damage as well. More than 230 firemen fought the three-alarm blaze before it was brought under control. The cause of the fire was never determined.

Also that year Fire Chief DeGraves launched an appeal through the city council to seek a deferment from active duty for city firefighters with more than one year's experience instead of three years, because one out of five men on the VFD had responded to the wartime call-up. The efficiency of the department was being compromised.

November 4, 1943 Ottawa announced that Vancouver was to have three new fireboats added to her Air Raid Precautions (ARP) fire fighting equipment. The fifty-three-foot (16-m) commando landing barges were the same as those that landed at Dieppe and Sicily. They would have a pumping capacity of about 3,000 gpm (13,500 liters). The *Vancouver Sun* had been waging a campaign for a firefighting fleet since the *Pluvius* was condemned February 28, 1937. The first of two barges was turned over to the city on September 21, 1944. The third barge went to North Vancouver, December 4, 1944. The city alderman who chaired the council's special fireboat committee said, "We realize they will not measure up as a permanent means of fire protection, but we feel that they will give some relief."

G.L. Pop fire.
Photo: Vancouver Sun/VFD Archives

Fire barge at No. 16 Fireboat Station. *Photo: CVA 354-147*

Crash Kills Fireman Wootton

"City Fireman Killed In False Alarm Crash," read the *Vancouver Daily Province* on Saturday, November 13, 1943. The day before at 8:30 p.m., while responding to Box 3719, at 6th Avenue and Blenheim Street, No. 19 Pump was in collision at the corner of 8th Avenue and Alma, with a police car that was also responding to the same location.

The police car struck the fire rig's left side sending it careening down the sidewalk, where Fireman William Wootton, age twenty-seven, was thrown off into a rock garden, striking his head. He was rushed to Vancouver General Hospital, where he died of his injuries. He left a wife and two young children. Wootton had been on the fire department for two years and was a relief fireboat pilot. He had been transferred to No. 19 Hall from No. 18, for the night shift.

CITY OF VANCOUVER
REWARD

The sum of **$500.00** will be paid by the City of Vancouver for information leading to the discovery, apprehension and conviction of the person or persons, who, on November 12, 1943, turned in the false fire alarm which resulted in the death of Fireman W. D. Wootton, and injuries to members of the Police and Fire Departments.

A standing reward of **$100.00** is offered for information leading to the discovery, apprehension and conviction of any person who subsequently to this date, wilfully turns in a false fire alarm in the City of Vancouver.

DATED the 30th day of November, 1943.

FRED HOWLETT, City Clerk.

1943 notice of reward regarding false alarm resulting in death of firefighter Wootten.

No. 19 Pump at the accident scene.
Photo: CVA 1184-642

Fire department records showed that six false alarms to that box had been received in the previous week. The fire chief suggested that in future, anyone caught pulling a box would be prosecuted under the criminal code instead of the lenient city bylaw, and the firemen's union urged that a standing reward of $100 be posted for the apprehension of those who pull boxes. The city posted a $500 reward for the conviction of the person responsible for Wootton's death, but no one was ever charged.

The last of the wartime deliveries of fire apparatus was the delivery of the 1943 Bickle-Seagrave Underwriter 625 gpm pump, which went into service at No. 2 Hall on February 10, 1944.

During 1944, two-way radios began to be installed in chief's cars and Rescue & Safety rigs. By 1948 pumps would begin to get radios, followed by trucks, and by 1949 all VFD vehicles were equipped.

Chief DeGraves accepting delivery of the new Bickle-Seagrave Pump.
Photo: Steffens-Colmer Ltd. #49016-2/VFD Archives

Stoker George Dean. *Photo: VFD Archives*

Twelve members of the VFD who served in the Canadian Fire Fighters Overseas Corps returned home to be discharged, and returned to duty on the VFD. Their leave-of-absence gave them the option of returning to duty if they wanted, although some opted out.

When they paraded in Britain for the last time, the Canadian High Commissioner to Britain, Vincent Massey, made the official farewell speech. They were thanked for their dedication to duty and praised for "a grand job, well done," and told, "No job was too hot, too smoky or too dirty for you; you tackled them all with true firemen's courage and enthusiasm." In total, 400 men from across Canada served in the badly bombed cities of Plymouth, Bristol, Southampton and others. (Vincent Massey later became the first Canadian-born Governor General of Canada, serving from 1952 to 1959.)

In August 1944 the last of the old steam-era engineers, George Dean, retired. He joined the VFD, August 2, 1910, as a stoker, an engineers assistant.

The Year 1945

Chief John DeGraves retired on January 28, 1945, and First Assistant Chief Edward L. Erratt, age fifty-seven, was named to succeed him on January 18.

Edward Landon Erratt

Born: May 17, 1887, Rockville, Ontario
Joined VFD: August 9, 1911
Appointed Chief: January 28, 1945
Retired: May 16, 1947
Died: November 6, 1970

Hugh Bird, president of Vancouver Fire Fighters' Local No. 1 urged a new pay deal for the city's firemen that averaged around 12 percent. He cited wage comparisons and stated that the firemen's sixty-hour workweek was a glaring exception to British Columbia's general eight-hour day and forty-eight-hour week. A district chief, for example, with thirty or more years service, made twelve and a half cents an hour less than a shipyard riveter and thirty five cents less than a plumber; a high school boy working as a whistle punk in a logging camp made five cents more an hour than a twenty-year lieutenant and it took up to five years to become a skilled fireman. Comparing Vancouver to other Canadian cities, Bird further stated that Vancouver firemen to district chiefs made from $506 to $1,020 a year less than their counterparts in Toronto and Montreal. Bird also said that, "Victoria firemen make more than Vancouver."

Negotiations continued into the spring, and on May 8 firemen ratified the city's wage offer that would give a first-class fireman a raise from $176.50 per month to $185, about $150 more per year than was estimated. City police at this time were making $190 per month.

The Capilano Stadium Fire

Cap Stadium, formerly Athletic Park, at West 5th Avenue and Hemlock Street, was built in 1913 and was one of Vancouver's major sports grounds for baseball, softball and soccer with basketball in the adjacent gymnasium. Emil Sick of Seattle, owner of Sick's Capilano Brewery, had purchased it in 1944 and the name was changed to Capilano Stadium.

On February 28, 1945 the first alarm was turned in at 6:07 p.m. and by 6:15 it was a third-alarm with men and apparatus responding from seven firehalls, including the fireboat, *J.H. Carlisle.* Many of the fire-fighters who responded were ballplayers who had played numerous games over the years in that facility, one of whom was Fireman Charlie Miron, a former ace outfielder and home run hitter, and father of Captain Alan Miron, who retired from the VFD in 1998.

The fire started near the main entrance at the northwest corner of the park and adjacent to the gym next door, where the smoke was first discovered in the change rooms and showers. Kids playing basketball in the facility ran to a nearby house to turn in the alarm. The fire moved so rapidly that by the time the fire department arrived on scene, the old wooden structure was doomed.

Capilano Stadium destroyed.
Photo: Vancouver Sun: Art Jones

Described as one of the best spectator fires since the Pier D fire in 1938, it attracted thousands of homeward-bound workers. Many stated that if the fire had started a few hours later it might have taken some lives; less than two hours after the fire started, a league basketball game was due to begin and likely the gym would have been packed. The only casualty of the fire was the building manager's wife who tripped over a hose line, injuring her ankle. Cause of the fire was never determined because of the total destruction.

The SS *Greenhill Park* Disaster

At noon on March 6, 1945 there was such a violent explosion that people thought that the city was under attack in an air raid. The No. 3 hold in the 10,000-ton SS *Greenhill Park*, tied up at CPR Pier B had blown, rocking the downtown area of the city, killing eight, injuring scores of people and causing widespread damage. The blast was so powerful that it blew the bridge of the ship 100 feet (30 m) into the air, dropping it as a shapeless mass into the water. A second blast blew out of No. 2 hold followed by a

Greenhill Park at Pier B following the explosion.
Photo: VPL 45869

Greenhill Park aground at Siwash Rock. *Photo: VFD Archives*

third, which was between Nos. 2 and 3, stopping short of No. 1 hold, which was full of explosives.

Within minutes of the first explosion (at least three others were heard, as many as ten were reported), every emergency agency and military installation was represented on the scene with the fire department rescuing and tending to the injured and getting water on the subsequent fires.

Before the blast, five gangs of longshoremen were working on the ship, then it blew. Men and debris were blown high into the air by the explosion, which caused damage for many blocks around. Tracer bullets and flares filled the air and small arms ammunition continued to explode at intervals.

Men escaped any way they could; diving and jumping over the side with some seen climbing and sliding down hawsers and cargo nets strung to the pier. Sailors on board, ignoring the dangers, ran to quarters to get their gear to help the others. A couple of them were seen swimming back to the ship after being blown overboard.

With debris flying everywhere, many different types of boats came into the area to help. The scene was likened to London after a bomb attack.

Many people on the street were cut by flying and falling glass. Some of it was blown out of office building windows as far as eight blocks away and along Georgia Street west of Burrard, then known as Automobile Row, where the large plate glass windows of some of the dealerships were blown out. Windows also blew out of the Post Office, Spencer's Department Store (site of today's Harbour Centre), the top floors of the Marine Building, the CPR Station and many small stores. One story was told about a large piece of glass that took the arm off a wooden chair at the Ration Office on Hastings Street, two blocks away, where one of the employees usually sat. Miraculously, most of the glass injuries were relatively minor.

Fireboat *J. H. Carlisle* and one of the fire barges fighting the *Greenhill Park* fire. *Photo: Courtesy Vancouver Maritime Museum*

When the ship was towed out of the harbor around 1:00 p.m., still on fire, it was met by *J.H. Carlisle* as it passed under the Lion's Gate Bridge. It was taken through the narrows and run aground at Siwash Rock in Stanley Park where the fire fighting continued with the fireboat and North Vancouver's fire barge. (Vancouver's fire barges weren't in service yet.)

Two days later, with the fire now knocked down, a team of firemen was overcome by fumes while searching for bodies. Once the fumes were cleared out, two more bodies were found bringing the total dead to eight.

In January 1980 Chuck Davis, local author, columnist and historian, received a letter from an old man who related the true story of the cause of the *Greenhill Park* disaster. In 1957, while in hospital, the old man related that his roommate, a man he called "Joe," told him a story in confidence about how some of the longshoremen that day, were siphoning liquor from a barrel in the ship's forward hold. After several had drawn off a drink or filled a container there was some spillage and the last man lit a match to see what he was doing and, with the fumes in the hold, the ship blew up. The man was killed instantly and the other man in the hold at the time was "Joe," who was blown out of the ship and across the dock. As he was only slightly injured, he brushed himself off and ran home. He never told anyone the story until he told it to the old man in the hospital in 1957. "Joe kept the story to himself all those years and must have felt his days were numbered," related the old man, "because a week later Joe was dead."

In June 1945 the firemen complained about the proposed manning and accommodations for the fire barges. The crew was to consist of a pilot and two men, and without a firehall the men would have to use the barge for quarters. The city said that there was no money available to build a firehall. After reviewing the firemen's complaint, the city agreed to build a hall and decided that each shift would consist of a fire captain, a pilot and crew of two men. The second barge would remain unmanned, but cared for and serviced by the barge crews for use as required. If a large fire occurred, the off-shift fire barge crew would be called in to man the second barge.

They went into service January 16, 1946 as No. 10 Firehall in leased space in the Immigration Building on the Vancouver waterfront.

During July 1945 firefighters had their hands full with four, three-alarm fires in two weeks. The first fire occurred July 16 at Mainland Transfer, south foot of Smithe Street, when a fire destroyed their warehouse and garage containing about fifty trucks. The first alarm came in at 11:28 p.m. and quickly became a third alarm and required calling out the off-duty men. Luckily, once again, there was little wind, as the fire was near the Cambie Bridge and the Imperial Oil Company facility where thousands of gallons of gasoline were stored. The only casualty was a horse that was stabled in the garage.

The Mainland Transfer fire showing the fully involved building.
Photo: news photographer unknown/VPL
Photo #84839A

The Moberly School Fire

On July 22 the worst school fire in the city's history occurred at Walter Moberly Elementary at East 59th Avenue and Ross Street, in South Vancouver. All that remained of the school was a brick shell when the roof fell into the building.

Moberly Blaze Toll Estimated $130,000

By RALPH DALY

Fire wardens and police are investigating the possibility of arson in a three-alarm $130,000 fire which almost completely destroyed Walter Moberly grade school, 59th Ave. and Ross St., early Sunday.

Raging for nearly two hours before being brought under control by firemen from six halls, the blaze was the worst school fire in city history.

Even while firemen battled the flames from a dozen ladders on all sides of the brick building, police uncovered five silver sport trophies, battered flat with rocks, 150 yards south of the building.

Trophies were stolen and similarly destroyed when an attempt was made to burn Mackenzie school, E. 39th and

Marauders May Have Used Matches

Windsor, four months ago.

Though no direct evidence of arson was uncovered by fire wardens probing the blackened shell of the building Sunday, arson was a "possibility," Chief Fire Warden Harvey G. Bowering said Sunday night.

QUIZZ YOUTHS

"There is no doubt the school was broken into, but whether the fire was set deliberately or accidentally has not been determined," Chief Bowering said. "They might have been using matches to see and dropped some."

Several youthful suspects were questioned Sunday, but no action taken, he revealed.

Fire officials estimated that the building and contents were about 90 per cent destroyed. The building is insured for its replacement cost and furniture at cost less depreciation.

The blaze originated in the heart of the three-storey building, on either the main or top floor, and broke through the roof around a central bell tower.

ROOF CRASHES

Burning furiously, it spread over the entire tar and gravel roof, which collapsed about 30 minutes after firemen arrived.

The Vancouver N

VOL. 13, NO. 76   VANCOUVER, B.C., MONDAY, JULY

FIRE-RENT SCHOOL: The bell tower that loomed behind burned completely away, along with the roof, the outline of the main peak of Moberly School, 59th and Ross, remained only on the brick front wall Sunday. The peak is over the main entrance on 59th. The $130,000 blaze, largest school fire in city history, originated somewhere back of the peak.

News clipping of Moberly School fire aftermath. *Photo: Vancouver News Herald*

The first alarm was turned in from Box 8544 at 4:23 a.m. with the second shortly after the arrival of No. 17 Company, then to a third at 4:47 a.m. by Chief Erratt, which also brought the off-shift out. It was brought under control at 6:13 a.m.

There was an earlier fire at Sir Alexander Mackenzie School at East 39th Avenue and Windsor Street in February. Both school fires were similar in that the schools were broken into and sports trophies had been taken outside and destroyed. The school fires were definitely arson and a number of boys were questioned, but because of a lack of evidence no charges were laid.

The Terminal City Ironworks Fire

The next three-alarm fire was one with fire department connections; it was in the only foundry in British Columbia manufacturing fire hydrants and secondly, part of the full-block complex contained the old No. 8 Firehall.

The first alarm to Terminal City Ironworks, Victoria and Pandora Street, came in at 5:47 p.m. on July 24, and when the fire quickly spread throughout the foundry, it soon became a third alarm, using Box 2318. By 6:20 the building began to collapse and crews were quickly removed from the front of the building where they had ladders up. The old firehall and several houses exposed to the fire, as well as a dairy's stable across the street, were saved. (Old No.8 is still part of the foundry in 2006). The biggest loss, said the president of the foundry, were the many wooden patterns collected over the years, since 1906, when the foundry began operations. Ten firemen were injured and the estimated loss was about $100,000.

Terminal City Ironworks July 24 1945. Fireman Frank Bain & unfinished fire hydrant.
Photo: VFD Archives

Fire at Happyland

The fourth, three-alarm of the month occurred at 6:55 p.m. July 31 when the Giant Dipper rollercoaster at Happyland amusement park at Renfrew and Cambridge Streets, at Exhibition Park, mysteriously caught fire. On Chief Erratt's arrival he immediately made it a third alarm, because of the extremely dry, wood construction in the area. The first-in company, No. 14, confined the fire to the Dipper, a shed, and a warehouse, and it would have had far-reaching consequences if it had gotten away. Most of the old, wooden buildings on the exhibition grounds were at risk.

It appeared that the fire had started in the oil room near the Dipper's loading platform, and in thirty minutes the fire was out. One fireman suffered burns waiting for the line to be charged when flames shot about twenty feet (6 m) from the platform, burning his hands. Total damage was estimated at about $5,000 to part of the framework and the two trains, which would be rebuilt by the staff. In the evening, as the crowds began to arrive, the firemen were still hosing the site down, but the park was open for business.

Also on July 31, No. 12 Truck collided with a wood truck at West 6th Avenue and Balaclava Street, seriously injuring one of the men. He suffered a fractured leg, lacerations and acid burns to his hands, face and body from the rig's damaged soda/acid extinguisher. The other two members escaped injury and the driver of the wood truck only sustained a bruised leg.

It began to look like August was going to be a repeat of July. On August 7, a spectacular fire struck two businesses and a private dwelling at West Third and Pine Street. The fire that burned from roof to roof, was quickly brought under control and damage to the cottage, trucking company, and construction company amounted to less than $20,000.

Then the next evening at 7:10 p.m. in Coal Harbour, at the Chappell Brothers Shipyard, 1779 West Georgia St., a fire was reported that could have been disastrous, as a short time before an American Army transport barge with 1,800 empty gasoline drums on its deck was launched at the Crane's Shipyards Ltd. next door.

Minutes later, a sheet of flame shot out of the machine shop at Chappell's and struck the ship's ways at Crane's, igniting everything. The first alarm brought all of the downtown halls, Nos. 1, 2 and 6, and the first-in

Total destruction of Chappell Brothers Shipyard.
Photo: Vancouver Sun

121

Fire fighters on aerial ladder fighting McMaster Building fire.
Photo: Vancouver Sun

chief immediately made it a third alarm. With a lot of luck, in spite of the oil, gas and tar barrels burning furiously, not to mention the buildings, by 7:53 p.m. the fire was brought under control and, a short time later, struck out. Chappell's was completely destroyed, including a small yacht still on their ways, and Crane's was badly damaged.

The McMaster Building Fire

At 6:38 a.m. on the morning of September 14, an alarm came in from Box 1421, Homer and Helmcken Streets, for a fire in the McMaster Building, 1176 Homer Street. The fire was in the premises of an envelope and card printing company, Pioneer Envelope Limited. By 6:50 a.m. a second alarm was put in, then it was made a third alarm at 7:00 a.m.

The fire was spreading and the crews were warned to stay out of the building and not to go in beyond the windowsills. At 8:30, a backdraft blew out the rear doors and windows with smoke and flames coming out of every opening with a roar. The walls seemed to tremble and then, with a thunderlike clap, the second and third floors of the five-story building collapsed, carrying with them tons of heavy machinery, paper and printing materials.

Fifteen firemen were injured, being thrown off ladders or struck by flying and falling debris. Others suffered from smoke and dust inhalation, heat and stress. One near-miss was related by a Kingsway Ambulance attendant who happened to be watching Vancouver's wrestling fireman, Jack Forsgren, who was working on a hose line from a ladder at a window

Vancouver Sun

VANCOUVER, BRITISH COLUMBIA, FRIDAY, SEPTEMBER 14, 1945 ★★★ PRICE 5 CENTS BY CARRIER $1.00 Per Month

3 Firemen Killed, 15 Hurt In Homer Street Blaze
Men Trapped As Floors Collapse

when the collapse occurred. He described the building falling "within a cigarette paper thickness" of hitting Forsgren, and was certain he would be a victim. Out of the smoke and dust walked Forsgren, uninjured, who coolly said, "That was close!"

Then it was learned that some members had been buried under the rubble. Fireman Reg Hill, age forty-four, of No. 2 Hall and Lieutenant Jim Hunt, fifty, of No. 13 Hall, were victims of the falling floors and suc-

Firefighters recovering body of comrade lost in McMaster Building fire.
Photo: Vancouver Province

cumbed to their injuries. The third victim was Captain Bill Barnett, fifty, of No. 2, who collapsed and died of smoke inhalation and stress on the roof of the building next door while pulling a hose line. Another firefighter, buried up to his armpits, was rescued.

The three-alarm fire created a heart-breaking battle for the shocked and weary firemen who tried to rescue their brother firefighters. Lieutenant Hunt was recovered at 12:20 p.m. and Reg Hill was brought out about 1:30 p.m. after a crane was brought in to remove the heavy machinery from the scene. The Rescue & Safety crew worked on Captain Barnett but were unsuccessful in reviving him; he died in St. Paul's Hospital.

Funeral of three fire fighters killed in McMaster Building fire.
Photo: VFD Archives

The fire was struck out at 1:55 p.m.

"I won't be home for breakfast," was the prophetic farewell Bill Barnett gave his wife when he left for work the night before. He and several of his crew were going to help in the search of a missing little girl, who was later found murdered.

Lieutenant Jim Hunt's son-in-law kept watch at the scene and when Hunt's body was recovered, he made the telephone call to his home to inform his wife and mother-in-law.

Many firemen, including Fire Chief Erratt, continued digging for their brother firefighter, Reg Hill. Hill's brother, RCAF Sgt. Roy Hill, stood nearby, keeping a lonely vigil until Reg's body was found. Roy joined the VFD the following year, on August 1, 1946, and served until retirement in 1975.

Two days later the cause of the fire was still unknown, and four of the nine firemen who were hospitalized were still there but reported to be in good condition.

On Wednesday, September 19 a civic funeral was held at Christ Church Cathedral in downtown Vancouver, and was attended by hundreds of citizens. The service was conducted by Reverend Dean Cecil Swanson, assisted by Lt. Colonel the Reverend C.C. Owen, unofficial VFD padre; Reverend H.J. Greig, St. Philip's Anglican; and Reverend Lorne McKay, Vancouver Heights Presbyterian.

The three coffins, each covered with a Union Jack, white ribbons with each man's name and surrounded by a mass of floral tributes, were borne to the cemetery on the 1941 LaFrance city service truck. The procession traveled east along Georgia Street, across the Georgia viaduct, then to the Firemen's Plot in Mountain View Cemetery.

The following, titled "The Last Alarm," was penned by A. McNeill, of Vancouver.

> For them it was the last, God's will be done:
> In death they did not see the battle won:
> They did not know, nor did they fear
> The end, its presence ever near.
> Often had they heard the Siren's mournful wail,
> Fight fire they must, they dared not fail;
> The dreaded flame their glory and desire,
> As it reaches upwards, ever higher.
> Hose on hydrants fastly filled,
> Tons of water soon are spilled,
> The pumps in rhythm help to quell
> The seething flame they know so well.
> Their last alarm, the fire struck out;
> Our hearts in sympathy beyond a doubt,
> We mourn with those they leave behind;
> For them we pray that God be kind.
> By yonder road where silence reigns supreme,
> These brave men lie, their face unseen;
> At rest in peace, until the dawn,
> Always in memory, not just gone.

A veteran who had served on the VFD from 1927 to 1930 was thinking that he should re-join the fire department and resume a career begun several years earlier. He telephoned the administration office at No. 2 Hall, the headquarters, and asked, "Are you hiring today?" The terse reply was, "You sure aren't wasting any time, are you," and the phone was hung up. The veteran thought this was strange, but didn't call back. Later in the day, he learned of the deaths of the three firefighters that morning in the McMaster Building fire and now understood the response he had received when he called the VFD office. He never rejoined the fire department.

After an extensive investigation, the cause of the fire could not be determined.

In early November, the first, hydraulic, all-metal 100-foot (30-m) aerial ladder on the VFD was received and demonstrated for interested spectators at the City Hall. It was a 1945 LaFrance Type JOX-M5-100 service aerial and it was placed in service November 11 at No. 2 Firehall.

With the war over and the prospect of better times ahead, it was hoped that much of the old fire apparatus — some of it over thirty years old — would be replaced with new, modern pieces.

The VFD Annual Report for 1945 showed that the fire department had a record 3,954 calls, 550 more than in 1944. There had been sixteen major fires, including five two-alarm and eleven three-alarm fires with the deaths of ten citizens.

Delivery photo of the 1945 LaFrance 100-foot service aerial. *Photo: VFD Archives*

Post-War Progress

(1946–1949)

News clipping of Broadway Hotel fire.
Photo: Vancouver Sun

On April 17, 1946 a potentially tragic hotel fire occurred. The first alarm came in at 9:00 a.m. for a fire on the top floor of the Broadway Hotel at Columbia and Hastings Streets. Forty-nine guests, many still in nightclothes, escaped relatively unharmed from the smoky fire. Sixteen of them on the top floor ran to the elevator, which fortunately was put out of order by the fire, forcing them to escape down the stairway.

Two chambermaids discovered the fire on top of the elevator shaft as they prepared to begin their daily chores. They alerted the guests and the front desk, and the alarm was turned in.

Cause of the fire was unknown, but it appeared to have started on a stairway leading to the roof, where it spread to the top of the elevator shaft. Damage was estimated to be about $15,000.

A Plaque to Fallen Fire Fighters

On Wednesday, May 8 a simple and solemn ceremony was held at No. 2 Firehall to dedicate a bronze plaque to the sixteen Vancouver firefighters who had lost their lives in the line of duty. Retired Assistant Chief William Plumsteel, with the invocation and blessing by Reverend Dean Cecil Swanson, unveiled the plaque.

Lieutenant Art Wilkins Dies

Fire destroyed two houses and damaged two others in the 2200 block East 24th Avenue, May 19 and claimed the life of a veteran fireman. Lieutenant Arthur Wilkins, age fifty-nine, collapsed due to smoke inhalation and exhaustion and died at the scene. The fire started in the basement suite of one of the houses, quickly spread throughout the house and then extended to the one next door.

Lieutenant Wilkins joined the department in 1918 after military service in the Great War. He was an original member of the VFD band and a goalkeeper on the firemen's soccer team. His wife, daughter and four sons, one of whom joined the department in 1944, survived him.

As a result of Wilkins death, and based upon the recommendations of the coroner's inquest, the Vancouver Fire Fighters Union set up a committee to look into firefighting conditions in the city. The jury asked for an increase in the number of men on fire crews, updating of firefighting equipment, and periodical medical examinations. The committee, made up of VFD members, represented The Firemen's Benefit Association (FBA), the British Columbia Professional Fire Fighters Association (BCPFFA), and the City Fire Fighter's Union, Local No. 1. Mayor Cornett warned the union that sizes of crews and equipment was the sole responsibility of the city council and didn't want them to do a survey. "We have a perfect right to conduct our own survey...and no outside body has the right to interfere with the activities of this union," said the union president, Hugh Bird.

It was recommended that three-man companies be increased to five and four-man to six, at all times. Further testimony revealed to the public that many of the old, out-dated fire apparatus had to be cranked to start them, and it was felt that this physical exertion along with stress at the fire scene contributed to Wilkins' death. He had cranked No. 20's old 1912 American-LaFrance hose wagon/chemical that day because it had

10   THE VANCOUVER SUN: Thursday, May 9, 1946

IN MEMORY OF CITY FIREMEN who have lost their lives in the line of duty, the fire department Wednesday unveiled a bronze plaque in Nnumber 2 firehall inscribed with the names of the 16 fire heroes. Retired assistant fire chief I. W. Plumsteel here unveils the plaque during the simple ceremony attended by members of the force, relatives and friends. On the left is Dean Cecil Swanson, who offered the invocation and blessed the memorial.

PLAQUE TO HEROIC FIREMEN STIRS SON OF ONE TO TEARS
Gordon Hill Weeps, Unashamed, As Dad Honored With Other Dead

News photo of plaque dedication ceremony.
Photo: Vancouver Sun

Lieutenant Arthur Wilkins is pictured second from left in front of the hose wagon he helped to start. *Photo: VFD Archives*

been difficult to start, and then he helped the crew drag hose lines at the fire scene. Over the next five years, nine of the old, obsolete pieces of fire apparatus would be replaced.

The 48-Hour Work Week Begins

Effective August 1 the provincial government legislated the reduction of all paid firefighters' hours of work from sixty hours per week to forty-eight hours per week, and this required the city to increase the fire department strength by about 25 percent, or 132 men. The two-platoon system would remain, with two shifts working an average of forty-eight hours per week with a twenty-four-hour Saturday shift. The proclamation was not an option and many municipalities, Vancouver included, had made no provisions for additional firemen, which resulted in budget deficits. Cost of the increase in staff would be about $165,000 and increased the number of district chiefs by three, to twelve, added one more to the office staff, and raised the department strength to 537 men. The fire department entered a new era with these new men, called the "Forty-Sixers," and many of them would contribute to the future success of the department.

On October 8, 1946 the downtown fire-halls rushed to the Courthouse for a reported container fire that the companies found to be out on arrival. The fire had been in a spittoon!

On November 3 the old gabled, weather-blackened Spratt's Oilery cottage, the oldest building in the city and a survivor of the Great Fire, was destroyed by fire. The alarm came in at 9:45 p.m. and the building was pretty well gone by the time the first companies arrived. The cause was believed to have been faulty wiring.

Another school fire of note was at Vancouver College on December 6. Seventy-nine resident students of the Catholic school were forced into the street shortly after midnight when the fire broke out, and fast action by the fire department was credited with saving the building, even though the damage to a three-story wing amounted to more than $200,000. No. 18 Firehall was directly across the street from the school and within minutes of the call the crew had several hose lines on the fire. No. 18's district chief immediately made it a third alarm, which brought help from six other fire companies.

The Vancouver College fire.
Photo: Vancouver Sun:
Art Jones/VPL #84835

The old, outdated rigs of the VFD put firefighters in danger every day. While responding to a fire along Commercial Drive on December 9, one of the front wheel spindles on No. 9's old 1913 truck broke off and with all the weight now on the remaining front wheel, that spindle also broke, sending both wheels rolling down the street ahead of the now-stationary rig. Luckily, no one was injured. Repaired, the old truck remained in service until March 1955.

Fire Equipment Unsafe, Says Union Leader

Another serious breakdown involving a Vancouver Fire Department truck may occur at any time because the city fire committee more than a year ago ignored a warning from the firefighters' union to order new equipment, Hugh Bird, union business agent, said today.

Mr. Bird said he has been informed that deliveries on new equipment, if placed now, will not be made for at least three years. Two new pumpers have been ordered by the city, he added, but there is no sign of delivery.

Commenting on the breakdown of a fire truck in the 800 block Commercial on Monday morning, the union leader said: "It should be a well-known fact that the equipment here has gradually deteriorated to the point beyond the safety measure.

"Most of the rigs are 34 years old. The steel in them just crystalizes. You're liable to have an accident at any time with that kind of equipment."

Mr. Bird said the union made the recommendations for new equipment at the time of the Mc-Master Building fire in September, 1945.

LIVES OF VANCOUVER FIRE-FIGHTERS were endangered Monday when the axel of this 34-year-old truck snapped and it skidded 60 feet after losing both its wheels in the 800 block Commercial. Three firemen escaped injury.

Old rigs were endangering firefighters' lives as shown in this *Vancouver Sun* news photo.

"Jack of all Trades, Fireman Betts, to Retire," read the *Vancouver Sun* on February 24, 1947. It described the "smoke-eater" as a violinist, watchmaker, ironworker, farmer, knife grinder, wrestler, acrobat and mathematician. He used a team of oxen when he came west from his native Prince Edward Island in 1904 and farmed on Vancouver Island. He also drove the last rivet into the Connaught Bridge on its completion in 1912 and knew exactly where it was because he put a washer on it. Soon after, on June 13, 1912, he joined the VFD. In 1922, he transferred into the machine shop as a mechanic, from where he retired on his 60th birthday in 1947 as chief mechanic.

Collision of BC Electric Interurban Train and two streetcars, pinned the auto beneath them.
Photo: Vancouver News Herald

One of the most spectacular accidents in the city's history also happened on February 24. Shortly after 10:00 p.m. in front of the BC Electric Interurban Train Depot at Carrall and Hastings Streets, a runaway two-car interurban train left the station and headed for the street, where it struck two passing streetcars and an automobile, pinning the auto between the lead interurban tram and a streetcar.

The VFD responded to rescue the four people who were trapped in the auto and, on arrival, it was feared that their attempts would be in vain. Working with axes, hacksaws, an acetylene torch and Lieutenant (Peggy) Duff's penknife, No. 1 Hose Wagon crew and the Rescue & Safety crew from No. 3 worked for over an hour to free the trapped victims — a man and three women. Despite being dragged 200 feet (60 m) and sandwiched between street railway equipment weighing more than fifty tons (50.8 tonnes) there were no injuries.

Four trapped people miraculously survived this accident.
Photo: Vancouver News Herald

Ambrose Elmer Condon

A.E. Condon.

Born: December 7, 1890,
Trenton, Nova Scotia
Joined VFD: August 29, 1911
Appointed Chief: May 17, 1947
Retired: October 31, 1949
Died: December 3, 1952

Over the years, Peggy Duff, who joined the department in 1925, was involved in many rescues and first aid incidents with the R&S. He retired as the Chief Rescue & Safety Officer in 1962.

A.E. Condon Succeeds Chief Erratt

It was announced, after much debate by city council, that District Chief Ambrose E. (Con) Condon, age fifty-six, would succeed retiring Fire Chief Ed Erratt, effective May 17. First Assistant Chief Cy Ruddock, who some felt should get the chief's job, was due to retire in November and the Fire, Police and Traffic Committee recommended that the successful candidate should have at least three years as chief.

Alderman Miller wanted to make the job open to every man, but the union pressed for the promotion of the senior men, as any other terms penalized good and dedicated men who had grown with the service.

IAFF Returns Local 18 Charter

On May 1 the IAFF reinstated Local 18 and, a formal announcement and ceremony was held on May 22 to return the Vancouver Fire Fighter's Union to the International Association of Fire Fighters, from which it was expelled in 1936 for taking a strike vote contrary to the IAFF constitution. George J. Richardson, secretary-treasurer of the International, who officiated in the reinstatement of Local 18, found the occasion most gratifying, as he was the member of the VFD who was involved in the formation of the IAFF in 1918. An American by birth, George was always proud of his membership in Local 18.

Local 18 International Charter.
Photo: VFD Archives

On June 24 Chief Condon asked council for four more firemen to replace the four men sent to bolster the Fire Warden's Branch. Alderman Miller, the finance chairman, warned him that the fire department must economize in view of the recent pay boost for firemen and should defer hiring more men at that time.

Fire Truck Collision Kills Lieutenant Eric Robinson

At 10:45 p.m. June 26, No. 17 Truck collided with a car at East 33rd Avenue and Knight Street, while responding to a chimney fire. On impact it careened into a hydro pole and flipped on its side, pinning Lieutenant Eric R. Robinson, age forty-nine, under the right side of the truck. Eight people lifted the truck to free the trapped firefighter and he was rushed to hospital, where he succumbed to head and chest injuries five days later.

Aftermath of the collision of No. 17 Truck.
Photo: VPL Photo #84848

He left a wife and three children. (His nephew, Harry, served on the VFD from 1950 to 1984.)

A civic funeral was held for Lieutenant Robinson, under the auspices of the Firemen's Benefit Association. Traditionally, his casket would have been carried to the cemetery on a fire truck because of his line-of-duty death, but because of the shortage of fire apparatus in the city he was carried in a hearse.

The ex-South Vancouver truck in which he had been killed, a 1927 Studebaker city service truck, had became part of the city's apparatus roster following amalgamation and had a notorious reputation for being mechanically inadequate, with some of the complaints being: top-heavy, light on the front end causing poor steering, poor brakes, and insufficient power with a top speed of 25 mph (45 km/h). It also had an inefficient electric siren and poor warning lights. It was damaged beyond repair and the city was now without any spare equipment. At that time the delivery of any new equipment was taking more than eighteen months from the time of placing the orders.

July 19, 1947 marked the retirement of two veteran district chiefs, Torquil Campbell, the sole survivor of the streetcar and No. 11 Hose Wagon crash in 1918, and Angus McLeod, who was followed on the VFD by his son, Alex, (1946) and grandson, Alan (1971). The two old Scots joined the VFD in 1911 and 1912 respectively and were the first fire captains to serve on the fireboat *J.H. Carlisle.*

On October 3, 1947 Chief Condon was instructed by the city council to call for estimates for the replacement of city firehalls and a suitable location for a new headquarters firehall to replace the old No. 2 Hall on Seymour Street. With the city building inspector, a committee was to secure property for a new firehall at East 54th Avenue and Kerr Street (No. 5 opened there April 2, 1952); property for a new No. 17 Hall at East 50th and Fraser Street (No. 17 opened at East 55th and Knight, June 3, 1955); and property at Taunton and Kingsway to replace No. 20, Wales Road and Kingsway (the new No. 20 didn't open until 1962).

New Apparatus Begins to Arrive

In early October two new 1947 LaFrance 1,000-gpm (4,500-liter) pumps arrived in the city, were tested and placed in service. The first of the new Spartan 700 series pumps went to No. 2 and the other to No. 19 Firehall.

In November the city council agreed to pay the funeral expenses of Lieutenant Eric Robinson after it was noted that the funeral costs for two

No. 19 Hall's new
LaFrance Pump.
Photo: VFD Archives

policemen killed in a gun battle in February 1947 were paid for by the city. Finance Chairman Alderman George Miller warned council that this would not establish a precedent as the city pays a Workmen's Compensation Board (WCB) assessment on all employees and shouldn't have these expenses.

The city council also announced that in 1950, on the retirement of Fire Chief Ambrose Condon, his successor would be Captain R.H. Pinkerton of No. 2 Hall, who was promoted to Third Assistant Chief, effective November 27. This was an attempt by the city to get a much younger man into the top job. At this time Malcolm (Roary) MacDonald was First Assistant Chief and Allan R. Murray was Second Assistant Chief, and both were senior to Pinkerton.

An early morning fire in Chinatown took the lives of four elderly men on November 29. The three-alarm fire in a Shanghai Alley rooming house was discovered shortly after 2:00 a.m. and had a good hold by the time Nos. 1 and 2 Companies arrived. A cabinet shop on the main floor suffered huge losses to machinery and lumber stock; also lost were many costumes, masks and props used by a touring Chinese theatrical company that was stranded in Vancouver when World War II broke out.

Cause of the fire was never determined, but it started on the second floor of the old three-story building and quickly spread throughout because of the tinder-dry wood and papered walls and ceilings.

The year 1948 began with a three-alarm fire on January 8 in the Oscar Brown & Company Limited warehouse at 163 Water Street, a wholesale fruit and vegetable merchant. The first alarm came in at 7:15 a.m. with heavy smoke reported coming from the basement of the building and later filling the entire three-story building. Once the smoke was cleared the cause was found to be a stubborn oil furnace used to dry stalks of bananas in the basement. Smoke damage was estimated at $15,000.

This was one of the last fires at which the old-style smoke masks were used. The new MSA self-contained breathing air (SCBA) masks were beginning to replace the old model masks.

On March 2, 1948 Lieutenant George H. (Lefty) Kaye, age forty-seven, died in hospital following a lengthy illness that had forced his early retirement. Lefty earned his nickname as an outstanding pitcher in the Senior League for many years; he was also the pitcher for the firemen the night of Canada's first-ever night baseball game. He was buried in the Firemen's Plot at Mountain View Cemetery.

New masks are used at a multi-alarm fire at the Oscar Brown warehouse.
Photo: Vancouver Sun: Gordie Sedawie

Captain Jack Anderson Dies

On March 15 Captain Jack Anderson, aged fifty-six, VFD first-aid instructor, founder of the Inhalator Squad, and the originator of the diaphragm method of artificial respiration, passed away. "Captain Jack" had been in ill health for some time, a direct result of his attempt to save a man and his son who were trapped in an eighty-foot (24-m) well on their Surrey farm, in July 1938. The well was filled with methane gas. He was lowered into the well three times, while wearing an air mask, and successfully brought the bodies out, but later collapsed from the effects.

He was awarded the Bravery Medal, the civilian equivalent of the Victoria Cross.

Forced into early retirement after thirty-three years' service, he collapsed at his home earlier in January 1947, and was revived by the crew from No. 14, men that he had trained. A Guard of Honor consisting of fellow firefighters, Red Cross, and St. John Ambulance personnel attended his burial at Forest Lawn Cemetery. It was said that his memorial was the Rescue & Safety Company's "Men in White" (a nickname coined by the local press because of their white coveralls).

In April the firefighters' union won a wage increase of $40 per month, and within hours Alderman Miller, the finance committee chairman, suggested that with this increase the fire department might have to lay off some men. Union President Hugh Bird reminded council that the arbitration board decision was binding on both sides and if talk of cutting staff was attempted, "we will fight it," he said. Even with their wages raised from $195 to $235, many firefighters across Canada in much smaller cities were still making more money.

Evans Products Sawmill Fire

A three-alarm fire at 10:00 a.m. on May 2 totally destroyed the Evans Products sawmill and destroyed thousands of board feet of lumber at the neighboring Northern Timber Company at the south foot of Main Street, on the Fraser River. The Evans sawmill produced more than 50,000,000 board feet of lumber a year, and employed 180 men.

First-in companies were Nos. 17 and 22, followed by Nos. 3, 13 and 20, and No. 2 and No. 12 came in on the third. Water pressure was poor and drafting from the river was necessary. The damage exceeded $400,000, and the cause was found to be electrical.

On June 30 the engine of No. 4 Truck blew up at Broadway and Larch Street as it responded to a fire alarm on West Third Avenue. The truck was the department's 1928 American-LaFrance Type 31, eighty-five-foot (26 m) aerial. Pedestrians and several parked cars narrowly escaped injuries and damage as the old rig weaved down the street, out of control. Because it was a front-wheel-drive vehicle, all control was lost when the engine blew; fortunately, it came to a safe stop on its own.

The machine shop was notified of the breakdown and when the mechanic turned the corner of Broadway and Granville Street, he just followed the ten-block-long trail of oil on the roadway until he came to the dilapidated truck.

Now the department was short an aerial and the next new one was not due to arrive until the following spring. It was decided that rather than scrap the twenty-year-old rig, it would be returned to service as a tractor-aerial. Pacific Truck and Trailer of Vancouver built an open-cab tractor for $13,000 and the rig was returned to service in November 1948. With the new tractor, the rig was now longer than before and wouldn't fit into the firehall. To solve the problem, a "dog-house" was built into the rear wall of the hall to accommodate the overhang of

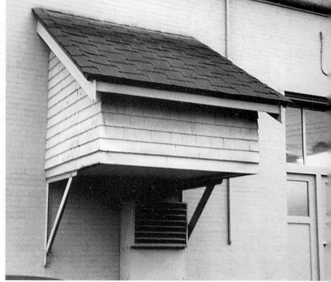

Rear of No. 4 Firehall had to be extended with a "dog-house" to accommodate the aerial ladder.
Photo: VFD Archives

the aerial ladder.

In the days before Vancouver had public bars and lounges it had many private, members-only licensed clubs. A call to one such club sent a truck to the scene for a small outside fire. With the small fire taken care of, the men attempted to get into the club to ensure that fire hadn't spread to the inside. They were only able to enter after the watchful, overzealous doorman made them sign in. (From the *Vancouver Sun,* August 6, 1948.)

In July of 1948 the VFD took delivery of three LaFrance-Dodge 500-gpm (2,250-liter) pumps costing $7,000 each, which helped to speed up replacement of old apparatus, giving the department three rigs for the price of one custom model. The first-ever, closed-cab apparatus on the VFD, they went into service at Firehall Nos. 14, 15 and 20, replacing three 1912-vintage hose wagons.

Newly built Pacific tractor on No. 4 Aerial. *Photo: Jack B. Thompson*

Copp Shoe Store Fire

At 11:00 p.m. August 23 a passing pedestrian noticed smoke coming from the Copp Shoe Store in the 300-block West Hastings Street. Within fifteen minutes of the fire department's arrival, the fire had gone to a third alarm. The fire burned out of control for almost three hours because the thick, blinding smoke and scalding heat of the burning leather products and packaging prevented the crews from entering the building. Seven firefighters were injured during the course of fighting the blaze.

Before the fire was struck out, police arrested two suspects on a nearby street, both of whom were wearing new shoes and carrying several other pairs. One of them was an employee of Copp's. The $275,000 loss included 15,000 pairs of shoes in the Copp warehouse and all the

Copp Shoe Store three-alarm arson fire.
Photo: Vancouver Sun:
Harry Filion/VPL #84855

Three-Alarm Blaze Causes $275,000 Damage

OLICE SUSPECT ARSON in three-alarm fire at Copp's 10e store, 339 West Hastings, near Homer, at 11 p.m., :onday. Eight firemen were injured, and one was sent hospital in $275,000 blaze. Sixteen pieces of equipent clogged Hastings and 120 firemen poured water on fire for three hours. Fifteen thousand pairs of shoes were lost. Arson angle was probed when two men were found with new shoes on Cordova. Hundreds watched fire, attracted by plumes of white smoke which spread over city. By Harry Filion Photos.

assets of the National Dress Company, that occupied the entire top floor. Originating in the basement, the arson fire was set by the petty thieves in an attempt to cover their theft of a half dozen pair of shoes.

Ralph Ravey Retires

The longest-serving member of the fire department retired in mid-October 1948 after some forty-five years, three months and a few days. He was W. Ralph Ravey, who joined the VFD July 1, 1903 at age twenty. In 1910 he became one of the city's fire alarm operators.

Ralph liked to tell the story of a new fireman, an ex-sailor. Late one night the bell sounded at No. 2 and the new man was told to "stand by," which meant he was to harness the horses when they came to the front of the rig. As he stood by, one of the horses sped by him and ran off down the street. Ravey then mounted the remaining horse, chased after and caught the first horse, returned it and hitched up the team. The crew then responded to what turned out to be a serious fire.

"Firemen Fight Use of Tugs as Fireboats," read the *Vancouver Sun* headline on November 9. The firefighters' union president said the tugs would be inefficient and, unless used properly, would be a danger to fire-

Drivers shown with the children are: *J.H. Carlisle,* fireboat Pilot Hector Wright (left) and Pilot George McInnis (right), who was also chairman of the union's Children's Hospital Committee.
Photo: Local 18, IAFF

men working on the shore. Union President Hugh Bird was supported by Alderman Gervin, against Acting Mayor George Miller, who protested that the city must think of saving money. The on-going argument of cost and necessity continued. In January 1949 a plan to purchase a fireboat was put before council by Alderman Gervin, on the recommendation of the Fire, Police and Traffic Committee.

On January 11, 1949 the members of the firefighters' union began what was to become a daily occurrence on the VFD by driving physically handicapped children to school at the Vancouver Children's Hospital. The firemen did this every school day of the year and would sign up as the list rotated around to every firehall in the city. This tradition continued for thirty years, until it was no longer necessary and was discontinued in 1978.

Airport Terminal Destroyed

On February 19 the administration building at Vancouver International Airport caught fire, and before the fire could be reported, all the phone lines were severed. The radio operator at Vancouver called the Calgary airport operator, 640 air miles away, who then reported the fire to the VFD about 8:00 p.m. through the local Trans-Canada Airlines office in

The airport terminal before the fire.

BEFORE THE BLAZE, administration building looked like his except for minor changes made since the picture was taken. Airline executives who work on the premises seem not too unhappy about the prospect of a new building

MONDAY, FEBRUARY 21, 1949

After the $150,000 fire at the airport terminal.
Photo: Vancouver Sun

'There'll Be Some Changes Made' at Fire-Swept Airport

AFTER THE FIRE did $150,000 in damage to the administration building of Vancouver's International Airport, Saturday night. Picture shows front of the 18-year-old structure a few hours after firemen battled flames.

the Hotel Vancouver. It was a record 1,280-mile call for a fire less than three miles away.

The weather was below freezing and the road conditions were very slippery, and by the time the city fire department arrived, there wasn't much left to save. They fought the fire with the Richmond volunteer firemen and an RCAF fire crew that was stationed across the field.

At the same time that crews were dealing with the airport fire, a three-alarm fire was being fought downtown in the six-story Coast Warehouse. Damage at this fire exceeded $425,000.

The first of two 1949 LaFrance, 100-foot (30-m) service aerials was put in service on March 10 at No. 3 Firehall. And on April 5 it was announced that the new fire department headquarters would be built in the 700-block Hamilton Street, at a cost of $250,000 to $300,000. When the Hamilton Street site, described as one of the quietest downtown streets, was chosen, the expenditure for the new firehall at East 54th Avenue and Kerr Road was postponed.

Fireman Malcolm W. McPhatter, age twenty-nine, fell three stories from the drill tower at No. 2 Hall on April 7. He was climbing the vertical ladder on the tower when he fell, struck the safety net and plunged to the pavement striking his head. He was rushed to Vancouver General Hospital where he died of his injuries. A member of the fire department for three years, he was married but had no children. He was buried in Mountain View Cemetery.

Men working on the aerial at the Coast Warehouse fire. *Photo: VFD Archives*

No. 3 Hall's new aerial. *Photo: Frank Degruchy*

Firemen accepted a $16 per month wage increase for 1949, bringing a first-class firefighter to $256 per month. Recruits started at $190.

On April 28 a chimney fire and small roof fire broke out at a girls' club in the 600-block Vernon Drive. The building was being remodeled

—Brian Kent Photo

COMMUNITY EFFORT helps to polish up one of Vancouver's oldest firehalls for use as Kiwassa Neighborhood House at 600 Vernon Drive. Vancouver firemen apply fresh green paint donated by local merchants. Brushes and scaffolding are also provided without charge. Firehall was built in 1905 but closed down in 1918 when fire service changed from horse-drawn to motorized vehicles.

and was due to open within days. The fire was quickly extinguished in the old building that had originally opened in 1905 as No. 5 Firehall. Over the years, whenever the old place needed a new paint job, the firemen would get together and do the job for the Kiwanis, the sponsors of the club.

On June 8 Allan Murray was promoted to first assistant chief with Robert Pinkerton to second assistant chief.

The first four-alarm fire in many years occurred at 2:00 a.m. June 15 just west of Main Street on False Creek. where it burned violently for the next three hours, fanned by winds that reached 25 mph (40 km/h). With the fireboat *J.H. Carlisle* on scene in minutes, it pointed out the value of a fireboat and, once again, it was thought wise to place one in Burrard Inlet.

The $1 million blaze swept through a dozen sawmills, lumber yards,

Off-duty fire fighters repainting old No. 5 Firehall for the Kiwanis Club.
Photo: Vancouver Sun: Brian Kent

steel fabricating plants, fuel lots, assorted small businesses, and house-boats. All available off-duty firefighters were called in, and at one point more than 400 men were on the scene. It was described as the worst fire since the Pier D fire of 1938.

Witnesses said the area where the fire seemed to start "went up like a matchbox" and quickly spread. The fire consumed buildings, lumber, paper products and more than 8,000 tons of coal being stockpiled for the winter trade.

Ten firefighters suffered the effects of coal fumes, and one stepped on a nail that went through his foot. Firefighters remained on the scene for days, putting out small fires after the conflagration was struck out. The cause was never determined.

First Assistant Fire Chief Allan R. Murray was appointed deputy fire chief, effective August 23, replacing retiring Deputy Chief Malcolm (Roary) MacDonald.

Chief Condon Resigns

On October 11 Fire Chief Condon, age fifty-eight, submitted his resignation to the city council citing poor health as the reason. Allan Murray became acting fire chief, November 1.

Also in October, eleven ladies of the Red Cross were made honorary members of the International Association of Fire Fighters (IAFF) by the Vancouver Fire Fighter's Union, Local 18, in recognition of their volunteer work with their canteen wagon at all major fires of three alarms or greater. At a ceremony held at the Beatty Street Armouries, each was thanked and presented with an IAFF pin by Acting Mayor R. K. Gervin, on behalf of the union.

In November Vancouver's long-standing regulation requiring city employees to live in the city was removed, allowing them to live in any of the adjacent municipalities.

It was also announced by the city council that retiring Fire Chief Condon would received three months pay as a retirement gift and that two dozen applications have been received for his vacant position. The selection committee hinted that the new chief would be "young — probably under fifty years of age."

The *Vancouver Sun* headline on November 24 read, "Battle Looms Over Fire Chief Job" and the news item went on to say that the "not over fifty" provision (by the city's personnel committee) barred many good firefighters from applying, including some thirty-five extremely qualified men. Acting Fire Chief Allan Murray was not being considered, even though he came highly recommended by Chief Condon in his retirement letter.

Alderman Miller, chairman of the personnel committee, declared that the union was advised that applications were open to all members of the department. As a result, Hugh Bird resigned his position as president of the union and applied for the job, too, and was reported to be one of five men under consideration. He was about 100 names and fourteen years junior to Chief Murray.

The eighty or so senior officers from assistant chiefs down to captains formally protested that after many years of loyal service and practical experience they would be denied promotions and, further, that this situation was detrimental to the morale of the department.

"Union Boycotts Fire Chief's Job," read the headline in the *Vancouver Sun* on December 6. The union protested the ten-year clause as it eliminated too many good men, and stated that it endorsed Allan Murray as fire chief. Previous applications were then discarded and when new applications were called for, none were submitted as all members refused to apply. One alderman suggested that a solution to the problem would be to seek legislation to extend the retirement age of a fire chief to age sixty-five and blamed union interference in past appointments for the current impasse.

A statement by a union executive stated "members feel very strongly that the 10-year clause amounts to interference in a department which is operating efficiently." They further reaffirmed the union principle of seniority with ability and, further, endorsed the competency of the acting fire chief in the best interest of the city, the department, and the morale of the members of the fire department.

The debate continued as to who the next fire chief would be and everyone hoped that "seniority with ability," the long-held and proven principle, would prevail.

The memorial plaque to fallen firefighters in 2006 at No.1 Firehall.
Photo: Author's collection

New Chiefs, New Fireboat and New Goals

(1950–1955)

Allan Robert Murray

The year 1950 saw a number of changes. January 10, Allan R. Murray, age fifty-seven, became the seventh fire chief of the City of Vancouver Fire Department. On the new chief's recommendations, Mayor Charles Thompson announced the following promotions: Assistant Chief Lorne Foley to deputy chief; second assistant chief, Robert H. Pinkerton; District Chief Syd Gillies to third assistant chief. Murray was described as fearless, a strict disciplinarian, and a man whose firefighting ability was respected by the firemen.

City aldermen decided that all future promotions would be won on a competitive examination basis with seniority as a secondary consideration.

Born: June 6, 1892, Peterhead, Scotland
Joined VFD: June 14, 1913
Appointed Chief: January 10, 1950
Retired: June 5, 1952
Died: May 3, 1975

On January 26 the *Vancouver News-Herald* reported that tenders for a new fireboat were about to be called. The boat, which would cost about $300,000, would become the most powerful on the continent.

Vancouver firefighters presented their sole demand for 1950 — an increase in future pensions. At that time the average pension was about $42 per month. The demand went to binding arbitration, and in April a decision was handed down setting the average fireman's pension between $94.48 and $134.35, with special superannuation contributions of 2 percent and 2.5 percent to be paid by the firemen and city, respectively.

Waterfront Fires

The largest fire of the year was at Celtic Shipyards at the foot of Blenheim Street on the Fraser River, on May 22. It came in shortly after 1:00 a.m. and quickly went to a third alarm. Exploding drums of gasoline and burning oil drums quickly spread the fire to a storehouse where it destroyed engines, batteries, rigging, booms and masts, and tons of marine supplies.

Also destroyed was *Dominion,* a $50,000 seiner that burned to the waterline; three other fishing vessels were heavily damaged, saved only when tugs pulled them out into the river. A fish packer, *Kimsquit,* on the ways, was badly scorched, and only a west wind saved the BC Packers fishing fleet that numbered about 100 boats. Employees turned on the sprinkler system, an action that saved the machine shop. Several employees were burned trying to save boats. The damage was estimated at $500,000.

Another large fire of 1950 was, once again, a waterfront fire that destroyed the Columbia Grain wharf, a seed mill and nine boxcars, on July 22. The fire momentarily spread across the tracks into the Columbia Grain Elevators on Commissioner Street, near the north end of Renfrew Street, but it was quickly extinguished.

The three-alarm fire consumed the two conveyor loading ramps and by 11:55 a.m., both had collapsed. Within an hour, with the help of the fire barge, the fire was knocked down and the many grueling hours of overhauling began. The cause was attributed to spontaneous combustion. Damage was estimated at over $150,000.

The *Vancouver Daily Province* Fire

The fire at the *Vancouver Daily Province* newspaper on Pender Street — described as the most difficult and smoky fire in the city's history — happened January 3, 1951. The first alarm came in at 1:05 p.m., and before it was brought under control at 5:24 p.m. fifteen men were hospitalized and twenty-five more treated at the scene for the effects of the smoke caused by the burning newsprint. Even though the firemen were wearing air masks, they were still affected. It was later learned that a broken gas main was adding its deadly fumes to the smoky fire and, miraculously, didn't ignite.

The stubborn, three-alarm fire started in the 500 tons of newsprint stored in the basement, and it could have been smoldering for hours before being noticed, investigators said. The smoke and fire extended throughout an adjoining cafe, gutting it and causing heavy damage to the adjacent Parks Hotel.

Hotel guests and cafe patrons had to flee for their lives; many of the people in the hotel were shift workers who were sleeping at the time. The fire caused a major traffic jam by halting dozens of trolley buses and attracting thousands of people who gathered to watch the blaze.

The *Vancouver Daily Province* newspaper fire. *Photo: CVA 354-134*

After the fire, the crews received praise from Mayor Fred Hume and from Fire Chief Murray, who directed fire operations. The chief said that this fire was much worse than the G.L. Pop Furrier fire of 1943 that felled thirty-six firemen.

Both the *Vancouver Sun* and *News-Herald* offered the *Daily Province* the use of their presses for its evening edition, an offer that was gratefully accepted.

"We Salute the Smoke Eaters," read the editorial in the next day's edition of the *Vancouver Sun:*

"Business routine in The Sun Tower was disrupted yesterday afternoon, when staff members crowded the windows to watch Vancouver's efficient firefighters in action.

We had box seats across Pender Street for a basement blaze,

which horribly demonstrated how the smoke-eaters got their nickname. They had a treacherous task and they knew it, but they went at it as calmly as if it had been a backyard bonfire. Spectators had no inkling of the real dangers until ambulance sirens in monotonous rotation screamed away to hospital with men overcome by the billowing fumes. Other men took their places and the work went on without interruption, hour after hour, with no confusion and no heroics. It was an object lesson in how bravely the public's protectors respond to the call to duty.

We regret the outbreak and wish the injured firemen swift and full recovery. We're also sorry that more members of the public couldn't have seen such an inspiring example of selfless courage. Because the men themselves treated it as a matter of routine only heightens our admiration."

The *Daily Province* editorial the next day said, "We Say Thank You!"

"Because of the courage and skill of Vancouver firefighters the *Daily Province* presses are rolling again today after an interruption on Wednesday afternoon.

In fighting the newsprint fire that started in the building next to the *Province* plant many firemen suffered from the effects of the acrid smoke. They fought one of the most stubborn fires in the history of Vancouver. They overcame great difficulties and it was only by their persistence and skill that the *Province* plant itself was not damaged. The *Province* is grateful to the men of the department."

Western Canadian Softball Champions, 1950.

Back, left to right: Charlie Miron, Herb Baillie, Bill Heath, Alex Turkington, Butch Gilmore, Clare Foster, Frank Ambler. **Front**: Jim Miller, Roy Riley, Peter Proctor, Laurie Shuttleworth, Dick Pickering, Norm Warren, Norm Ford. **Bat boys**: Alan Miron and Gary Gilmore.
Photo: VFD Archives

February 1950

On the third of the month, firemen accepted a $30 per month wage increase that gave a first-class fireman $280 and a captain $325 per month from $291, effective March 1.

February 26, Alderman Anna Sprott launched Vancouver's new *Fireboat No. 2* at Yarrow's Shipyards in Victoria. Alderman R.K. Gervin and Fire Chief Allan Murray were also at the ceremony. The $300,000 vessel was expected to arrive in Vancouver within a month.

Firemen at No. 17 Hall laughed when a citizen came into the hall to report, "The cemetery is burning down!" The firehall was directly across the street from the Mountain View Cemetery where the Vancouver crematorium is located. The firemen looked out and discovered that the crematorium was indeed on fire. The blaze in the electrical/furnace room was brought under control in twenty minutes, after causing about $10,000 damage.

On February 27, firemen waged a desperate battle against a fire that threatened to destroy the BC Packers million-dollar fish plant at the foot of Campbell Avenue. The fire burned through the roof of the 300-foot-long building, and became a second alarm shortly after 7:00 p.m. Two Vancouver fire barges contained the fire from the waterside, with the land companies attacking from the shore along the narrow docks, which at

Rendering of new fireboat. *Photo: VFD Archives*

times hampered their progress. After two and a half hours, the fire was brought under control. One small fish boat had to leave the dock because of the fire, but the 10,000-ton British freighter, *Thistledale,* tied up nearby, remained at dockside with her crew playing their hose lines on the ship to prevent any radiant heat damage.

Fire wardens found the cause of the fire to be loose sacks that ignited in the fish smokehouse. Other goods destroyed were assorted fishing equipment and thousands of cardboard cartons used in packaging fish products. Damage was estimated at $75,000.

BC Packers fish dock fire. *Photo: VFD Archives*

The New Fire Department Headquarters Opens

The new No. 1 Firehall and Headquarters at 729 Hamilton Street went into operation at 10:30 a.m., Wednesday, March 7, 1951. The first run from the new hall was to a chimney fire at Thurlow and Haro Streets at 7:35 that night. The official opening ceremony, with Mayor Fred Hume officiating, was held March 16. With this opening, the old No. 1 at Gore

VFD Band playing at official opening of No. 1 Hall.
Photo: Stuart Thomson #6337/VFD Archives

The new No. 1 Firehall.
Photo: Local 18, IAFF

and Cordova Street became known as No. 2.

Within days of the opening of No. 1, Chief Murray recommended that a new firehall be built at West 49th Avenue and Cambie Street, following a recent survey on fire protection in the rapidly growing Oakridge area. It was never built.

And on March 8, every fireman's nightmare: the fireman on watch at No. 17 answered the phone and discovered that the call was for a fire in his own home, in which all the living room furnishings were destroyed.

Disaster Unit Introduced

In October, Local 18 firefighters introduced their new disaster unit to the people of Vancouver. The Vancouver-built, 1937 Hayes-Anderson, ex-Pacific Stage Lines bus was donated by the BC Electric, the parent com-

Program of dedication and official opening of No. 1 Hall.
Photo: VFD Archives

pany of PSL. The firefighters converted it into a canteen wagon for use at major alarms, and it could also be used as a field hospital in the event of a large disaster emergency.

The unit went into service on November 14 at No. 2 Hall, and was later stationed at No. 18 Hall. When it was needed at large, multiple-alarm fires, it would be brought to the scene and the ladies of the Red Cross would be called out to prepare and serve hot drinks and sandwiches to the firefighters. On cold, rainy, wintry nights the warm and dry bus was very much appreciated at a fire scene.

Vancouver Fire Fighters'
new canteen wagon.
Photo: VFD Archives

Red Cross ladies making
sandwiches and serving
coffee and soup in the
canteen wagon at a
multiple alarm.
*Photo: A. Duplissie:
VFD Archives*

This service continued until the early 1980s when it was decided by the union members that the old bus should be replaced, either by a newer bus, a smaller coffee-wagon unit, or a call-out service on a contract basis. In 1985, the bus was returned to the transit authority in exchange for recently retired bus #5501, a 1965 Prevost. The canteen wagon concept did not work out in the 1980s, so the bus was sold and the project was abandoned.

The employees of BC Transit restored this last-remaining Hayes-Anderson bus to original condition for their centennial celebrations in June 1990, and today it is part of their collection of historic transit vehicles.

The Fireboat Arrives

On April 5, 1951, after an overnight trip from Victoria where she was built, the new *Vancouver Fireboat No. 2* was towed into Vancouver Harbor by the steam tug *Burnaby*. A civic welcome was extended to her on her arrival and she was declared "the most powerful fireboat in the world" by marine underwriters, after her trials. She pumped 25,074 US (20,895 Imperial) gallons per minute, 3,000 gallons more than the old record holder, Seattle's fireboat, *Duwamish*.

Vancouver Fireboat No. 2 demonstration. *Photo: Vancouver Sun: Dave Buchan*

Designed in Vancouver by Milne, Gilmore and German, naval architects and marine surveyors, the vessel was said to be "an almost all-Canadian product."

With an overall length of 87 feet (28 m), beam of 21.5 feet (7 m) and a draft of 9 feet (3 m), the boat had a top speed of 12 knots (15 statute miles per hour). Powered by five 12-cylinder, 525-horsepower Kermath gas engines, two for propulsion and three for pumping, with the propulsion engines also able to be used for pumping, if necessary. There were four monitors, three of 3,000 gpm (13,500 liters) and one of 2,000 gpm (9,000 liters) and four manifolds each fitted with six 3½-inch (87-mm) discharge ports for hose lines. The crew consisted of a pilot, fire captain, engineer, pump operator, and up to three firefighters.

After ten days of crew familiarization, *Fireboat No. 2* went into service on April 14, at No. 10 Hall, quartered at the Immigration Dock at the foot of Burrard Street.

The first fire that the new fireboat attended was at the Coal Harbour Boat Rentals at Chilco and Georgia Street, on June 12. The firm lost

Fire Chief A. Murray accepting delivery of *Vancouver Fireboat No. 2*. *Photo: VFD Archives*

twenty, small, twelve-foot (3.5-m) inboard motorboats and suffered damage to two larger pleasure craft and a houseboat when boys playing with firecrackers threw one of them into the bottom of one of the small boats. The firecracker ignited some spilled gas in the boat's bilge, causing the total destruction of the rental business in less than five minutes. By the time No. 6 Company and the fireboat arrived on scene, the boats and the office containing all company records were destroyed.

More Replacement Apparatus Arrives

Several new fire apparatus were received to replace some of the vintage rigs still in service: a 1950 Bickle-Seagrave eighty-five-foot aerial went to No. 18; a 1951 Bickle-Seagrave 1,200-gpm (5,400-liter) pump went to No. 17 (it was nicknamed Rudolph because of the red light mounted in the center of its hood); a 1951 Bickle-Seagrave city service ladder truck for No. 9 Hall, and two 1951 LaFrance, Type 715, Dominion 1,250-gpm (5,625-liter) pumps, delivered to Nos. 2 and 6 Firehalls.

Bickle-Seagrave aerial shown at No. 21 Hall, 1970s. *Photo: VFD Archives*

Bickle-Seagrave city service truck. *Photo: Courtesy Patrick L. St. James: VFD Archives*

On Monday, December 31, firemen fought a losing battle in a spectacular early morning blaze that destroyed a Fraser River sawmill and caused $100,000 damage. Freezing cold weather hampered firefighting efforts to save the mill in the fire that was first reported at 6:35 a.m. When crews first arrived, the mill was fully involved so exposures were covered, including several piles of lumber, and more than 200,000 board feet of lumber was saved. The two-alarm fire was at the foot of Argyle Street and the hose lines had to be laid across the BCE interurban tracks, causing hours of inconvenience to rush-hour travelers. A shuttle service had to be used with passengers walking through the deep snow from one tram to the other because of the hose lines.

Then downtown, an overheated furnace at Jermaine's Limited, a fashionable women's clothing store at Granville and Smithe Streets, caused many thousands of dollars of smoke damage to all the clothing, including many expensive fur coats. The first alarm for this fire was at 8:30 a.m., just shortly after the change of shift.

Retired Captain Arthur Clegg, age ninety-one, an original member of the paid department in 1889, died on December 31, 1951. He originally held Badge No. 2, and when Captain John W. Campbell died in 1910, he was given Badge No. 1, which he held until his retirement in 1929. He continued to take an active interest in the Firemen's Benefit Association and in 1949 the association presented him with a gold medal for over fifty years of service.

On January 18, 1952, Hugh Bird was re-elected president of Vancouver Fire Fighters Union, Local 18, IAFF.

Arthur Clegg.
Photo: VFD Archives

No. 5 "Fort Apache" Opens

At 8:00 a.m. April 2, 1952, No. 5 Hall at East 54th Avenue and Kerr Road opened, the first No. 5 Firehall since the original closed in April 1918.

Firehall No. 5, "Fort Apache," opens in southeast Vancouver. The hall's first pump is the 1929 Bickle.
Photo: Stuart Thomson #7755/VFD Archives

Bowman Storage fire. *Photo: VFD Archives*

The firehall was nicknamed "Fort Apache" because it was surrounded by bush and undeveloped land. The crew answered their first alarm at 8:15 a.m. to a chimney fire in the 2200-block Uplands Drive.

Sunday, April 20, saw one of the worst warehouse fires in the city, and after a thirteen-hour fight all that was left was a blackened shell of brick and burned timber that had been the Bowman Storage and Cartage building at 829 Powell Street.

A cab driver turned in the first alarm at 3:01 a.m. to two police officers on the beat, who relayed the alarm to fire dispatch. Eight minutes later, with No. 2 Hall on scene, the district chief put in a "six-six" alarm, which was a second-alarm response, using on-shift personnel without calling the off-duty members. Then at 3:24 a.m., it became a third alarm, with the off-shift men called in. The fireboat was also called and lines were stretched from the boat to the fire scene, under railroad tracks where possible so as not to tie up the CPR trains.

The Red Cross ladies came to work in the canteen wagon that arrived about 5:00 a.m. and remained until late in the afternoon, serving coffee and sandwiches to the tired firefighters. The four-story building suffered $500,000 damage, including all the contents. Firemen said the public was very cooperative in keeping back a safe distance and it was reported, "Only the very curious got splashed by water from hoses." The cause of the fire, the biggest in two years, which also caused damage to four other businesses, was not determined. Ten firefighters were treated for injuries.

The next evening at 10:30 p.m. a two-alarm apartment fire, the second multiple alarm in two days, claimed the lives of a young mother and her seven-year-old daughter.

Hugh Bird Named Fire Chief

In a "sharp divergence" from past policy of appointing senior men, the city council selected the next fire chief of the VFD on April 22, 1952. At the special council meeting, Hugh Bird won the top job on the first ballot. The next day, Captain Hugh S. Bird, age fifty, was named fire chief. Immediately, Chief Murray was openly critical of council's choice.

Both the union and Chief Murray cried foul when it was learned that the council, rather than call for applications, took only the five successful candidates from two years previous when a successor to Chief Condon was to be selected. Chief Murray and the union also believed in the basic tenet of "seniority with proven ability".

Chief Murray was also critical because, in spite of recent commendations given the department's senior officers on an outstanding record for 1951, he and his senior officers hadn't been approached regarding the applicants for the position, and he made it clear that his choice was his deputy, Assistant Chief Lorne Foley. Unfortunately, Foley had only two years to go before he reached retirement age.

Despite union meetings, accusations, newspaper editorials and innuendo, the decision was final, and Hugh Bird, who had vaulted over some eighty senior men, would become the department's eighth fire chief,

Hugh Stuart Bird

Born: March 25, 1902, Broadview,
Saskatchewan
Joined VFD: July 11, 1927
Appointed Chief: June 6, 1952
Retired: March 14, 1962
Died: October 21, 1981

effective June 6, 1952. Because of his age — and if the city's ten-year clause prevailed — it was said that many good men over forty years old had lost all chance of serving as fire chief and other senior positions. With the announcement, Bird once again resigned as president of the Fire Fighter's Union, and the morale of the VFD was at an all-time low.

The United Grain Growers Elevator Fire

On the same day as the Bird announcement was made, a multimillion-dollar, four-alarm fire hit one of the largest wheat pool elevators in the city.

The fire alarm to the United Grain Growers' dock came in at 11:34 a.m. and in thirty minutes it had wiped out a 700-foot-long (210-m) loading jetty. The dock, on the waterfront at the foot of Vernon Drive, was destroyed along with hundreds of feet of track, a loading platform and fourteen boxcars. Every available man on the department was called in to fight it. At 12:30 p.m., an offensive fight against the dock was abandoned and resources were put to saving the elevators and the exposures. The nearby Buckerfield's Limited warehouse, which contained four tons of fertilizer, was in danger of blowing up, as well as having the fire extending into straw, hay, grain and other flammables. The marine equipment of Vancouver Barge and Transportation was also threatened. Two ships tied up at the dock were pulled clear by Cates Towing tugs and both suffered scorched paint from the intense heat and were covered with soot and flying debris.

Vancouver Fireboat No. 2 at United Grain Growers' Dock fire.
Photo: VFD Archives

Chief Murray said this fire was the worst waterfront blaze since the Pier D fire fourteen years earlier and could have been the worst in the city's history. As it was, VFD resources were stretched so thin that if another major fire had occurred, men and equipment would have had to be taken from this fire.

The new fireboat was on scene in nine minutes, and paid for itself that day. Only the fireboat could have stopped the fire's advance before it could extend to the massive concrete elevators. At the height of the fire, men taken from the fire scene and brought around into the harbor to assist the new fireboat manned the False Creek fire barge.

Considered under control at 2:00 p.m., it was struck out at 4:00 p.m., but crews remained on scene until early the next day.

The fire was caused when a small, portable welding generator on a trailer, being used for repairs on a nearby ship, was accidentally tipped when struck by a passing line of grain cars backing along the track to the unloading site. When it tipped over, gasoline was spilled out and then ignited by an electrical short, and the fire started. The fire raced up the supports of the loading platform and quickly spread along the dock. The men working the loading platform scrambled to safety.

Original damage was estimated at $3 million but later was set at just under $2 million. The fireboat was credited with saving the day. It was estimated that without the fireboat, the property damage on the waterfront could have easily been $8 to 10 million.

The MV *Dongedyk* Fire

The MV *Dongedyk,* a 12,000-ton, luxury cargo-liner of the Holland-America Line, arrived in Vancouver early in the morning of April 28 on her maiden voyage from the Netherlands. She tied up at Ballantyne Pier No. 1 and at 10:41 a.m. a fire alarm was received for a fire in her No. 3 hold. A stubborn three-alarm fire followed and, after a three-hour fight, the fire was brought under control using hose lines supplied by the new fireboat. As the fire was being fought, three Cates Towing tugs stood by in case the ship had to be towed out into the harbor.

The fire began in the engine room when fumes were ignited and spread to the No. 3 hold, which contained twenty British cars, ten of which were destroyed. The hold also contained steel, lumber and a number of 950-liter fuel tanks that were to be off-loaded. Tons of water and foam were used to put out the fire and then a dangerous list-to-port developed that, if not checked, would have resulted in the ship capsizing. The order to flood the starboard tanks righted the ship, and once the fire was out the water was pumped out, leveling her once again.

Ten of the more than 200 firefighters were overcome by smoke and

MV *Dongedyk* fire. *Photo: Courtesy Patrick L. St. James: VFD Archives*

twelve others suffered minor injuries, including Assistant Chief Foley, one of the officers in charge of the fire. All were treated at dockside by No. 3 R&S and the ship's doctor. Damage was estimated at about $100,000 and repairs took ten days.

The ship originally was the *Deldftdyk,* and on her maiden voyage after the war she struck a mine and was badly damaged. Completely rebuilt, the ultramodern ship began her new service on the Netherlands–Vancouver run, via the UK, Panama Canal, and west coast US ports.

The following day, April 29, the sixth major fire in nine days occurred at 6:00 p.m. at the National Shingle Mill on Taylor Street, on False Creek under the Cambie Bridge. The small mill was completely destroyed in the two-alarm fire, which was attended by twelve companies, including the fireboat *J.H. Carlisle.* A small mill with a staff of only ten men, it suffered a loss of $35,000 damage, but would be rebuilt.

On May 10 it was announced that First Assistant Chief Lorne Foley would be acting fire chief until Hugh Bird took over June 6. Various groups spent the entire month of May trying to get city council to rescind the appointment of Bird, but to no avail.

Veteran Fire Captain Dies in Squad Wagon MVA

On June 3, 1952 the Rescue & Safety Squad Wagon from No. 3 Hall was in collision with a car at East Broadway and Clark Drive, struck a second car, flipped on its side and then struck three parked cars and burst into flames. The two men riding in the back of the wagon were thrown out, one suffering painful pelvis and back injuries. Seven people in the cars involved were also injured.

The driver and the captain were trapped in the burning cab, but pulled out by the uninjured crewmember and citizens. Their clothes were on fire and once they were extinguished, the men were treated and taken to hospital, both suffering burns to various parts of their bodies.

No. 3 Pump responded with Exclusive Ambulance to treat the injured

R&S Squad Wagon after the crash and fire.
Photo: VFD Archives

and extinguish the burning squad wagon. It turned out that the run the wagon was responding to was of a minor nature and not life-threatening.

Captain Morley Oswald (Ozzie) Howell, age forty-six, suffered burns to 65 percent of his body. He died of his injuries on June 14 in Vancouver General Hospital.

The civic funeral was held at Christ Church Cathedral in downtown Vancouver on June 21, with the Reverend Dean Cecil Swanson conducting the service. He mentioned that six years earlier he had been present at the unveiling of the plaque dedicated to sixteen Vancouver firefighters who had died in the line of duty; Captain Howell's would become the twenty-second name on the memorial.

Ozzie Howell had been a member of the fire department for twenty-five years and was a former secretary of the union, as well as secretary of the provincial association of firefighters. He was buried in Forest Lawn Cemetery, and left a wife and three daughters. In 1993, Judy, one of Ozzie's daughters, said that on the day of the accident her mother had passed by the accident scene on a bus, only to learn later that her husband was one of the victims. On April 5, 2003, the new No. 13 Firehall was dedicated to Ozzie.

On March 25, 1954 the *Vancouver Sun* reported that the driver of the car involved in the collision with No. 3's Inhalator Squad Wagon was 85 percent to blame for the accident. The driver of the rig was 15 percent to blame for not slowing down enough for the intersection. The Supreme Court decision awarded the city $3,900 damages, and then the court learned that the driver, a mechanic, was currently serving a year in jail for burglary.

Captain Ozzie Howell is the officer on far right in this group picture of No. 3 R&S standing in front of the squad wagon door.
Photo: VFD Archives

Chief Bird Takes Over

At 9:05 a.m. June 6, 1952 Fire Chief High Bird walked into his new office and started work. He said, "the trouble is over, as of right now," and with his two assistant chiefs, Robert Pinkerton and Syd Gillies, he presented the following three-point program of reforms in the department that he proposed to carry out:

> Training: Familiarization of all firefighters to the dangers of fires in warehouses and major buildings, ships, and more knowledge of oil and chemical fires.

> Fire Warden's Branch: All firefighters will be trained as fire inspectors for the suburban areas, which will allow the wardens to inspect larger downtown buildings and other facilities.

> Inhalator Services: More trucks and pumps will carry first aid and inhalator equipment to give faster service, and a second squad wagon will be added to the downtown area.

At a news conference, Bird said that he had no hard feelings against those who tried to block his appointment.

New squad wagon at No. 3 Firehall.
Photo: Frank Degruchy

VFD band playing for March of Dimes drive.
Photo: Vancouver Sun

Two new squad wagons were purchased, one to replace the one destroyed in the accident. The second one was for service downtown at No. 2 Hall; designated No. 2 R&S (Rescue and Safety) it went into service in February 1953.

The Vancouver Firemen's Band and other members of the VFD marched for the *Vancouver Sun's* March of Dimes as they had done for a number of years, and in 1952 they collected over $4,000 in three hours of canvassing, a record amount raised by any group in such a short period of time. Over the years the fire department members have helped in many ways to raise money for the Children's Hospital and have long been considered "big brothers" by the toddlers and children at the hospital.

Ex-Fire Chief Condon Drowns

On December 3, 1952, while a passenger on a North Vancouver-bound ferry, retired Fire Chief A. E. Condon, age sixty-one, fell overboard and drowned. He was quickly recovered from the water but all attempts to revive him were in vain. He had been in ill health for some time.

Simma Holt Named Honorary Fire Chief

In February 1953, Simma Holt became Canada's first honorary woman fire chief in a ceremony held at No. 1 Firehall. Fire Chief Bird presented Mrs. Holt, a *Vancouver Sun* columnist and a good friend of the fire department for many years, with a white chief's helmet. The chief paid tribute to her cooperation with the department, her reporting of the Rescue & Safety Inhalator Squad activities, and "our co-operation in the great *Vancouver Sun* March of Dimes program." In the 1970s she served as a member of parliament, and in 2006 she is still writing.

Honorary Fire Chief Simma Holt with retired fire chiefs Ed Erratt, Jack DeGraves, Charlie Thompson, Archie McDiarmid and Allan Murray. Back row are retired assistant and district chiefs.
Photo: CVA 354-363

On June 3, 1953 Deputy Chief Lorne Foley retired after thirty-nine years and was replaced by Assistant Chief Robert H. Pinkerton, fifty-seven, a veteran of thirty-seven years. Second assistant was Syd Gillies and the new third assistant chief was Elmer Sly, who returned to the VFD from a secondment with the Civil Defense.

One of Foley's favorite stories was about the time he rescued District Chief William Eaton from a smoke-filled furrier's shop fire. Eaton had apparently dashed into the smoke and was overcome by some tanning material fumes. Foley crawled into the smoke when Eaton failed to come out. He found the unconscious Eaton, and as there was no time to carry him out, he simply dragged him out by the feet, facedown, to safety. Next day, Eaton, "peering through a bloody eye with a swollen and skinned nose and forehead," said Foley, tried to find out who it was that had dragged him across twenty feet of cement floor. For more than a week nobody would tell him what happened. On July 1, 1907, Eaton became the first, and only, lieutenant on the fire department.

In June 1954, District Chief W. (Bill) Scott, who joined the VFD, April 1, 1932, became the first fire chief of the District of North Vancouver Fire Department.

The *Vancouver News-Herald* headline of November 30, 1954 read, "Fire Rages on 3 Floors High In Marine Building." The first alarm for the fire in Vancouver's (then) tallest building, the twenty-one-story Marine Building, was turned in by the janitor at 12:31 a.m. for a fire burning in the elevator shaft on the fifteenth floor, and on arrival of the first on-scene company a second alarm was immediately put in. With the fire floor well beyond the reach of the tallest aerial ladder, firefighters had to climb the stairs and carry hose up fifteen stories and haul hose lines up the outside of the building by rope. It took some time before firemen had water on the fire.

The fire, which was mainly confined within the elevator shaft, spread smoke throughout the upper floors, and after a short time the fire weakened the elevator cables and one fell to the main floor. A hole had to be broken through the wall to the burning elevator to completely extinguish the fire. The fire was struck out about 2:00 a.m. and the leg-weary firemen began picking up their hoses.

All of the elevators remained out-of-service until the cause, determined to be electrical in nature, was repaired. The renowned Guinness Brewery family of Ireland owned the building, that had opened in 1930.

Budget Cutbacks

For the 1955 city budget, Chief Bird was told that he would have to cut back $20,000 from his budget. Despite his objections, the cuts remained, so he informed city council that the only way he could make cuts was to remove the service of No. 21 Truck, its crew and inhalator service. Some aldermen cried foul and said the reason that the Kerrisdale truck was chosen was that a number of aldermen lived in the Kerrisdale area. Bird replied that he didn't know where the city's aldermen lived.

He then announced that on June 1, eleven firemen would be given thirty days' notice and the last eleven hired would be gone by July 1. The union then advised that the previous year, thirty-five inhalator calls were answered by No. 21 Truck resulting in two-dozen lives saved and President Hector Wright of Local 18 said that the city was playing with twenty-four lives for the sake of $20,000. The residents got into the fight and put on a drive with handbills to save the truck and inhalator.

Editorials began to appear and feature writers for the newspapers began filling their columns with support for the firemen. One columnist reported that the acting mayor hoped to offset $20,000 in concessions granted the firefighters because of the $40-a-month pay boost awarded them by an arbitration board, which was binding on both sides. He went on to say that the fire department "only answered 664 more alarms than the previous year; had saved 250 lives with the inhalator and first aid service and treated 115 others, with various ailments and injuries."

THIS IS NOT
AN ADVERTISEMENT

AND HAS NOTHING TO DO WITH POLITICS OR RELIGION

(It will only take 95 seconds to read!)

TO SAVE TAXES IS NECESSARY

BUT . . .

NOT AT THE PRICE OF OUR LIVES

The population in our district, together with Vancouver, is growing, but due to a cut in manpower, ordered by the City Council

> THE FIRE DEPARTMENT LADDER AND INHALATOR TRUCK
> IN OUR DISTRICT HAS TO BE TAKEN OUT OF SERVICE
> **24 LIVES were saved by the crew of this truck last year**

Do you know what this means?
If you are trapped in a burning building, rescue may be delayed by precious minutes which may mean the

DIFFERENCE BETWEEN LIFE AND DEATH
FOR YOU AND YOUR LOVED ONES

This truck also answers all the INHALATOR calls for our area. Last year alone 24 LIVES WERE SAVED by the crew of THIS ONE TRUCK in our district. This can be checked with records at the Fire Department. In future it may take twice as long to get emergency calls answered. Doctors explain that the FIRST THREE MINUTES are the MOST IMPORTANT ones and THESE WILL BE LOST in future, because this equipment will be decommissioned.

Do not permit this to happen and

Write to the Vancouver City Council today
City Hall, 453 W. 12th Ave., Vancouver 10, B.C.

It took your fellow citizens five years to get this ladder truck, for which the insurance underwriters had asked, stationed in our fire hall. And now, all this may be wiped out again, in spite of growing population figures and during prosperous years.

It is irresponsible that for a saving of LESS THAN 1/20 OF 1% of the budget, scores of citizens' LIVES should be RISKED.

IF YOU LIKE TO LIVE IN SAFETY
YOU MUST ACT WITHOUT DELAY

DO NOT RELY ON THE NEXT DOOR FELLOW TO DO IT FOR YOU!
PLEASE TALK IT OVER WITH YOUR FRIENDS AND NEIGHBORS.

Vancouver Citizens' Fire Protection Committee
1042 W. Hastings Street.

Citizens' handbill protesting removal of
No. 21 Truck and Inhalator.
Photo: VFD Archives

While fire and police department cuts were being proposed, aldermen had to wrestle with a call for an increase of city hall staff. Eventually, the budgets were reconsidered and the firemen kept their jobs.

In February 1955 three more of the new LaFrance Type 715, 4,730-liter Dominion pumps were placed in service at Nos. 1, 4 and 22 Halls.

On March 29, No. 22 received a new truck, a LaFrance Veteran city service truck, which had a 300-gallon (1,350-liter) booster tank and a high pressure, dual booster hose reel arrangement, which was specified for the truck's response to the Vancouver International Airport.

No. 22 Hall's new LaFrance 1250-gpm pump.
Photo: Walt McCall: VFD Archives

No. 22 Hall's new 1955 LaFrance Veteran city service truck.
Photo: Frank Degruchy: VFD Archives

On June 3 the new No. 17 Firehall opened at East 55th Avenue and Knight Street with pump, truck and district chief.
Photo: VFD Archives

All ranks, firefighter to captain, began wearing the new Maltese cross-style cap badges in place of the laurel leaf and numeral badges that had been worn since the days of the volunteers. The significance of the laurel badge relates to the ancient laurel leaf victory wreath of Roman times. The Maltese cross, long-associated with the fire service, dates back to the Crusades of the Middle Ages. Saint Florian, the most well known patron saint of firefighters, was a Roman centurion and fire brigade officer. St. Florian was martyred in 301 A.D. and his feast day is May 4.

New Maltese cross-style cap badges. *Photo: VFD Archives*

June 10 saw yet another tradition of the VFD changed to suit the times. Departmental Directive #11/55 stated, "Suspenders will no longer be worn with trousers," (only belts) under signature of H. S. Bird, Fire Chief.

Arson Fires Keep the VFD Busy

The evening of Saturday, September 10, 1955 began with a fire in the 1200-block Granville Street involving a deliberately set fire in a store. Within an hour, the fire scene had moved to the second floor above four stores in the 1100-block Granville Street, which went to a second alarm; again, a deliberately set fire. All available fire wardens were called in to investigate these fires, including some who were called away from the BC Lion's football game.

Before these fires were out, another fire was reported in the 100-block of West First Avenue, on False Creek, at 10:51 p.m. and by 11:37 p.m. it had gone to a third alarm. The fire in two Arrow Transfer Company warehouses caused $1 million damage to the buildings and contents, a sawmill had $70,000 damage, a cartage company lost eleven trucks and three lumber carriers for another $70,000, and another warehouse and a steel company's storage shed were lost. Fed by oil, paint and other flammable materials, the area literally exploded. Eight firefighters were injured and one was hospitalized with severe head lacerations and a possible fractured skull, when a falling beam struck him.

Investigators believed that a firebug had set the warehouse fires. Earlier, an apartment owner had surprised a man on the main floor of her building and he ran off. Then she found paper burning in the hallway.

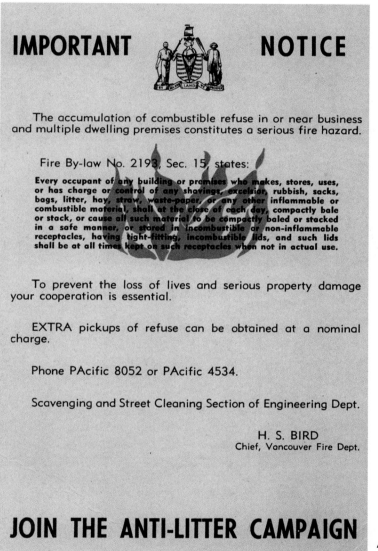

Anti-litter notice.
Photo: VFD Archives

The total fire damage just for the weekend had exceeded the entire previous year's $1.1 million total by almost $100,000.

Then on September 14, a house being demolished about three blocks away from the Saturday night False Creek fire was torched, extensively damaging neighboring houses.

A drive was now on to encourage citizens to keep the areas around their homes, garages and yards free of litter and to be aware and keep their eyes open to anything that might be suspicious. By September 19, after five more arson fires in the downtown area and West End of the city, the total stood at eleven and the investigators had a description of the suspect.

On the morning of October 4, tragedy struck the city's West End when an arsonist's fire in an unfinished apartment building spread into three adjacent rooming houses, and by the time it was struck out, four people had perished.

Arson fire in the West End.
Photo: VFD Archives

Two policemen on patrol turned in the first alarm at about 6:50 a.m. and nearby No. 6 Company arrived on scene within minutes. By 7:02 a.m., the second alarm was turned in, but the fire spread so quickly that the only people who could be saved were the first ones awakened by the two policemen. These rescued people reported that the fire seemed to spread following a small explosion and "whooshing sound," and before the tenants in the upper floor of the Comox Street house could be warned, the top of the house was already on fire. The victims of the fire were an 81-year-old woman, an 80-year-old man and his wife, aged 79, and a 47-year-old woman. This fire was the forty-seventh fire of the year to be suspected as arson, with twenty-one of them occurring since August 1.

A suspect was picked up at a nearby cafe and charges were laid but a conviction wasn't forthcoming as he was found to be mentally unstable.

Shortly before midnight, Tuesday, November 8, the BC Sash & Door Company warehouse at 533 West Broadway was destroyed, fed by a large stock of dry wood and paint, varnish and finishing products and nothing could be saved. When firemen arrived, the fire was through the roof. The first alarm was sounded at 11:10 p.m. by two police officers noticing smoke coming from the building. They noted three youths in the area, but when one of officers went to question them, they had disappeared.

The three-alarm fire threatened nearby houses but they were saved, although with some scorching and water damage. The factory behind the warehouse was also saved. By 1:00 a.m., the fire was under control, struck out at 1:37 a.m. and crews "overhauled" it (removed rubble to ensure the fire would not rekindle) throughout the night to ensure that the fire was completely out. The damage was over $100,000 and the cause undetermined, but suspicious.

A two-alarm fire at St. Joseph's Catholic Church and School destroyed part of the roof on November 21, 1955 but "it was quickly put out in minutes after the call went in," said the ecstatic priest, "and they had four hoses laid out in no time flat!" He also praised the firefighters for saving the altar from water damage by covering it with a tarpaulin. The alarm, coming in at 2:27 p.m., was struck out at 2:45. Cause of the fire was unknown, but it appeared to have started behind the electric organ. Years later, at 6:00 a.m. on March 3, 1981, the new St. Joseph's Church was destroyed in a spectacular, two-alarm arson fire.

Deputy Chief Robert Pinkerton retired on November 21, 1955 and was succeeded by Chief Syd Gillies. Elmer Sly became second assistant chief, with Walter A. Carlisle, son of the late Chief Carlisle, promoted to third assistant chief.

It was hoped that the many arson fires would cease in the coming new year, but it was not to be.

St. Joseph's Catholic Church destroyed in early morning fire.
Photo: VFD Archives

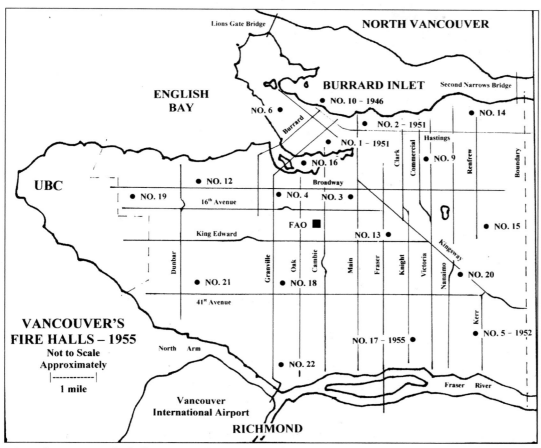

Map of Vancouver firehalls - 1955. *Photo: VFD Archives*

Fire Prevention Awards and the Big Fire

(1956–1964)

An early morning fire destroyed the top floor of a downtown card club on January 8, 1956. The fire alarm came in at 10:40 a.m. and immediately went to a third-alarm, with the off-shift members called out. The blaze, having a good hold before it was reported, went through the roof, causing dense smoke from the burning roofing material, and making the fight more difficult, but by 11:18 it was struck out. The store and restaurant on the ground floor suffered smoke and water damage but there were no injuries. Damage was estimated at over $40,000 and the cause was believed to be smoking materials improperly disposed of during the previous evening's activities.

On January 11, 1956 another three-alarm fire, this time in the warehouse district at 1100 Hamilton Street, destroyed several thousand dollars worth of tires, carpeting and stored goods at the Coast Warehouse. As well as the contents, the top floor of the three-story building was badly damaged. The first alarm was turned in at 8:00 p.m. and an accountant working on the main floor didn't realize that the building was on fire until firefighters found him on their arrival from the nearby No. 1 Hall, four blocks away. Damage was estimated to be about $150,000; the cause was unknown.

Downtown Arson Fires

Between 8:00 and 9:30 p.m. on May 25 fifty firefighters were kept busy answering seven fire alarms in the downtown area. At the scene of a 4:00 a.m. house fire, a thirty-five-year-old man was arrested on a tip given police by a cab driver who had seen him at one of the earlier fires.

One of the fires resulted in damage to two occupied houses. Another fire was set in the same block and three others were set within a four-block area. When the arrested man, known to police, was found to have pockets full of paper napkins, he told the officers that he "didn't like sirens." The man was also questioned regarding seven fires set eight months previously and charges were laid following further police and fire department investigations.

An early morning four-alarm fire on May 12 caused $450,000 damage to Pacific Bolt Manufacturing on Granville Island. The alarm came in at 4:45 a.m. from Box 1862.
Photo: Vancouver Sun: Dan Scott

A Third Generation Carlisle Joins the VFD

On May 28 nine probationary firefighters started their training with the fire department and one of them was Ronald H. Carlisle, the grandson of the first chief of the VFD, John H. Carlisle and the son of Assistant Chief Walter A. Carlisle. Ron served to retirement in 1987 as a captain. Coincidentally, he started on the seventieth anniversary of the founding of the fire department.

On September 10, 1956 the entire front page of the *Vancouver Sun* — with a headline that read, "275 Fire Chiefs Arrive for National Meeting" — was devoted to the chief's conven-

Ron Carlisle, grandson of J.H. Carlisle.
Photo: Local 18, IAFF

tion. Chief Hugh Bird was pictured as the host of the forty-eighth annual conference of the Canadian Association of Fire Chiefs (CAFC) with articles on special visitors, delegates and table officers, a display by the newly formed scuba team, a mention of the early days of the band, a Len Norris cartoon and the convention's planned wrap-up dinner at the Forbidden City, for an exotic Chinese buffet. At the convention's conclusion, the members made Mayor Fred Hume and Attorney-General Robert Bonner honorary fire chiefs.

The last delivery of the LaFrance Type 715 Series, 4,730-liter pumps was made in December, when the two 1956 Quads went in service at Nos. 12 and 17 Halls, respectively. In the US, the Model 715 series were rated at 1,500 gpm (US) and only twenty-four units with this pumping capacity were built*, comprised of twenty-two pumps and two quadruple combinations. (*This was an unusual size in 1956; however, 1.500-gpm capacity is more common now.) Of these twenty-four units, nine were delivered to Vancouver, three went to Mexico City and the remainder went to US cities.

1956 LaFrance quad shown at No. 17 Hall, c. 1973. *Photo: VFD Archives*

Chief Hugh Bird accepts International Association of Fire Chief's Fire Prevention Award. *Photo: Villy Svarre: VFD Archives*

The VFD Wins Fire Prevention Awards

In January 1957 the fire department was notified that it had won the National Fire Protection Association (NFPA) Grand Award for the most outstanding fire prevention program for 1956 and first place for Class A Canadian cities (over 100,000 population). It was the first time that a Canadian city had won the coveted Grand Award and the first time a western Canadian city won the Class A award.

Seattle Fire Chief William Fitzgerald, past president of the International Association of Fire Chiefs (IAFC) said, "Vancouver proved in 1956 that it had the best fire department on the continent. And this trophy is not merely awarded; it has to be won against keen competition from every city in Canada and the US."

There were 696 entries in the competition and Vancouver's organizer was Lieutenant (later captain) Jim MacBeth of the Fire Warden's Branch. His son, Jim MacBeth, Jr. joined the VFD in 1965, retiring as a captain in 1996.

The Butler Hotel Fire

The first alarm for the Butler Hotel fire on Water Street came in at 5:05 p.m. on March 29, 1957, a time of day when most people are awake; but the smoke from this fire, which started in the basement, shot up the elevator shaft so quickly that many people were trapped. It was made a second-alarm at 5:15 p.m., and then a third-alarm at 5:23 p.m. At that point there was a call-out, because the off-shift men were needed to help put this fire out.

The heavy smoke throughout the building and the dozens of overhead wires in the front and rear of the building hampered the firefighters' efforts. There were many daring rescues, a few involving unruly drunks. Two blind men lived in the building. One of them, a forty-two-year old, stumbled around in the smoke until a fireman brought him out. The other, an eighty-one-year-old was found in the smoke and safely taken down an aerial ladder, along with six other men. In the end, three male tenants lost their lives and ten other people were injured.

The fire broke out in a basement room of the six-story, skid road building and was caused by careless smoking. Damage was estimated at $100,000.

A dramatic ladder rescue
at the Butler Hotel fire.
Photo: Vancouver Sun: Dan Scott

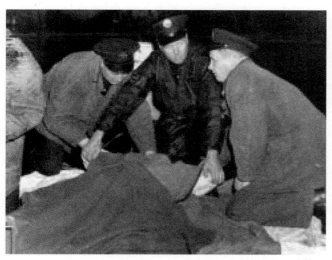

Tending to victim of Butler Hotel fire.
Photo: Vancouver Sun: Dan Scott

Aerial ladder rescue at the Butler Hotel fire.
Photo: Vancouver Province: Eric Cable

A three-alarm blaze at 1701 West Broadway at 9:45 p.m. April 9 destroyed the building and injured fifteen firefighters, mostly from smoke inhalation and minor cuts from breaking glass. One was almost killed in a fall. He and another firefighter were taking a hose line up a ladder to the roof and as the first man stepped onto the roof, the ladder slipped sideways and crashed to the ground. The second firefighter was left dangling three stories up on the hose line, on which he lowered himself safely to the ground, to the cheers of about 3,000 spectators.

After burning out of control for some time, the thick smoke chased the firefighters out of the building and off the roof. And shortly after 11:00 p.m. it got away and burned through the roof. With large volume master streams on the now-open roof, the fire was knocked down. The fire chief called the fire one of the toughest he had to fight.

Several doctors and an engineering firm occupied the building. Damage was estimated at $500,000 and cause was believed to be electrical.

Fire damage at 1701 West Broadway.
Photo: VFD Archives

Granville Mansions three-alarm fire. *Photo: VFD Archives*

Fire broke out in Suite 302, of the Granville Mansions, 715 Robson Street at Granville, at 6:00 a.m. on April 15, 1957. An eighty-one-year-old retired dentist was rescued from his smoke-filled rooms by firefighters and taken by ambulance to the hospital where he later died of smoke inhalation. A resident informed the janitor, who was just starting his work for the day, that there was smoke on the third floor and the alarm was turned in. One hundred and fifty people who lived in the building were able to get out safely, three by coming down one of the aerials at the front of the building from the third floor.

There was considerable fire, smoke and water damage to two suites on the third floor and water damage to the stores below. Cause was suspected to be careless smoking and damage exceeded $60,000.

At 7:47 p.m. on June 13, the Pacific GMC dealership at 2410 East Broadway was destroyed in a $400,000 fire.
Photo: Vancouver Sun: Ray Allan

A memo from the British Columbia Highways and Bridges Authority to the Sea Island and Lulu Island Toll Plazas, dated September 23, 1957, said:

> To: All Sergeants, Sea Island & Lulu Island Plazas: In future, Vancouver Fire Department equipment going to and from the Vancouver Airport, will pass through free. The trips will be recorded on the pass sheet, same as MP's, etc. There should be no problem regarding identity, as the vehicles are clearly marked V.F.D. and are painted the usual red. However, units from Richmond Fire Department or any other local municipality must pay.
>
> (Signed): Toll Captain.

It's not recorded if a previous incident had occurred or not. In 1961 the city of Vancouver sold the airport to the federal government for $2.75 million, and after the provincial government took over the bridges and highways in 1963, all tolls were removed.

At No. 1 Firehall on September 25, Firefighter William (Bill) Jenner collapsed and died while playing handball. He joined the VFD August 1, 1946 as one of the Forty-Sixers.

The year 1958 began with a two-alarm waterfront fire at the Pacific Elevators, Loading Jetty No. 1, at the LaPointe Pier. The fire, reported at 8:15 p.m. January 9, destroyed about 650 feet (195 m) of the 900-foot-long (270-m) wooden grain gallery house that enclosed the conveyor belts used to load the ships. Ten companies and the fireboat fought the fire for two hours before it was struck out and, once again, the fireboat and crew were given credit for saving what could have been another multimillion-dollar fire. There were no injuries in the $200,000 fire.

Also in January, Captain R.J. (Dick) Sowden, age forty-five, became the fire chief of the Richmond Fire Department. Following in the footsteps of his father Robert, who joined the VFD in 1910, he joined the department in 1934 and rose through the ranks. He retired in 1973.

In 1958 Firefighter Aubrey (Bear) Neff, another of the Forty-Sixers, formed the British Columbia Provincial Fire Fighters' Curling Association, and then the Dominion Fire Fighters' Curling Associations. By 1960 the latter grew to five teams and by 1983, when Newfoundland joined, it encompassed all of Canada and was known as the Canadian Fire Fighters' Curling Association.

He was a dedicated curler and epitomized fair play and good sportsmanship. In 1962 Aubrey's team won the BC Provincial Championship then went on to win the Canadian Championship in Edmonton, where he was selected as the All Star Skip. He also served as president of the Vancouver Curling Club. The "Bear" died of cancer in retirement in 1997, at age seventy-four. The fiftieth anniversary of the Canadian Fire Fighters' Curling Association is slated to be held in the Greater Vancouver area in 2008.

Pacific Elevators fire.
Photo: Vancouver Sun: Dan Scott

Aubrey Neff's 1962 Dominion championship curling team. Back row, left to right: Al Heaven, third; Chief Ralph Jacks; Aubrey Neff, skip. Front row: Bob Lynch, second; Ollie Ballam, lead. *Photo: CVA 354-265*

The Second Narrows Bridge Collapse

The main span of the new Second Narrows Bridge was inching its way toward the south shore, and people often commented on the crane working on the cantilevered north end and thought it looked too heavy. Then, with little warning, on the hot afternoon of June 17 in "fifteen seconds of thunder, waves and death," as one construction worker later described it, the leading section of the bridge with the crane on it collapsed into the inlet. When this first section, that was on temporary steelwork, fell, it pulled the next pier forward, which caused the second section to also collapse and fall into the inlet. The collapse killed eighteen workers.

The Second Narrows Bridge collapse, looking toward North Vancouver. *Photo: VFD Archives*

Second Narrows Bridge collapse with the old bridge and city in the background. *Photo: CVA 354-186*

Within minutes the call went out to fire departments and emergency personnel on both sides of the inlet, sending Vancouver's fireboat and scuba team to the scene where they spent many hours helping survivors and searching for victims, some of whom were trapped among the twisted steel and cables of the project. Many pleasure craft also responded to help search for and give aid to survivors.

One young employee of Dominion Bridge had taken a Steel Company of Canada employee to the bridge site to talk to the engineers about some bolts being used in the construction of the bridge. After leaving him with the project engineers, he began walking back down the bridge toward the North Shore when he felt a violent shuddering; as he was stepping from the second section to the third section, at the expansion joint, the bridge suddenly fell away beneath his feet. Alone, and in a state of shock and disbelief, he continued walking off the bridge towards the North Shore and into the nearest beer parlor.

VFD divers, Jack Bridge and Len Erlendson, searching for victims at Second Narrows Bridge collapse. *Photo: CVA 354-190*

Days later the search for bodies continued, and tragedy struck when one of the civilian divers lost his life in the search, bringing the total lives lost to nineteen. Overall, during the construction of the bridge, twenty-three men died. The cause of the collapse was said to be a mathematical miscalculation by an inexperienced junior engineer.

On a wet August 26, 1960 the $19-million, six-lane, two-mile-long Second Narrows Bridge system was officially opened, and in 1994 it was rededicated as The Ironworkers Memorial Second Narrows Bridge.

The 1958 fire of the year was the four-alarm at the E.L. Sauder Lumber Company on West First Avenue, on July 1, Canada Day.

The first alarm came in at 10:13 p.m. and upon the arrival of the first-in district chief, a second alarm was put in at 10:21 p.m. Soon the entire lumber warehouse was fully involved and when Chief Bird arrived, it quickly went to a third- then a fourth-alarm at 10:33 p.m. The $550,000 fire included the destruction of the warehouse and damage to three freight cars, including one loaded with lumber. Several nearby buildings and hydro and telephone poles were damaged but saved by firefighters. Sauder's main plant across the street was also saved. Damage to hydro lines blacked out nearby Vancouver General Hospital that had to go to auxiliary power.

Major traffic jams were caused when the fire attracted many holiday weekend people to the scene, and it was estimated that about 30,000 people watched firemen battle what was described as the hottest blaze in many years and the first four-alarm fire in nine years.

It took more than four hours to contain the fire, and at 2:30 a.m. it was struck out. Three of the 150 firefighters were slightly injured.

Fourteen pieces of equipment attended, including the fireboat *J.H.*

Sauder Lumber four-alarm fire. *Photo: Vancouver Sun: Ray Allen*

Carlisle, leaving five firehalls in the city without equipment. Cause of the fire was undetermined but arson was not suspected.

The 1959 annual report noted a 39 percent reduction in the number of fires since 1955 with the introduction of fire company inspections and the inspection of private homes, but it reported that commercial and industrial fires had not been reduced. There were twenty-two exceptionally bad fires during the year involving everything from boats, to large furniture and department stores to schools.

Lions Gate Bridge Jumpers

The Rescue & Safety Branch had a record year, attending to 2,222 inhalator alarms and 630 fire alarms, involving nearly 3,000 patients. Two of the runs described as "beyond the call of duty" concerned rescues of would-be suicides from the Lions Gate Bridge.

The first occurred on June 30 when a man left his hotel telling the desk clerk that he was "going to jump off a bridge." He took a cab to the middle of the bridge, made the driver stop, got out and climbed up a cable ladder on the bridge. Police were alerted that a man was going to jump off a bridge and learned that he was on the Lions Gate Bridge. The fire department was called and No. 2 R&S responded with No. 6 Truck at about 4:30 a.m.

Firefighters Len Erlendson and Ralph (Buck) Rogers with Lieutenant Fin Thomson in charge of No. 2 R&S arrived on scene. The 100-foot (30-m) aerial that also responded was put up and Erlendson and Rogers climbed up the ladder toward the man. As Erlendson reached him, the climber slipped and the firefighter grabbed him by the belt and called for Rogers to help him. Just as the rescuers got a rope on the would-be jumper, he suddenly began fighting and screaming. With the help of a third firefighter, they were able to subdue him enough to get him down the ladder to the ground after the fifty-minute incident. Erlendson said that it was a good job that the man's belt held or he would have fallen either to the bridge deck or the 300 feet (90 m) into the treacherous, swirling water of the narrows.

The second drama on the bridge unfolded on a cold, foggy November 24, when a slightly built forty-five-year-old man climbed to the top of one of the cable ladders, 165 feet (50 m) above the bridge deck and threatened to jump. Buck Rogers once again responded with the R&S, but this time an aerial couldn't be used because the climber was beyond its reach. Rogers climbed the cable ladder to within two or three rungs and attempted to talk the man down. Two priests on the scene also tried to talk him down from his perch near the top of the bridge tower.

Rescue of Lions Gate Bridge jumper.
Photo: CVA 354-312

During the course of the three-and-one-half-hour ordeal that began about 2:15 p.m., the largest traffic jam in the city's history was caused because of the bridge closure and the inability of the old Second Narrows Bridge to handle the re-routed North Shore traffic.

Finally, Rogers and the priests were able to talk the man into coming down, but all the way down he attempted to kick Rogers in the head and still threatened to jump. Rogers managed to stay a couple of rungs beneath him and when they reached a point about twelve feet from the bridge deck, the man suddenly lunged toward the waterside of the ladder. Rogers, alert to any sudden moves, was able to grab the man and deflect his jump causing him to fall to the bridge deck where he was grabbed by city police and taken to hospital by ambulance.

A few years later Rogers left the fire department. (He died in retirement, January 2001.) Captain Fin Thomson died before reaching retirement age, in July 1964, and Len Erlendson was killed in a motor vehicle accident in October 1973. Fin's son, Brian, joined the VFD in May of 1975 and was one of the first captains at the new No. 13 Firehall.

Two large multiple alarm fires of note in 1959 were the Wosk's Department Store fire on West Hastings Street and the John Oliver High School fire. The Wosk's fire came in at 10:25 a.m. July 17 and caused $350,000 damage to the building and three floors of household furnishings and appliances. Cause was unknown.

John Oliver High School Destroyed

The old John Oliver High School fire was reported just after 3:00 a.m. December 10, and by the time fire companies arrived the old wooden building was fully involved. It quickly progressed to a third-alarm, and then a four-alarm at 3:27. All that could be done was to protect other school buildings and the nearby homes. About 750 students were displaced when the sixteen classrooms were lost in the $230,000 fire, and arson was the suspected cause.

John Oliver High School prior to fire. *Photo: J.O. Grad Class website*

One of the oldest high schools in the city, it was attended by dozens of members of the fire department, including Chief Ralph Jacks, Union President Gordon Anderson, Deputy Chief Jim Tuning and Chief Training Officer Bud Kellett, to name a few. Jacks, in his early days on the job, worked at the nearby No. 17 Hall that sat adjacent to the school grounds when it was South Vancouver's No. 3 Hall. In the mid-1950s, the old firehall was torn down to make room for the new school expansion. When completed, the school was said to be the largest high school in Canada.

John Oliver High School after fire. *Photo: J.O. Grad Class website*

FIRE ALARM SYSTEM

FIRE ALARM SYSTEM LOG SHEET

DATE Wednesday October 14th 1959.

DAILY OCCURANCES REPORT

SOURCE	LOCATION OF FIRE	TIME	HALLS ANSWERING	FIRE OUT	COMPANIES O.K. AT QUARTERS	REMARKS
Telephone.	Columbia Hotel 303 Columbia St.	2.35 AM	2 Co.-2R&S		No. 2 Co. 2.46 AM No. 2 R&S 2.43 AM (From Police).	Considerable damage to mattress, possible smoke damage to room #110.
Telephone.	1720 Trafalgar St.	1.18 PM.	12-H.		O.K. 1.36 PM.	Fire damage to electric dryer & contents, possible smoke damage throughout house & contents.
Box.694.	Vancouver Auditorium.200.W.Georgia St.	1.29 PM.	1-2-2-R&S.	1.34 PM.	No.1 Coy.1.37.PM. No.2 Pump & Truck.1.42.PM. No. 2 R&S.1.44.PM.	Mechanical False Alarm
Telephone	447 Aubrey Place	5:40 PM	13		No 13 Co.- 5:55 PM	Short in wiring of basement suite. Slight damage wiring
Telephone	1230 Cardero St.	5:54 PM	1 - 6 3 R&S	6:03 PM	No 1 Co.- 6:09 PM No 6 Co.- 6:25 PM No 2 R. & S. Co.- 6:13 PM	Slight damage to ceiling of suite #6 wiring and attic. Slight smoke damage suite 6 and contents 1237 Cardero.
Telephone	Templeton Dr. and Cambridge St.	5:55 PM	9 - 14	6:00 PM	No 9 Co.- 6:05 PM No 14 Co.- 6:09 PM	Rubbish fire under porch at rear 2194 Cambridge St. no damage.
Telephone	1310 Burnaby St. #201	6:47 PM	2 R&S		No 2 R. & S. Co.- 7:42 PM	Errol Flynn 48 yrs. collapse no response to VGH Dr. G. Gould.
Telephone	1127 Beach Ave.	9:09 PM	1 - 6 J	9:14 PM	No 1 Co.-9:20 PM No 6 Co.-9:27 PM No 2 R. & S. Co.-9:22 PM	Coffee pot burned dry slight smoke damage suite #7
Telephone	65 East Hastings St.	11:36 PM	2 R&S		No 2 R. & S. Co.- 11:48 PM	Services not required.

Fire Alarm Office log entry of Errol Flynn's death. *Photo: VFD Archives*

Errol Flynn arriving in Vancouver, October 1959. *Photo: Deni Eagland: Vancouver Sun*

Errol Flynn Dies of Heart Attack

A telephone alarm on October 14, 1959 sent No. 2 R&S to the 1300-block Burnaby Street in the West End, for a collapse. The patient was movie actor Errol Flynn who was visiting the city. En route to the airport he complained of feeling ill and his host took him to the apartment of a doctor friend. After a visit, he said if he could lie down for a short time he was sure he'd feel better and when he was later checked on, he was dead. The rescue crew worked on the fifty-year-old Australian-born star for almost an hour without success.

With a career that spanned over twenty-five years Flynn starred in forty movies and was remembered as the ultimate swashbuckler in such movies as *Captain Blood, The Adventures of Robin Hood* and *The Sea Hawk,* as well as several notable westerns. He was the matinee idol of the thirties and forties.

Again, as in 1956, the VFD took the NFPA Grand Award for the most outstanding fire prevention program and first place, Class "A" Canadian cities over 100,000 population and first place, Class "A", Province of BC. Chief Bird also accepted a special award from the International Association of Fire Chiefs.

New No. 9 Firehall Opens

On February 12, 1960 the new No. 9 Firehall opened at 2nd Avenue and Victoria Drive. A big improvement over the old hall that was built during the horse days, the new hall was modern and spacious and had a much needed drill area and tower. Mayor Tom Alsbury (1959–1962) and Fire Chief Hugh Bird officially opened it on March 11, 1960, with many friends and neighbors in attendance.

The last-surviving, original volunteer fireman died on March 3, 1960 in Vancouver. He was Gabriel William (Gabe) Thomas. Born in Virginia City, Nevada, in the 1860s, he arrived in Vancouver in 1884. A printer by trade, he had lost everything in the Great Fire of 1886.

The Stanley Park Armory Fire

Vancouverites have always compared big fires to others that occurred over the years and the Stanley Park Armory fire of March 18, 1960, was no exception. It was described by Chief Bird as the most spectacular fire since the 1936 blaze that destroyed the ice arena that was located just across the street. Others compared it to the CPR Pier D fire of 1938.

Vol. 10, No. 11   VANCOUVER, B.C., WEDNESDAY, MARCH 23, 1960

Photo by H. Dube, General Photo Service

Fire rages through Stanley Park Armories

The first alarm came in at 4:00 a.m., and within minutes of the arrival of No. 6 and No. 1 Companies the roof had fallen in and the walls began to sag and fall. The second and third alarms were quickly called for to protect the surrounding buildings. Telephone and hydro poles and lines were badly damaged by the heat from the huge bonfire. In spite of the falling wires and the exploding small arms ammunition, the fire was struck out in an hour and a half.

Also known as the Horse Show Building, the fifty-two-year-old, three-story-high wooden building was the home of the Irish Fusiliers

Photo: Vancouver Sun

Regiment and was leased by the Department of National Defence. It covered almost the entire 1900-block West Georgia Street and much credit was given the firefighters for saving the small homes and businesses located at the west end of the block. The destruction of the building was so complete that no cause was ever found and the building was never rebuilt, with the property sitting vacant for the next thirty-five years.

The first new aerial to be received in ten years was put in service at No. 1 Firehall during March. The 100-foot (30-m) LaFrance service aerial was the first aerial purchased during Chief Bird's term and one of a set of five new rigs, which included three pumps and a quad.

The BC Forest Products Fire

Fire Alarm Office with Laurie Shuttleworth dispatching. *Photo: VFD Archives*

The captain of the fireboat *J.H. Carlisle* reported what turned out to be the first five-alarm fire in the history of the VFD at 5:24 p.m. on a very hot Sunday afternoon, July 3, 1960. The captain had noticed smoke coming from under the lumber storage and dock area of the BC Forest Products property on the south shore of False Creek. The fireboat was located directly across False Creek from BC Forest Products and was the first company to respond.

Within a very short time, with the fire rapidly spreading, it soon became a multiple alarm with the off-duty members being called out. The first-in companies had laid lines into the large lumber storage area that covered six to eight city blocks in size. The fire kept growing as it was fed by creosoted pilings and many piles of dry lumber. The hot weather and increasing winds created a firestorm.

Firefighter Bud Black was the driver of No. 12 pump that day, and he and his crew soon found themselves in an

BC Forest Products fire as seen from Oak Street. *Photo: VFD Archives*

area quickly being overtaken by the spreading fire. With several hose lines in operation, Bud yelled at his brother firefighters that they'd better get out of the area because he didn't want to be "the second generation Black to lose a rig in a fire," referring to Fireman George Black (no relation), also of No. 12 Company, who lost his rig at the Pier D fire in 1938. They shut down and abandoned their lines and got themselves and the rig out of the growing inferno.

The fire continued to spread and by the time it had been made a five-

Fire Alarm Operator Mark Reed reading the tapes. *Photo: VFD Archives*

The fire as seen from the north shore of False Creek. *Photo: VFD Archives*

alarm fire, many piles of lumber, the general office, storage sheds and several box cars were burning furiously, and with the westerly winds it continued to spread eastward. Years later, Fire Alarm Operator Mark Reed related comments made by Deputy Chief Elmer Sly after he requested the response of *Vancouver Fireboat No. 2* from the harbor. When reminded that General Orders require the boat to stay in the harbor, the chief

The BCFP lumberyard, fully involved. *Photo: VFD Archives*

replied, "If we don't stop this damn fire before it gets to Cambie Street, it's gonna burn clear to Nova Scotia!"

The fireboat was then brought around from Burrard Inlet to help fight the fire. Over 350 men and all but a handful of the VFD's equipment fought the blaze. Many of the firefighters suffered the effects of the heat; twelve were injured and three were hospitalized. And it didn't cost the city a nickel's worth of overtime because at that time the city paid for the firefighters' telephones, and this agreement allowed it to call men out to big fires as needed. The damage was estimated at over $3 million including damage to adjacent lumber and equipment firms. The mills were never rebuilt and the area along the south shore of the creek was cleaned up and redeveloped as a residential community.

On September 26 a three-alarm fire destroyed the Hamilton-Harvey Department Store at 2015 Main Street, which resulted in almost $900,000 damage to the store and contents. This fire, along with the armory fire and BC Forest Products accounted for 84 percent of the fire losses for 1960, and the annual report also noted that none of the premises that had major fires during the year were equipped with sprinkler systems.

Santa's helpers, Bill Metcalfe and Ian Anderson at No. 6 Hall in 1961. *Photo: Vancouver Sun: George Diack*

Santa's helpers at No. 13 Hall, 1938 with toys just built.
Photo: VFD Archives

Christmas Toys

For more than thirty-five years, between the mid 1930s and late 1960s, city firefighters built, repaired, and accepted donations of toys for needy kids for distribution throughout the province. When plastic toys that were not repairable began to appear in the late sixties, the program ended.

Between April and June 1961, Chief Bird's order for three LaFrance 4,730-liter pumps and a 4,730-liter quad arrived in the city and were placed in service. The three pumps went to downtown Firehall No. 1, 2 and 6, and the quad went to No. 17 Hall. The previous year, a LaFrance aerial that was part of this order had been placed in service as No. 1 Truck.

On March 25, 1962 Fire Chief Bird officially retired. In spite of the criticism regarding his appointment he was remembered as a progressive chief who was responsible for many changes in the department. On retire-

ment, he successfully ran for city council as an alderman, where he served for twelve years and worked on many committees. He died October 21, 1981 at seventy-nine years of age.

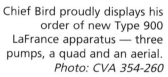

Chief Bird proudly displays his order of new Type 900 LaFrance apparatus — three pumps, a quad and an aerial.
Photo: CVA 354-260

Retired Fire Chief Hugh Bird as Acting Mayor with Local 18 executive: President Gordon Anderson, John Bunyan and Alex Turkington.
Photo: VFD Archives

Ralph Raymond Jacks

M. R. Jacks

Born: September 9, 1909, Vancouver, BC
Joined SVFD: May 11, 1928
Appointed Chief: March 15, 1962
Retired VFD: June 17, 1969
Died: March 7, 1976

Ralph Jacks is Appointed the New Fire Chief

Hugh Bird was succeeded by Rescue & Safety Chief Ralph Jacks, age fifty-two. He joined the South Vancouver Fire Department in May 1928, and became a member of the VFD on amalgamation in 1929.

Also in 1962 the Vancouver firefighters' soccer team took part in the North American Soccer Championships in Los Angeles and defeated Mexico in the finals to capture the Kennedy Cup, named for US President John F. Kennedy. The team of '62 also took the Provincial Cup and the Pacific Coast Soccer League Championship. Over the years the firefighters' soccer team won the Dominion Cup a record nine times, and in 1957 was the first Canadian team to have a chance to play in the early rounds of the World Cup.

In November the new No. 20 Firehall opened, replacing the old, ex-SVFD firehall at Wales and Kingsway.

The annual report for 1962 reported the following recruit statistic averages: Height: 5 ft.10 in. (179 cm); Weight: 178 lbs (81 kg); Age: 24 years; Education: slightly under Grade 12.

There were 6,685 alarms with the lowest fire loss in the last two years at just under $2 million. Careless smoking accounted for the highest number of fires at 333, and there were sixteen fire deaths. Four large fires during 1963 in a school, two hotels and a cleaning establishment, accounted for more than a third of the year's total fire losses as well as three fire deaths.

Fire fighters' soccer team with the Kennedy Cup, 1962.
Back row, left to right: Norm McLeod, Dave Hutton, Art Bennett, Bud Walton, Gordie Nordby, Lou Trischuk, Terry McKibbon, Dave Sullivan, Jim McGuckin, Ken Pears, Ed Bak, Ernie Durante.
Front row, left to right: Bob Mills, Art Hughes, Gary Stevens, Al Lenfesty, Bob MacKay, Deryl Hughes.
Photo: CVA 354-291

The Spotless Cleaners and Dyers plant at 5th Avenue and Main Street was destroyed in a three-alarm fire on March 3, causing almost $375,000 damage. The call came in 6:34 p.m. on a Sunday afternoon, and by the time No. 3 Company arrived on scene, it was fully involved. Cause of the fire was electrical; it was said to be a faulty light fixture.

Then on Sunday, March 31 at 3:31 a.m., a three-alarm fire in the Austin Hotel, 1221 Granville Street claimed three lives in a fire caused by a cigarette. No. 6 Truck crew made many rescues by aerial ladder, particularly at the rear of the hotel. They plucked three people to safety when they were trapped on two fourth-floor windowsills while flames licked out at them from their rooms. Mere seconds made the difference between life and death. One hotel tenant was seen climbing down a small diameter pipe on the side of the building in an area almost inaccessible by ladder.

On November 10 at 2:39 a.m. Queensbury School in No. 5's district was badly damaged by a fire—deemed to have been set—that caused $250,000 damage.

Finally, a small hotel over some stores on East Pender Street sustained $245,000 in damage but, luckily, no loss of life, as the call came in at 9:00 a.m.

Captain Watson Dies

The morning of January 1, 1964 began on a sad note with the death of Captain Gordon B. Watson. No. 18 Company was returning to the hall from a run and as the pump was rounding the corner at West 41st Avenue and Oak Street, the captain collapsed in the front seat of the rig. Efforts failed to revive him. He was born February 4, 1908 and joined the department February 4, 1929, his twenty-first birthday. In his younger days he was a well-known local boxer.

Gordon Watson on left, center Hugh Coleman, Jim MacBeth on right at No. 4 Firehall in 1943.
Photo: VFD Archives

The Plimley Motors Fire

The first alarm to the fire at Plimley's auto agency came in at 3:05 a.m. on May 29 and by 3:17 a.m. it was a third-alarm. It took nearly 150 fire-fighters, including eighty-three off-duty men, about two hours to bring the fire under control; it was finally struck out at 5:30 a.m.

The blaze tore through the North American-built car department of the firm, but the neighboring British car section at the end of the block was spared. Destroyed were thirty cars, the showroom, repair shop, parts department and staff facilities. Damage estimates were placed at $500,000. One firefighter, Bill (Honest & Fearless) Johnstone, suffered head and arm injuries when a hose coupling burst, throwing him to the pavement. The cause of the fire was never determined but was believed to have started in the repair shop.

When the chief interviewed Bill Johnstone for the department in 1934, he asked him what he had to contribute to the job and Bill replied, "I'm big and I'm strong and I'm honest and fearless." Bill retired as a captain in 1971 and died in 1994 at eighty-three years.

The 42-Hour Workweek Begins

The reduced hours of work for firefighters became effective August 1, 1964. The new contract saw the average hours of work per week reduced to forty-two hours from forty-eight hours, but the city wanted the hours to be reduced from forty-eight to forty-six for six months, then down to forty-four hours and over a period of months eventually reduced to forty-two hours. Union President Gordon Anderson told the city that it could not be done that way and said it would be easier and more cost-efficient to make the change in one move rather than over a longer period of time. The city didn't agree and after some calculations on the University of British Columbia's computer it was discovered that the union's sugges-

tion was valid.

With the reduction of hours it was necessary to hire eighty more men, who later became known as the "Sixty-Fourers" (or "backwards Forty-Sixers" after the large group hired in 1946). Like the Forty-Sixers, this group also influenced many changes in the department. The men were hired on June 22, were trained in eight groups of ten men by senior captains who were seconded to the Training Branch, and were ready to be placed in the firehalls for the start of the shorter workweek on August 1.

On November 5 retired assistant chief William Plumsteel died at age eighty-three in St. Paul's Hospital. The well-known chief joined the department in 1906 and retired in 1941. The year 1964 also saw the death of William D. Frost, age eighty-seven, the last of the nineteenth-century firefighters. He joined the department as a driver on October 1, 1899, and was the brother of Captain Richard Frost who was killed when No. 11 Hose Wagon collided with the streetcar in May 1918.

Christmas Eve warehouse fire behind No. 4 Firehall. *Photo: VFD Archives*

The year ended with the largest warehouse fire of the year occurring December 24 across the lane from No. 4 Hall. The first alarm for the furniture warehouse and dance studio came in at 8:54 a.m., for a fire that was caused by an overheated incinerator being used to burn rubbish in the building. Before it was knocked down, the smoky three-alarm fire caused more than $170,000 in damage, but no one was hurt.

The 1964 group of recruits at No. 1 Firehall. *Photo: CVA 354-267*

187

Major Hotel and Apartment Fires (1965–1969)

On January 20, 1965 it was announced that because of the new forty-two-hour workweek, the strength of the department would be increased by one assistant chief to four, one more district chief to twelve, twelve more captains to around ninety, and two more lieutenants.

Many of the rookies would be present at one of the dirtiest fires in memory. The two-alarm "peanut fire" was on the top floor of the six-story Pacific Cold Storage Company warehouse at 21 Water Street on January 27. The thick, gray-green, smoke prevented firefighters from easily gaining access to the building to clear the smoke and fight the fire. Eventually, jackhammers had to be used to punch holes in the concrete roof of the old Swift's meat-packing and cold storage plant to clear out the smoke in order to get at the fire.

The fire itself involved more than 700 bags of raw peanuts, and the firefighters had to haul the 100-pound sacks away from the main pile and throw them out the windows at the rear of the building. No. 1 Truck was sent along the tracks at the rear to assist in reaching the fire, but got stuck in the soft ground between the rails. A CPR engine later freed it.

The first alarm came in around 1:25 p.m. and the four-hour battle resulted in some smoke injuries with relatively little fire damage, but the firefighters and equipment were covered in an oily residue from the burning peanuts, resulting in many hours of equipment clean-up.

On February 18 Chief Jacks announced that the fire department would purchase its first tractor-drawn, 100-foot (30-m) tiller aerial since the *Queen Mary* was received in 1938. Hub Fire Engines and Equipment in nearby Abbotsford, BC would build it, with a trailer built by Coast Apparatus Trailer of California, and a ladder built by Grove Manufacturing Limited of Shady Grove, Pennsylvania. Not only would it be the first aerial of this type built by Hub, it would also have the first aerial ladder built by Grove and be the first VFD rig with an automatic transmission and a canopy-cab for the crew. Chief Jacks had tried for two years to convince the city of the need of a tiller-aerial to better maneuver the city's congested West End streets and lanes.

Deputy Fire Chief Elmer Sly retired from the fire department in March and on April 2 he collapsed and died of a heart attack, one week before his sixtieth birthday and official retirement. He joined the Point Grey Fire Department on May 1, 1926, and later served as Vancouver's union president from 1938 to 1941.

On April 30 a conciliation board awarded Vancouver firefighters a 6 percent wage increase, bringing the wage for a first class firefighter to $524 per month.

A CPR diesel engine came to the rescue of No. 1's aerial when it became stuck on the tracks while fighting the peanut fire on Water Street.
Photo: Vancouver Province

THE PROVINCE, Thursday, January 28, 1965 ★★★★27

The Shanghai Alley Fire

Firefighters were always concerned about fires in certain parts of China-town because many of the old wooden buildings dated back to the 1880s. At 10:38 p.m. December 22 an alarm was received for a fire on Carrall Street, and No. 1 and No. 2 Companies responded. By 11:01 it was turned into a two-six alarm (second-alarm assignment).

The fire was found in one of the old tenements in Shanghai Alley that housed single, elderly Chinese men who lived in the building's small cell-like rooms. Shortly after the first-in company located it, the fire began to spread and help was needed. The buildings involved were 509, 517, 517½, 529, and 531 Carrall Street, with the fire starting in 517 Carrall.

Before the fire was knocked down it was learned that two men had died on the top floor of one of the buildings. Four others were treated at St. Paul's Hospital for smoke inhalation and one for injuries suffered when he jumped from a third-floor window. The cause was not deter-mined but thought to be electrical in nature, and with the poor housekeep-ing in such crowded conditions the rapid spread destroyed needed evidence. Fifty elderly men were left homeless.

On the second alarm, No. 4 and No. 6 Pumps responded to the fire, according to the control card (see chapter 3 on Vancouver's fire alarm system). En route, at Seymour and Dunsmuir Streets, they nearly col-lided. No. 4 Pump, northbound on Seymour, turned right onto Dunsmuir and cut off No. 6 Pump, which was eastbound on Dunsmuir. They were so close that a car couldn't have parked in the space between them and the two men on the tailboard of No. 4 Pump tried to pull themselves up onto the hose bed to avoid being hit by No. 6 Pump. Needless to say, both crews were somewhat shaken.

Shortly after arrival on the scene of the fire, the district chief asked what No. 6 was doing there and the captain said that the control card indi-cated that No. 6 Pump responded. It turned out that all the control cards in the city for multiple alarms in that area, using Box No. 1264, Shanghai Alley and Pender, had all been updated, except the one at No. 6 Hall.

Picture shows the aftermath of the Shanghai Alley fire where two people died. *Photo: VFD Archives*

Two Die in Hose Wagon/Car Collision

Response to another 2-6 alarm at 11:30 p.m. March 10, 1966 had dire consequences when No. 2 Hose Wagon collided with a car while respond-ing to a lumberyard fire at East 20th Avenue and Commercial Drive. With lights and siren operating, the hose wagon was eastbound on Hastings Street when at Clark Drive it struck a northbound car, killing its driver. The two men on the tailboard of the rig were thrown off on impact and slid along the roadway, striking the curb. Firefighter Don McCavour, age forty-four suffered head injuries and died five days later in St. Paul's Hos-pital. The lieutenant, the driver, and the other firefighter suffered rela-tively minor injuries and were released from hospital after treatment.

The hose wagon was one of the old spare 1948 LaFrance-Dodge pumps. It was felt that if the crew had their regular rig, which was one of the new, larger 4,725-liter Hub-International pumps, all of the firefighters would have been better protected in a collision such as this one.

A civic funeral was conducted on March 17, and Donald Harry McCavour, an eighteen-year veteran, was buried in the Firemen's Plot in Mountain View Cemetery. He was survived by his wife and two sons.

The funeral procession of Fireman Don McCavour traveling down Burrard Street.
Photo: VFD Archives

The New Aerial Arrives

On March 28 the new Hub-International 100-foot (30-m) tractor-aerial was put in service at No. 6, and members were cautioned not to make any disparaging remarks about it. In spite of some of its good points, like maneuverability and easy handling, it had some engineering flaws: a turntable leveling device that never worked properly and was finally welded in place, and manual jacks on the tractor that were removed after one fell off. The fenders over the tractor's dual tires were torn off when the trailer and tractor flexed in different directions on uneven ground, and a booster hose-reel nozzle was cut off when it got caught between the turntable and the water tank when someone failed to roll the booster hose tightly enough onto the reel.

Many agreed that the city's decision to buy cheaply — which was necessitated by budget constraints — was wrong. It was felt that the money would have been better spent on a tried-and-true custom tractor-aerial that would have cost more but would have lasted many more years than the ten-year life this truck had. It was placed out of service in August 1976, and disposed of at the city auction.

In May, city council relaxed the living boundaries when union President Gordon Anderson convinced them that the old limits were now outdated as all firemen had cars and getting to a large fire call-out was not like it was when the men had to walk or ride a streetcar or bus to work. The new extended area included New Westminster, Port Coquitlam and Port Moody, or roughly ten to fifteen miles outside the city limits.

No. 6's new aerial was previewed for the crew before being put in service. This is the only picture showing the rig prior to "VFD" being painted on the cab. *Photo: Author's collection*

The Washington Court Fire

The fire of the year was the Washington Court fire at Nelson and Thurlow Streets, November 12. The three-alarm fire in the six-story apartment building came in at 6:47 p.m., and had it been a few hours later there would likely have been loss of life. Even at the early hour of the fire, many in the all-woman apartment had to be rescued from balconies and assisted down fire escapes.

The cause of the fire was suspicious and believed to have been caused by a disgruntled suitor. The damage to the building and contents was over $400,000. In spite of the damage to the top floors of the building it was rebuilt, but the sixth floor was eliminated leaving it a five-story building.

This was the first Vancouver fire at which a "platform" was used. At the height of the fire a city electrical department platform was jerry-rigged with a line of hose and a couple of men and pressed into service as an additional elevated turret. It would be March 1973, with the arrival of the Firebird 125, before the department put its first fire-fighting platform in service.

At a December city council meeting, Chief Jacks recommended that all future fire apparatus be delivered with covered seating behind the cab

The first use of a "platform" on the VFD. Firefighters are shown using an electrical department giraffe to fight the fire.
Photo: A. Duplissie

for the men, reducing the need for men to ride on the tail-board of the rigs. The recommendation came following the investigation into the accident that caused Firefighter Don McCavour's death. Subsequently, two extended-cab equipped pumps were ordered from the Hub Fire Engine Company at a price of more than $60,000.

On January 1, 1967, the Vancouver Fire Department began using the twenty-four-hour military time.

Fire Apparatus Colors Change

The city council approved Chief Jack's recommendation that all future fire apparatus deliveries have a new, two-toned white-over-red paint job. Two Hub-International 1,050-gpm (4,725-liter) pumps scheduled for delivery at the end of the year would have the new color scheme.

Hose Wagon No. 1 is shown in the new red and white livery.
Photo: Author's collection

The largest fire of 1967 was the four-alarm warehouse fire at 1060 Homer Street that caused damages totaling $750,000 to stereos, television sets and other electrical equipment. The fire was reported just after midnight on August 29 and required the off-shift be called out. Soon, the 150 men on scene were able to confine the fire to its building of origin.

The Washington Court, shown the day after the fire.
Photo: Vancouver Province: Ross Kenward

A four-alarm warehouse fire on Homer Street.
Photo: Vancouver Sun

193

Fire Dispatcher George Shields showing his medals. *Photo: CVA 354-324*

Chief Ralph Jacks announced January 18, 1968 that Rescue & Safety Chief Marvin Wortman, age fifty-four, was promoted to assistant chief to fill the vacancy left by the death of Assistant Chief Charles Sutherland, who had succumbed to cancer in October at age fifty-three.

On April 11, there was a routine fire, a burning mattress in a second floor apartment, but this time something went wrong. As two firefighters were attempting to throw the smoldering mattress to the ground from the balcony, the railing gave away and the firefighters fell about ten feet, one striking his back on the first-floor porch railing. Firefighter George Shields' back was broken and he was partially paralyzed.

Fourteen months later, after ten months of treatment and rehabilitation, George returned to work on the VFD as a fire alarm operator, a high-stress position at which he excelled. Over the years, he competed in various wheelchair sporting events, winning many major medals. He retired in 1991.

The Beacon Hotel Fire

At 2105 hours May 25, the downtown eastside Beacon Hotel, at 7A West Hastings, was the scene of a two-alarm fire that took the lives of four men. The squalid living conditions contributed to the deaths and damage to the building — a four-story, turn-of-the-century hotel used in the early days by various vaudeville acts playing the Pantages Theatre that had been directly across the street. Some firefighters, disturbed by the poor housekeeping conditions of the fire scene, stated that the building was cleaner after the fire because of all the garbage they had shoveled out of it. Cause of the fire was believed to be alcohol-related careless smoking.

The front of the Beacon Hotel following the fire. *Photo: Vancouver Sun*

The Clarence Hotel Fire

Then a month to the day later, another hotel fire six blocks away, at Pender and Seymour Streets, claimed the lives of five men, four in the fire and one later in hospital. The fire alarm came in at 2120 hours for the Clarence Hotel, a seventy-five-year-old building housing primarily older men. When the first fire companies arrived, many people were driven to the windowsills by the thick smoke and their rescue was the first priority. Several were sent to the hospital suffering from smoke inhalation.

The cause of the third-alarm fire, was determined to be careless smoking in one of the second-floor rooms. Both the second and third floors were gutted. Chief Jacks said the conditions in the hotel were worse than those encountered in the Beacon Hotel fire the previous month because the rooms were much smaller and space was very limited.

Election day fire at the Clarence Hotel. *Photo: Vancouver Sun*

Smoke and water damage was extensive down to the main floor, containing the lobby, stores and the beer parlor, which normally would have been full of people, including most of the tenants. Many people felt that had the beer parlor been open, most, if not all of the dead would have survived; the beer parlor was closed because June 25 was the day of the federal election.

On September 9 of that year, the roller coaster at the Pacific National Exhibition's Playland was damaged and several midway buildings were destroyed. The fire was set by a group of boys who were seen fleeing the scene and the damage was set at $175,000. Luckily, the fire never got a good hold on the roller coaster or the spindly, all-wooden structure would also have been lost.

In 2006 the coaster is still operating and considered by international roller coaster enthusiasts to be one of North America's best rides.

Members and residents were evacuated from the Terminal City Club on West Hastings Street at about 2330 hours December 19 when a fire on the fifth floor filled the building with smoke. The lower floors of the private businessmen's club suffered water damage while the offices on the fifth floor were gutted. Three firefighters were injured and were treated

The PNE's roller coaster was saved from destruction by fire. *Photo: Vancouver Sun*

Terminal City Club fire at
837 West Hastings.
Photo: Vancouver Sun

Firefighters aiding rescued tenants of the
Stratford Hotel overcome by smoke.
Photo: Vancouver Sun

St. John Ambulance Association fire
at Granville and Davie Streets.
Photo: Vancouver Province

on scene and at the hospital and released. Damage was estimated at $95,000. The cause could not be determined.

Eight persons escaped unharmed at 0600 hours February 6, 1969, when fire raced through an apartment at 28th Avenue and Main Street. The three-alarm fire damaged six top floor suites and three stores below received extensive smoke and water damage. Officially there were no injuries but, during the course of the fire, Chief Jacks and a group of firefighters with hand lines were in the smoked-filled drug store when a large fluorescent light fixture came crashing down on their heads. The cause of the fire was undetermined and damage was estimated at over $150,000.

The costliest fire of the year was yet another warehouse fire. With damage exceeding $800,000, the three-alarm blaze came in at 0515 hours but had been burning for some time before it was turned in. Well over 200 firefighters, including the off-shift fought the paper-product-fed fire that kept the department busy for most of the day. Cause of the March 24 fire in the 1100 block Homer Street was not determined.

Another three-alarm fire in a Downtown Eastside skid road hotel

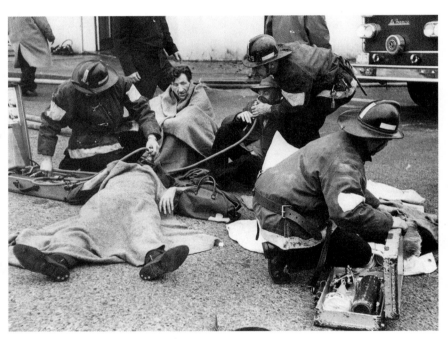

drove 200 people into the street, mostly elderly men on welfare, when the third floor of the Stratford Hotel was gutted at 1700 hours on April 3. Eleven of the tenants were taken to hospital for smoke inhalation and fractures from jumping to safety. Many of the tenants were handicapped and in wheelchairs, and firefighters had their hands full bringing them out to safety. Cause of the fire was said to be careless smoking, and damage totaled $50,000. The fact that the fire came in the late afternoon rather than the middle of the night, and that the stairway doors on the fire floor were closed, saved many lives.

A downtown landmark was destroyed by fire in the early hours of May 10. The old, vacant building that housed the St. John Ambulance Association at Davie and Granville Streets, was believed torched by transients. Damage to the three-story building, store and the hotel next store was estimated at $115,000.

A 3-6 alarm (a third alarm, without calling the off-duty men), at 2355 hours May 27 at the Pacific National Exhibition (PNE) stables killed twenty-one race horses and caused the stress-related death of a forty-eight-year-old trainer who was trying to free some of the trapped horses. The fire raced through the stables very quickly and firefighters had a difficult time laying hose lines into the area because the freed horses were galloping wildly around the track and many were trying to get back to their stalls. The fire was called suspicious and some juveniles were thought to be to blame. The dollar loss was in excess of $115,000.

PNE stable three-alarm fire, May 27, 1969.
Photo: Vancouver Sun

Armand Konig Becomes the City's Tenth Fire Chief

In July Fire Chief Ralph Jacks retired after forty-one years' service and turned over the leadership of the VFD to Armand Konig, age forty-eight, an acting captain who had been named the city's new fire chief in June 1969.

Shortly after Chief Konig's appointment, his brother-in-law, Jim Tuning, a twenty-seven-year veteran, applied for and was selected by council to be Konig's deputy chief.

In 1965 Konig, a Forty-Sixer, was promoted to assistant chief from lieutenant, but the Union membership strongly voiced their objections at a special meeting because he was too junior at the time. The city then rescinded his promotion, but many believed that one day he would get the top job. Over his years as the fire chief he made many changes and improvements.

Armand Konig

Born: June 28, 1920, Vancouver, BC
Joined VFD: August 1, 1946
Appointed Chief: June 18, 1969
Retired: June 30, 1980

New Rigs, More Fires and Too Many Deaths

(1970–1974)

The Belmont Hotel fire of March 5, 1970 claimed the life of a young woman and sent eight people to hospital. The first alarm came in at 0600 hours and soon turned into a third alarm. The two top floors of the six-story building were involved and firefighters had it under control within an hour and a half. It was while they were overhauling the fire scene on the top floor that they discovered the woman's body; her husband had escaped but wasn't able to save his wife. (It was later learned that the woman was related to a Vancouver fire captain.)

Another young couple in a fifth-floor room was very lucky to get out. They woke up to the commotion of the fire with smoke in their room.

Finding the hallway full of smoke, the husband broke the window to let the smoke out and poked his head out just as a firefighter did the same thing at the next window, a half-story higher in a stairway. The firefighter had a rope, which he threw to the man, who then assisted his wife out the window on the small half-inch line. She swung out and was pulled into the stairway window with the help of three firefighters. Then the naked man, who had given some of his clothes to his wife to protect her modesty, got the rope back and tied a loop around his waist. He hesitated to jump until the heat from the fire and the hot water and burning roofing materials running down the wall forced him to leap. He swung out like a pendulum and four firefighters managed to pull him to safety. He told his rescuers that the smoke in his room was so thick he couldn't see to put any clothes on. "He was so glad to be alive after his ordeal, he vowed never to take drugs again!" said Firefighter Gerry Marleau, one of the rescuers.

Fighting the fire in the top-floor rooms of the Belmont Hotel.
Photo: VFD Archives

The newly married couple from Calgary lost everything they owned, including their welfare check. The firefighters were later commended for their action but "reminded" that such a small rope — normally used to raise and lower equipment — should not be used for a rescue. (Rope used for rescue is ¾ inch [20 mm].) The downtown hotel, at Granville and

Belmont Hotel three-alarm fire. *Photo: VFD Archives*

Nelson Streets, was repaired and reopened for business as the Nelson Place Hotel.

The Local 18 Public Relations Committee

On May 5, 1970 members of Local 18 formed a public relations committee to "promote good public relations and awareness through the press, radio and television." In the old days, only chiefs were authorized to give out fire information, which often made it difficult for the press to get what they wanted because some chiefs wouldn't take the time to deal courteously with them. It was learned that the media of the day didn't have any good sources except the Fire Alarm Office, which was often too busy to be of help.

With the help and advice of many media people, a plan was laid out and presented to Fire Chief Konig and, with his authorization, press cards were distributed to all media members, and lines of communication were opened up to satisfy their needs.

The firefighters became more involved in various civic and district parades, events, displays and neighborhood activities that needed some of their support, as well as their own Fire Prevention Week, Muscular Dystrophy, and UNICEF drives.

On July 24, 1970 the last telephone alarm (3-3) and the last box alarm — to Box 2891, Renfrew and Charles Streets — were transmitted on the Gamewell Fire Alarm tape system that had been serving the VFD since 1890, and the new tone-alert, verbal dispatch system began operations.

On September 23 Captain Lloyd Love of No. 3 Hall collapsed and died on a run to 13th Avenue and Main Street. The call was for "alarm bells ringing," (often a maliciously pulled alarm station) but on investigation no fire was found. Love was working as a substitute for one of the other shift captains in the firehall. One of the Forty-Sixers, he joined the VFD on August 1, 1946.

The Firefighters' Union and Chief Konig worked together to create new press cards. *Photo: VFD Archives*

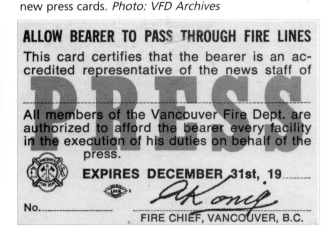

ALLOW BEARER TO PASS THROUGH FIRE LINES
This card certifies that the bearer is an accredited representative of the news staff of

PRESS

All members of the Vancouver Fire Dept. are authorized to afford the bearer every facility in the execution of his duties on behalf of the press.

EXPIRES DECEMBER 31st, 19____

No._____ FIRE CHIEF, VANCOUVER, B.C.

Crawford Storage Warehouse Fire

Once again fire struck in the warehouse district of the city and, like church fires, warehouse fires often became major fires due to the large spaces, lots of flammables, and the fact that the fire often has a good start before an alarm is turned in. The three-alarm fire on November 5 at the Crawford Storage building was no exception. By the time the fire companies arrived, the contents of the building, described as everything from drinking cups to insecticide, were burning with many aerosol cans exploding; it took four-and-a-half hours to bring it under control. The building and the exposures were saved, but the bulk of the contents were lost with damage estimated at $750,000. The cause of the fire in the two-story building in the 1000-block Hamilton Street was not determined.

Rookie firefighter Brian Singleton was injured and lucky to get out

The Crawford Storage warehouse three-alarm fire.
Photo: VFD Archives

Injured Firefighter Brian Singleton being assisted by Captain Jock Stewart and his crew. *Photo: Vancouver Sun*

alive. Struck by some falling live wires, he was knocked off his feet. When he fell, his air mask face-piece came off and he was overcome by smoke. Treated at the scene, he was taken to the hospital for the effects of smoke inhalation then sent home.

Before the fire companies cleared the scene, a second third-alarm warehouse fire occurred on Grant Street on the east side of town. The concrete block structure was razed. Cause was never determined and the contents, mostly grocery items, valued at about $250,000, were destroyed. With two simultaneous three-alarm fires in progress, the VFD resources were stretched very thin that night. Fire losses totaled $1 million for the night.

No. 9 Hall's new LaFrance pump.
Photo: Author's collection

In November and December the first two fire apparatus ordered by Chief Konig arrived in the city and were placed in service at No. 6 and No. 9 Halls. They were identical LaFrance-Ford C-Series, 1,050-gpm (4,725-liter) pumps in the new red and white color scheme and the first to have "Fire Dept." across the front of the cab and "Vancouver Fire Dept." on the cab doors.

The Stork Visits No. 9

On the evening of January 12, 1971 a family that was racing to the hospital decided that, because of the heavy snowstorm, it might be better to find an alternative venue for their impending birth. Just as they were passing No. 9 Firehall, the birth began to happen.

The driver ran into the firehall, explained the situation and the crew jumped into action. They moved the truck out of the hall and had the man back his station wagon into the now-empty truck bay. A phone call was made to advise fire dispatch that No. 9 Pump had a "verbal alarm" and to request that an ambulance be sent — fast!

Things began to happen quickly and, amid lots of excitement, a baby was born. The captain, right in the thick of things, got so excited he couldn't tell the crew, when they asked him, if it was a boy or a girl. Baby Berner, turned out to be a little girl, weighing in at 4 pounds 11 ounces (2.5 kg).

Thibault-Ford aerial shown at No. 3 Firehall. *Photo: Author's collection*

New Aerials and Super Pumps Arrive

In February the new equipment began to arrive and Nos. 1 and 2 put new aerials in service: No. 1, a Thibault-Ford 100-foot (30-m) service aerial, and No. 2 got the LaFrance-Ford 100-foot service aerial, not only the last LaFrance aerial built in Canada, but also the last gas-powered rig on the VFD.

No. 1 Hall's Super Pump.
Photo: Author's collection

After fifty-seven years of fire apparatus production in Canada, the American-LaFrance Fire Engine Company went out of business due to declining sales and competition from other Canadian builders. During its years in business in Toronto it produced 1,264 motor fire apparatus, according to American-LaFrance historian John Holden. In 1997 the re-structured company returned to Canada.

Around this time, the first two of the King-Seagrave "Super Pumps" arrived. The four identical pumps were ordered for the firehalls situated around False Creek and they were ordered as replacements for the forty-four-year-old fireboat, *J.H. Carlisle,* scheduled to be retired. When the fireboat was launched in 1928 for service in False Creek there were almost thirty mills that needed fire protection, but in 1971 only four remained, and the area could now be better served by land companies because the shoreline was easily accessible.

Each of the pumps, the department's first diesel-powered rigs, was

Four King-Seagrave Super Pumps
at the test demonstration.
Photo: VFD Archives

rated at 1,500 imperial gallons per minute, but when they were tested and displayed for the mayor, public and press, it was found that each could pump in excess of 1,750 gpm (7,950 liters). By the end of July all were in service at Firehalls No. 1, 2, 3 and 4. The total capacity of the four pumps was more than 7,000 gpm (31,500 liters) compared to the fireboat's 4,550 gpm (20,650 liter).

The fireboat was taken out of service August 5 and tied up at No. 10, the Burrard Inlet fireboat station, at the foot of Campbell Avenue until it

The fire scene involving the three houses on West 7th Avenue, showing No. 4 Pump.
Photo: Vancouver Province: Ken Allen

One of the fire dept. scuba dive teams. *Photo: VFD Archive*

The morning after the fire, revealing the damage to the three houses.
Photo: Author's collection

was sold for $13,333 to Clean Seas Canada Ltd., who used it to clean up spills.

In June the VFD scuba team was back in business after an absence of six months because of the lack of available divers. All volunteers who received no extra pay, these new divers completed a three-month course and were now available as required.

On the warm summer evening of July 14, 1971 a fire in the 1000-block West 7th Avenue destroyed three houses and badly damaged two others in another spectacular two-alarm blaze. Because of the steep hillside terrain in the area, fire crews had a difficult time getting to the rear of the houses; the houses were close together and there was no back lane. As a result, the middle house of the five was so consumed by fire that when it collapsed in upon itself it disappeared down into the back yard. It took two hours to bring the fires under control. Most of the houses were occupied by groups of young people, some squatters. There were no serious injuries and the cause was undetermined, but poor housekeeping and careless candle use was suspected.

Vancouver firefighters received a wage hike of 17.2 percent, bringing the wage of a first-class firefighter to $849 per month from $724, retroactive to January 1.

In the early afternoon of September 8 a large black column of smoke was seen from No. 1 Firehall and it looked like another nearby warehouse

THIRD SECTION — **The Sun** — LIVELY ARTS, LIVING TODAY
VANCOUVER, BRITISH COLUMBIA, THURSDAY, SEPT. 9, 1971 ****35

The CPR locomotive was extensively damaged.
Photo: Vancouver Sun: Dave Donnelly

was on fire. The alarm that came in directed firefighters to the CPR roundhouse in the area just south of the warehouse district.

The first-in crew found that one of the large CPR diesel locomotives had derailed going onto the roundhouse turntable and had ended up nose-first in the turntable pit. As it went off the tracks, one of the diesel tanks ruptured, the fuel ignited, and the engine was rapidly being destroyed. Lines were laid from a nearby CPR hydrant and the fire was extinguished; however, the damage to the engine and turntable was extensive.

Effective January 1, 1972 the position of president of Local 18 was made a full-time position and Gordon Anderson served in that capacity from 1972 until October 1977 when he retired from the VFD. In February 1978 he was appointed British Columbia's first fire commissioner, retiring in 1988.

During January *Vancouver Fireboat No. 2* was converted from gas to diesel power with the installation of two V16-71 Detroit diesel engines for propulsion and three V12-71 Detroit diesels for pumping. On February 1 the Canadian Underwriters determined her capacity was 18,400 gpm (83,500 liters) and anticipated this would exceed 20,000 gpm (91,000 liters) when the engines were broken in, which it did.

Arthur Fiedler Visits Vancouver

The world-renowned conductor of the Boston Pops and long-time fire buff, Arthur Fiedler, visited the city as guest conductor of the Vancouver Symphony Orchestra, April 17, 1972. On the evening of his opening, much to his delight the maestro was driven from his hotel to the Queen Elizabeth Theatre on the department's vintage 1912 American-LaFrance hose wagon.

At the intermission, Mr. Fiedler was presented with a white fire chief's helmet and declared an "Honorary Fire Chief of the Vancouver Fire Department" by Deputy Chief Jim Tuning. He accepted the helmet and the honor with, "What, no badge?" which he explained was the traditional American presentation. The Boston-born musician was a collector of fire memorabilia and was an honorary fire chief of many fire departments around the world.

Arthur Fiedler, guest conductor of the Vancouver Symphony Orchestra, with fire fighters Alex Matches and Al McLennan, is driven to his concert on Vancouver's 1912 hose wagon.
Photo: Vancouver Sun: Brian Kent

On a subsequent visit to the city in the mid-seventies, he was again driven by members of the VFD, this time on the 1926 LaFrance pump, to the "Topping-off" ceremony of the 700 West Pender Building. Arthur Fiedler died on July 10, 1979 at age eighty-four.

On June 4, 1972 another stable fire at the Exhibition Park racetrack occurred, causing tragic results. About 2030 hours a groom fell asleep in one of the tack rooms with a cigarette and started a smoldering fire in a sofa. Three other grooms discovered the fire, got the sleeping groom outside, removed the sofa, doused it with water and left it, thinking it was out. Then the driver of a passing pickup pushed the sofa up against the wall of a feed shed to get it off the narrow roadway. A short time later it burst into flame, ignited the shed, and quickly spread throughout the stable area, destroying two barns and killing twenty-five horses.

Upgrading of the buildings at the track was now called for, with the installation of sprinkler systems to help reduce the incidence of fires like this.

A new, two-year wage package was awarded the city's firefighters when they received a 9 percent increase for 1972 ($914 per month) and 6.5 percent for 1973, beginning January 1, which raised a first-class firefighter's wage to $1,069 per month.

Three Major Fires in Eighteen Hours

At 0730 hours August 3 a fire in progress was reported at 321 Water Street, and by the time firefighters arrived on scene, the fire, which started on the second floor of the six-story Mussenden Building, was well on its way to the roof and was quickly made a third alarm. The false ceilings in the building made it difficult to get at the fire and it took more than four hours to bring it under control.

Damage was estimated at $1.2 million to the building and to the seven shoe, textile and clothing businesses that occupied it. Five female employees were injured when they jumped to safety and a fire captain suffered cuts to his hands. The fire was described as another typical warehouse fire, and it was expressed once again that sprinklers would likely have helped save the building and some of the contents.

Then the next morning, at 0100 hours yet another fire at the Exhibition Park stables killed twelve horses and caused considerable damage to the stables. Horses that were released onto the track once again posed a hazard to firefighters trying to lay hose lines into the stable area. Soon after the arrival of the first-in chief, the fire was made a second-alarm. Some of the horses got out of the compound and were seen running down the nearby streets and along the freeway median.

Then at 0130 hours another second alarm came in for an apartment fire in the 3400-block Kingsway, which fortunately was quickly extinguished. The cause was electrical and forced people from thirty suites into the street, but there were no injuries.

It was a busy eighteen hours, with three multiple alarms in three corners of the city and, as it turned out, No. 1 Pump attended all three: the Water Street fire which was in No. 1's fire district; the PNE stable fire which it answered on the second alarm, and the apartment fire as a special assignment, because it was available and able to go as it hadn't laid any hose or committed any other equipment.

The Phoenix Rooms Revival

In late 1972 No. 2 R&S received an inhalator call to the Phoenix Rooms on Hastings Street, just behind the firehall. Upon arrival on the top floor, Firefighters "Ike" Robertson and Gerry Marleau found a big, burly thirty-year-old male lying on his back, without a pulse and not breathing. Suspecting a drug overdose, they inserted an airway (a curved, flat vinyl tube that, when inserted in the mouth, extends from the lips down the throat and allows passage of air into the lungs) and began cardio-pulmonary resuscitation (CPR).

The police and ambulance attendants arrived on the scene and instantly, as if on cue, the patient sat up and began talking to the police in a thick accent, completely oblivious to the airway in his mouth and throat. As he now appeared to be okay, the firefighters began gathering up their equipment and prepared to leave and Marleau reached around a police officer and gently removed the patient's airway.

The young man immediately accused Marleau of stealing his dentures and called him a "smart-ass" who would only do that with police protection. He then threatened to place a "Newfie stranglehold" on him if he didn't return his teeth!

While the argument continued the crew returned to their rig, but before they could pull away the young man was at Marleau again for the return of his teeth. Marleau then told him to talk to the ambulance crew about them and drove back to the firehall. Shortly after, they were told by the man on watch at the hall that they had a visitor. Sure enough, it was the "Phoenix Giant" once again demanding his teeth and threatening the dreaded Newfie stranglehold.

By this time, most of the fire crew had gathered around the commotion and were amused by the rescue crew's predicament. Marleau, on impulse, asked the man whether it was the uppers or the lowers that were missing. "The uppers, the uppers," he excitedly exclaimed. Marleau suggested he reach in and check. He did so and took his upper plate out. Then he said that it must have been his lower set that Marleau took. Once again he was told to check and again he found his plate intact.

Now he had both plates out of his mouth and began swinging his hands around in bewilderment, with everyone expecting to see them shattered on the floor. In a moment the young man settled down and then shouted at Marleau, "How'd ya do that? That's the greatest damn trick I ever seen!" The firefighters' laughter followed the big man all the way back to the Phoenix.

The new Vancouver Fire Department flag.
Photo: Author's collection

The First Departmental Flag

It was Chief Konig's idea to have a departmental flag, and he wanted one developed for the band in time for the annual Pacific National Exhibition Parade in August. He wanted something that would exemplify the department's pride in its history and traditions. After several weeks of trying different colors, styles, and various elements, a design was drawn up and presented to the chief for his comments. The colors and dimensions chosen were the same as for the Canadian flag. With white and red signifying peace and danger, the gold VFD badge and band were used to illustrate the winning of peace over danger. In addition to the elements of the department's Maltese cross badge, 1886, the year of the formation of the department, was added. The chief ordered two flags, one for the band and one for his office.

Since the originals were made in 1973, others were ordered for the city and department's centennial celebrations in 1986 and were flown daily, along with the Canadian flag, at each of the city's firehalls. With the name change to Vancouver Fire and Rescue Services (VF&RS), a new flag was designed in 1995.

The VGH Fire Zone Incident

Numerous alarms are received from the sprawling complex of buildings that make up the Vancouver General Hospital (VGH). Early one spring evening, No. 1 Pump with Nos. 3 and 4 Companies arrived at the VGH's Bamfield Pavilion. Each company, with the assistant chief from No. 1 in charge, responded to their predetermined locations.

The fire alarm was found to be of a minor nature and was struck out. When the chief and his driver returned to their car, a member of VGH security approached them and proceeded to give the chief a lecture for parking his car in the fire zone! It is said that Chief Dick Enman, always cool under fire, "politely" informed the security person what a fire zone was for, and left him standing on the sidewalk with a red face and ringing ears.

The First Firebird Platform Arrives

On March 23, 1973 the 125-foot (38-m) Calavar Firebird 125 aerial platform arrived at No. 1 Firehall from California. The senior chiefs arranged to meet the new rig at the US border and escorted it into the city, via the Cambie Street Bridge, to the firehall. Only afterwards did it occur to everyone that the load limit on the old bridge was sixteen tons and the new rig weighed in at thirty-two tons!

The new Firebird 125 at No. 1 Firehall. Note doorway has been raised to accommodate this new large rig. *Photo: Author's collection*

At the McCormick's warehouse fire, the new Firebird 125 platform operated nose-to-nose with the aerial it would replace in a few short weeks. *Photo: Vancouver Sun: Dan Scott*

Not only was the new Firebird the heaviest piece of apparatus, it was also the tallest, so the canopy on the front of No. 1 Hall had to be removed and the top of the doorway of the truck bay had to be cut out to fit the big rig into the firehall.

Many weeks were spent teaching crews how to drive and operate the new platform before it went into service. As it was the first platform and rig of its type on the VFD, and a complex piece of equipment, every crewmember learned about its operation and capabilities.

On May 1, although not officially in service, it attended its first multiple-alarm fire at the McCormick's Limited warehouse in the 1100-block Hamilton Street. The top floor of the warehouse was destroyed in the three-alarm fire and the lower floors suffered major water damage, totaling more than $300,000.

At the fire the new platform operated nose-to-nose with the piece of apparatus it would replace in a few short weeks. The old rig that would soon be replaced was the 1950 Bickle-Seagrave eighty-five-foot (26-m) service aerial, called Shop No. 100 (Shop No. is a numerical way of identifying apparatus), and had been in service since July 21, 1950. The "Bird" was numbered as Shop No. 100, then it was realized that the old 1941 LaFrance city service truck, Shop No. 51, was still being used as a spare. As it was more practical to have an aerial as a spare than a city service ladder truck, the Bickle-Seagrave aerial was recommissioned as Shop No. 51, replacing the old LaFrance truck, and the Firebird went in service as No. 1 Truck on June 8.

In 1976 when the second Calavar Firebird 125 was placed in service, it replaced the re-commissioned Bickle-Seagrave aerial, Shop No. 51, the only rig on the VFD to be replaced, twice, by new rigs.

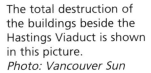

The total destruction of the buildings beside the Hastings Viaduct is shown in this picture.
Photo: Vancouver Sun

Hastings Viaduct Apartment Fire

A three-alarm fire at 0230 hours on June 1, at 1036 East Hastings Street claimed three lives and injured eight, including a firefighter who suffered burns dragging a physically disabled tenant to safety. The old, three-story wooden building at the east end of the Hastings Viaduct was a tinderbox, and within minutes the entire structure was burning. As No. 2 Company arrived on scene and headed to the nearest hydrant to hook up lines, people were scurrying from the building along the long porch that ran its length. By the time the pump got turned around and lines were laid from the nearest hydrant, most of the porch had collapsed along with the front part of the building.

The fire was determined to be arson and a suspect who had been seen recently at the racetrack fire was also seen at this fire. It was believed the fire was started at the rear of the grocery store. From the store, it spread throughout the forty-suite building then extended to the apartment building next door. Nothing remained of the store that was at the street and viaduct level. Only a gaping hole in the viaduct railing remained where the store had been before it collapsed to the ground level.

It took more than three hours to bring the fire under control, and crews spent all of the following day sifting through the completely collapsed buildings looking for more victims; none were found. The suspect was never caught.

The coroner, assisted by fire fighters, removing victims of the burned-out buildings.
Photo: Vancouver Sun

The Vancouver City College Fire

Vancouver City College fire.
Photo: Vancouver Sun

The dome just before it collapsed.
Photo: Frank Bain: VFD Archives

One of the most spectacular fires of the year was the Vancouver City College fire in the old King Edward High School building, at West 12th Avenue and Oak Street, on June 19. The first alarm came in at 1441 hours and very shortly went to a third alarm. The fire was believed to have started in the attic of the four-story high building, and within fifteen minutes the familiar dome was totally engulfed in flames and soon came crashing through the roof.

Over 1,000 students were in class when the fire broke out and all were evacuated to safety. The fire attracted many spectators, and with the heavy afternoon traffic the police were kept busy.

Opened in 1905 as Vancouver High School, in 1909 it became King Edward High, then in 1965 became Vancouver City College. The property was purchased by the Vancouver General Hospital for future expansion and the school was due for demolition in 1974. Cause of the total destruction was determined to be faulty wiring.

A hotel fire that could have had tragic results occurred on June 25 when an alarm was received for a fire on the fourth floor of the Georgia Hotel. The fire, involving two rooms, came in at 1315 hours and was soon a second-alarm. The crews attacking the fire attempted to use the house hose lines, but found that the hose cabinets were wired shut. Unable to easily access these, they used hose lines brought into the building via the Firebird from the pumps outside. The damage to the two rooms exceeded $50,000 and the cause was suspicious.

Charges were laid against the hotel for having wired the cabinets shut, but a provincial court judge dismissed the case, ruling that fire wardens as well as the hotel staff should have detected the wired cabinets during their inspections of the building. The cabinets had been secured to prevent misuse by vandals and pranksters attending the Grey Cup football festivities and final game eighteen months earlier.

The Nelson Street Fire

Three old two-and-a-half-story rooming houses in the West End were destroyed when a fire, believed started on the front porch of the middle house, spread to two adjacent houses. The first-alarm was reported by a passing cab driver at 0400 hours. He then found a man lying on the street with smoldering clothes and noticed a strong smell of gasoline in the area.

These men lived in the middle house and were lucky to get out alive.
Photo: Vancouver Province/ Ross Kenward

By the time nearby No. 6 Company arrived at the 1100-block Nelson Street, the three houses were described by the cab driver as "going like hell" and tenants were fleeing in all directions. The three-alarm fire killed three people and a fourth, the man lying in the street, died of burns later in hospital. Nineteen others were hospitalized with burns and injuries. Cause was determined to be suspicious.

The CIL Fire

Sunday, July 15, 1973, the five-alarm fire at the Canadian Industries Limited (CIL) was described by Fire Chief Armand Konig as a fire "that had the greatest potential for disaster of any we've ever had." The $2-million blaze in the CIL chemical warehouse on Terminal Avenue was turned in by a CNR train crew at 0812 hours and a city police officer at 0814 hours, and it quickly progressed to a multiple-alarm fire. At one point, consideration was given to evacuating the east end of the city because of the danger of cyanide and other toxic substances being dispersed into the air.

While firefighters were fighting the blaze that began with a loud explosion at the rear of the warehouse, many other explosions occurred within the building, some sending cylinders into the air like rockets. CNR train crews removed three tank cars and a boxcar from the site and workers and firefighters rolled many chlorine tanks to safety as the fire raged.

Fire companies took six hours to bring the fire under control in the 30,000-square-foot warehouse, and more than a little concern was voiced for the effects of the chemicals draining into the storm sewer system, even though it was well diluted.

The CIL fire scene on Terminal Avenue.
Photo: VFD Archives

One firefighter was knocked twenty feet off a ladder and suffered a cut forehead and hand injuries and four others were treated for smoke inhalation. The warehouse was destroyed but the general office was saved and sustained heavy smoke and water damage. It was not known how the fire started.

Earlier that morning, two firefighters went to the Royal Centre Mall to return the department's 1926 LaFrance pump that had been on display there and as they headed towards No. 1 they noticed the tall column of black smoke from the fire. As they came around the corner from Georgia onto Hamilton Street, one of the assistant chiefs standing in front of the empty firehall waved them down and began to direct them to respond to the fire before he realized that it was the antique rig that they were driving. The two firefighters spent the day manning the spare hose wagon at No. 13 Hall.

Effective September 1, firefighters began wearing yellow MSA helmets in place of the black ones they had worn for many years. Lieutenants and captains began wearing red helmets and chiefs continued to wear the traditional white helmets, with the addition of a leather front piece with VFD and the chief's rank on it. This feature of helmet fronts was eliminated about 1936 when the old leather helmets worn by all VFD ranks were replaced with composition helmets.

The new clothing issue was being worn at this time and had many new features: short sleeved summer shirts, work jackets and winter cold weather coats were issued. Shoulder patches were also introduced: red and white for all ranks, except chief ranks that wore patches of navy blue and gold. The dark blue shirts and red shoulder patches were selected so that firefighters would not be confused with policemen who, at that time, wore light blue shirts with blue and white shoulder patches.

The new VFD shoulder patches.
Photo: VFD Archives

Two Large Warehouse Fires

Large fires continued throughout 1973 with two large third-alarm warehouse fires. The first of the two alarms came in on the afternoon of October 17 when a four-story warehouse at 1072 Hamilton Street suffered $250,000 damage to paper products, paint, food and candy products. Five boxcar loads of toilet tissue had recently been unloaded and stored there, increasing the fire load significantly.

The warehouse fire at 1072 Hamilton Street October 17, 1973. Photo: *Vancouver Province: John Denniston*

The next day a second warehouse fire destroyed a fourth-floor business and included many bales of shredded paper and foam stuffing used in the manufacture of stuffed toys. Damage costs exceeded $200,000. This fire, at 780 Beatty Street, like the one the day before, could have had less damage if the sprinkler systems had been operating. Chief Konig stated that it was the owner's prerogative to have operating sprinkler systems, but he wasn't going to risk his men's lives by sending them into buildings such as these when no lives were at stake.

Firefighters at the rear of the three-alarm fire on Hamilton Street. The group on the left is four recruits getting on-the-job training with their training officer. Photo: *Vlad Keremidschieff: VFD Archives*

The Commercial Hotel Fire

In the early morning hours of October 21, a fire in the Commercial Hotel at 240 Cambie Street claimed the lives of five men; four died on the fire floor and one died falling or jumping from a window, narrowly missing a firefighter making up a hose line at the rear of No. 1 Hose Wagon.

Successful rescues of three other tenants from windowsills were made by firefighters using ladders, despite the wires and other obstacles they encountered in the lane behind the hotel. The entire top floor of the old four-story skid road building was gutted.

On investigation following the fire, it was found that the fire had started in the fourth floor washroom and a flammable liquid had been used. An empty paint thinner can was found at the scene. Three weeks later a coroner's jury ruled that the deaths were homicide and the work of an arsonist. They recommended that a new upgrading fire bylaw that included installing sprinkler systems, be instituted July 1, 1974, rather than the projected date in 1975. It was also recommended "a special emergency telephone number, such as 9-9-9 be developed." The now-familiar 9-1-1 had been in use in the US since 1969.

In November an apartment fire claimed two lives, then a house fire claimed another, and 1973 ended as the worst loss-of-life year in the history of the fire department, with the deaths of forty people in fires.

The Home Oil Barge Fire

On January 4, 1974 at 1420 hours an explosion ripped through the Home Oil marine fuel barge in Coal Harbor when a pleasure craft blew up while refueling. Three people were killed: two men and a seventeen-year-old boy. The men were career criminals and known drug traffickers, and one

Vancouver Fireboat No. 2 attacking the flaming Home Oil fuel barge. Photo: VFD Archives

of the ex-cons was believed to be involved in a Seattle bank robbery in March 1954 where a police officer was shot and killed. A definite cause was never determined, but careless smoking was probable. The fire was fed by 45,000 gallons (205,000 liters) of fuel, which threatened the three other nearby marine stations operated by Chevron, Shell, and Esso.

Vancouver Fireboat No. 2 was on scene in minutes and because of the extreme heat, heavy spray and fog nozzles were deployed; once cooled

A view of the fuel barges after the fire, with Gulf Oil replacing the Home Oil barge. *Photo: VFD Archives*

down, foam was used to extinguish it. Following the explosion, the barge began to drift toward the Esso barge and tugboat crews were credited with getting a towline on it and keeping it at a safe distance. This first towline was nylon and, as it burned through, a steel cable was attached.

The barge, still on fire, was eventually towed across the harbor to the North Shore where it was beached and the fire was left to burn itself out. Much credit was given the tugboat crews who did a tremendous job in spite of the dangers. Estimated loss was set at $150,000 and this was the first time that a fuel barge had been destroyed in more than thirty years of operation. The barge later reopened as a Gulf Oil barge.

A three-alarm fire in the 1400-block West Broadway on January 19, 1974 destroyed four buildings, including the carpet and flooring store in which it started, a pet store, a book and gift shop, and several apartments. The fire started at 1130 hours and was not brought under control until after 1345 hours when it spread through the old, wooden false-front buildings. The only casualties were fish in the pet store.

A $1 million fire at the Vancouver Sawmills Limited at 1111 East Seventh Avenue came in at 0230 hours on May 25 and totally destroyed the one-square-block complex. It was as if the entire block was a mound of kindling as the stockpiles of finished lumber burned like a bonfire. The heat was so intense that it cracked windows on a building across the street, and the side of No. 1 Pump facing the fire was badly scorched, with plastic parts being melted off. The revolving red light on the top of the cab also melted down onto the roof. Fifty off-duty firefighters were called out to fight the blaze that was struck out at 0528 hours, and the following day was spent overhauling the large site to ensure that the fire was completely out.

These firefighters were forced out of the basement by an explosion of flooring adhesives, and then the fire extended rapidly down the block to the apartment in the rear of the photo. *Photo: Vancouver Sun*

EHS Ambulance Service Begins

The British Columbia Emergency Health Services, Ambulance Service, more commonly referred to as EHS, began on July 1, giving the entire province the start of a uniform ambulance/paramedic service.

Effective October 1, 1974, new buildings in the city that were over seventy-five feet (24 m) were required to have a sprinkler system built in as part of their fire protection equipment.

The West Coast Transmission Building Fire

The West Coast Transmission Building in the 1300-block West Georgia was constructed from "the top down" in a design said to be earthquake-proof. A center pillar fifteen-stories tall, containing the elevators and stairways, was built first, then each floor was built on the ground, raised to the highest point and hung on huge cables and secured in place. This was repeated until all the floors were in place, then the walls and windows were installed, making it a truly unique building, suspended by cables and standing on a pedestal.

At 0215 hours on October 13 a spectacular and dangerous fire of unknown origin destroyed half of the fifteenth floor. Everything on the fire floor was burned beyond recognition. The heat caused the large windowpanes to fall to the ground in pieces; some of them floated down like playing cards. Luckily, no one was hit by these falling pieces.

With no sign of fire showing, the first-in captain and his "runner" (aide) took the elevator to the top floor. When the door opened, they met the heat of the fire head-on but, fortunately, were able to run across the hallway directly into the door leading to the inside stairway. The fire's extent was communicated to the ground, and then the fight began. All the hose lines and other firefighting equipment had to be carried up fifteen floors, and hoses were used from the standpipe system in the stairwell.

Damage was extensive and there was great concern for the integrity of the steel cables that held the building in place. There was also much concern for the computer room on the lower floors of the building, which, if damaged, could have had a significantly adverse effect on pipeline gas flows throughout the Greater Vancouver area. The next day, after the fire was struck out, the owners inspected the scene and could not believe the destructive result of a fire in what they mistakenly considered a fireproof building.

Many firefighters believe in the "rule of three," meaning that if one major fire occurs, often it is followed by two others.

The five-alarm St. John's United Church fire of October 29 was the first of three with two others to follow. The first alarm came in for smoke coming from the basement of the sixty-five-year-old church, which was reported by passersby at 0200 hours. No. 6 Firehall, a block away, responded its full company, but the fire could not be controlled and more alarms were called for. By 0300 hours the fire had spread throughout the basement and up the wood-paneled walls to the roof. Multiple alarms were sounded and the off-shift men were called out, and it wasn't until 0430 hours that the fire was brought under control.

Among the losses in the fire were a new $65,000 pipe organ, elaborate custom-detailed wrought iron railings, and stained glass windows. The Seaforth Highlanders Regiment, which used the church, lost two historic flags from the First and Second World Wars. Damage was estimated at $1 million and the cause was not determined.

St. John's United Church fire.
Photo: VFD Archives

The second fire was a third-alarm at the Georgia Rooms at Georgia and Main Street, in the early hours of October 31; the fire was burning on the top floor of the three-story building. When firefighters arrived on scene, ladders were raised and nearly fifty residents had to be helped out of the eighty-two-year-old skid road building that had been the original Woodward's Department Store. The fire safety upgrading had not yet been finished and the sprinkler system was not in operation. Three elderly men died in this fire, two on the fire floor and the third on the second floor in front of an exit door that had a one-story drop to the sidewalk below. Cause was determined to be careless smoking by a tenant who managed to get out safely.

The third major fire in eight days was a school fire. The third alarm on November 6 at Strathcona School, three blocks east of the Georgia Rooms, was turned in at 2330 hours for a fire on the top floor. At first only a small amount of smoke was showing, then it flashed over and the

Strathcona School fire.
Photo: VFD Archives

fire went through the roof. The Firebird was pressed into service and easily accessed the area. By 0130 hours the fire was under control, but the main wing of the school, built in 1891, suffered major damage. The cause of the fire was unknown but somewhat suspicious.

Georgia Rooms. *Photo: Vancouver Sun: George Diack*

New Firehalls Open

On December 5 the new No. 7 and No. 8 Firehalls opened at 0930 hours. Each opened as a one-piece firehall with No. 1 Hose Wagon becoming No. 7 Pump, and No. 2 R&S in new quarters at No. 8 Hall. When three of the four new firehalls were being built, No. 2 Hall was being built as No. 8, No. 7 as No. 2 and No. 8 as the new No. 7. Fire Chief Konig decided that Firehalls No. 6, 7 and 8 should be adjacent to each other and that the new No. 2 should serve the same area as the previous No. 2 Hall.

With the new fire bylaws regarding the upgrading of hotels and rooming houses well underway, the fire deaths in the city were reduced from a high in 1973 of forty, down to twenty-five deaths in 1974. Although still too high, steps were being taken to further reduce these numbers through a newly organized task force and more stringent fire company inspections.

Opening day at Firehall No. 7.
From left to right: Driver, Alex Matches, Bob Pachota, Captain Art Smith, Chris Stewart, and Senior Man Art Atkins. *Photo: Vancouver Sun: Brian Kent*

Canada's First "Class One" Fire Department

(1975–1985)

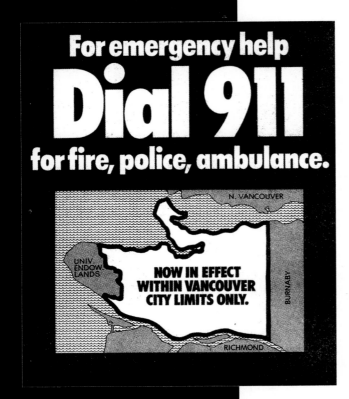

For emergency help
Dial 911
for fire, police, ambulance.

NOW IN EFFECT WITHIN VANCOUVER CITY LIMITS ONLY.

The year 1975 began with controversy when Fire Chief Konig, in his bid to increase the department's strength by sixteen men, made some remarks about the firefighters' absenteeism, and the press headlines reported that city firemen were goofing off.

The union president said that absenteeism was indeed excessive but blamed it partly on the fact that the average age of firefighters was between thirty and forty years and older men are more prone to sickness, that illnesses last longer, and that the increase in absenteeism was only just approaching that of other city departments.

The city doctor then cited the fact that a doctor must take the word of the patient and that anyone can read about the symptoms of everything from gastrointestinal problems to brain tumors!

The "malingering and goofing off" controversy died a slow death, and the fire department only got back to normal after many heated discussions.

The city council voted six to five in favor of hiring sixteen firefighters for the 760-man department (some of the fire department's "friends" on council voted against hiring more men). However, by the year's end seventy-four new firefighters would be hired, the largest number since hiring 108 men in 1964.

On March 10 at 1500 hours No. 1 Truck, the Firebird 125, moved to No. 7 Hall and became No. 7 Truck. At the same time, No. 1 Pump went in service as No. 8 Pump with No. 2 R&S, leaving only the administration office and machine shop at 729 Hamilton Street.

A $100,000 church fire in Kerrisdale destroyed the Knox United Church hall in the late afternoon of Sunday, March 23. The three-alarm fire necessitated the call-out of the off-duty firefighters who, after three hours, had the fire knocked down. Firefighters prevented its spread to the main building. Over 2,000 people watched the blaze, prompting the statement from a church member that "this is the best crowd we've had at the church in a long time."

Eighty-four year old wood frame building goes up in flames and smoke. The two-storey building had not been in use since Feb. 16 when an earlier fire ruined the interior.
—Peter Hulbert photo

Alexandra Neighborhood House fire.
Photo: Vancouver Province:
Peter Hulbert

Three-alarm fire razes city landmark

A three-alarm fire Monday afternoon destroyed Alexandra Neighborhood House, one of Vancouver's oldest buildings.

Hundreds of spectators and nine fire trucks clogged streets around the 84-year-old wood frame building at 1726 West Seventh after the 3 p.m. alarm.

Flames burst through the roof and lit up the interior while black clouds of smoke blew southeast across the city. When the blaze was brought under control about 4 p.m., the walls were still standing though the roof had fallen in.

The city fire warden's office is investigating the cause.

The two-storey building had been vacant since Feb. 16 when an earlier fire ruined the interior, said George Whiten, director of development for the Neighborhood Services Association which owns the building and the 1.5 acre lot.

Alexandra Neighborhood House as it was before fire destroyed it Monday.

"The building was due for demolition but we were trying to recycle useable material . . . hardwood floors and Douglas fir timbers . . . that's why it was still standing," he said.

It had housed offices of the Neighborhood Services Association but the bulk of the association programs now operate from Kits House, 2305 West Seventh.

"The building has one of the first cornerstones laid in the city . . . the historical society was quite interested in it and took three or four fireplaces and other artifacts out of it," Whiten said.

First the Alexandra Hospital for Women and Children, it then became the city orphanage and then a neighborhood house offering a variety of social service programs.

An eighty-four-year-old landmark was destroyed May 12 when a three-alarm fire razed the abandoned Alexandra Neighbourhood House in the 1700-block West Seventh Avenue. The fire alarm came in at 1500 hours for the old, wooden two-story building, and an hour later the roof caved in, leaving only parts of the walls standing. Originally opened as the Alexandra Hospital for Women and Children, it later became the city orphanage, then a neighborhood house offering a variety of social services. Earlier, in February, a suspicious fire severely damaged the old heritage building.

Hamilton Street Fire Headquarters Closed

On June 16 the administration office and the machine shop opened at the new No. 1 Firehall HQ at 900 Heatley Street, and on August 8 the fire companies went in service at No. 1 and No. 2. This final move into the new firehalls saw the end of the hose-wagon era, as there were no halls operating two pumps and the need to designate one of them a hose wagon.

The training department decided that the new No. 7 should be used as an interhall training site because of the large front apron and drill tower. On September 11, the first interhall drill was scheduled with No. 6, No. 7, No. 8, and No. 2 R&S. The companies arrived before 0900 hours for

New No. 1 Firehall. *Photo: Author's collection*

New No. 2 Firehall.
Photo: Author's collection

the start of the drill and when each piece of apparatus got parked it was found that with No. 6 Pump and Truck, No. 7 Pump and the Firebird, No. 8 Pump and the R&S rig, the district chief, the chief training officer and the assistant training officer's cars, plus the crews, which totaled over thirty men, there was no room to have a good drill! As a result, No. 7 Hall was removed from the interhall drill schedule. Then the hose tower was condemned as a drill site when it was learned that the safety-net hooks were unsafe for tower drills.

On September 21 at 2130 hours, fire raged through the Stanley Park Brockton Point sports field grandstand, leaving a pile of glowing embers behind. When No. 6 arrived on scene the structure was fully involved, with nothing left to save except the nearby large trees exposed to the flames. At the height of the fire an alarm was received reporting a fire at the park's miniature railway but on investigation this was found to be untrue. A second alarm was radioed in to get more firefighters on scene and to take care of any fires started by flying embers and brands. Cause of the fire was determined to be arson when it was found that gasoline had been used to start the fire in the women's washroom on the ground level.

The Burrard Grain Terminal Fire

One of the largest fires of the year for Vancouver firefighters occurred across the inlet on the North Vancouver waterfront, on October 3 when an explosion and fire at the Burrard Grain Terminal killed four workers. The fire was started when a stuck conveyor belt caused ignition of grain dust and North Vancouver City firefighters were soon fighting the worst fire in their history.

As there was no automatic mutual aid agreement system in place, if an adjoining municipality needed help from Vancouver their request had to be made through the mayor and/or fire chief and the aid would be sent. It was obvious from the Vancouver side that a large fire was in progress but it was about twenty-five minutes before North Vancouver City requested that *Vancouver Fireboat No. 2* respond to help with the blaze, and also to supply a badly needed water supply because of a broken water main. At the same time Chief Konig sent the Calavar Firebird 125-foot platform, increasing the firefighting strength by about ten men.

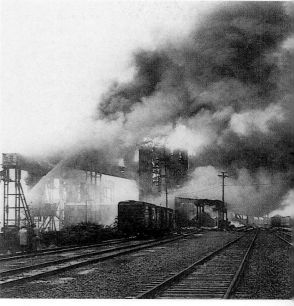

Burrard Grain Terminal fire.
Photo: NVMA #8512

One month later the City of Vancouver sent North Vancouver City a bill for services in the amount of $115,370. The North Vancouver mayor thought the bill showed a "lack of good-neighbor policy" and that Vancouver should bear some of the costs because "its facilities were the only ones available" to which Vancouver's council responded, "it was not our fire." Then it was suggested that the National Harbors Board should contribute and finally, almost two months later, after much political haggling, North Vancouver City accepted a bill for $16,000.

Chief Konig stated that, "the use of the fireboat on an "as used" basis is a cheap (way) out and is a bad policy." For years, the city had been trying, without success, to get the other municipalities to contribute to the fireboat's operating cost, which was more than $500,000 annually.

Now another battle was taking place. Because of additional vacation days negotiated and won by the firefighter's union, Chief Konig requested that an additional twenty-eight men be added to the department's strength. The council voted six to four not to increase the strength and also threatened the closure of No. 19 Firehall on the city's West Side, based upon a report by the city manager as the best way to reduce the manpower needs.

Konig responded that No. 19 was needed and the union's president, Gordon Anderson further stated that "fire fighting was the city's most vital service and it's ridiculous to eliminate a firehall." Then the neighbors got involved and one man interviewed pointed out that the hall was indispensable, and then told how these firefighters had saved his wife's life. The Vancouver Board of Trade wrote to council telling them it was folly to think of closing a firehall and a local community resources board sent a telegram to council stating that, "there was a greater need for inhalator services in the area now because of the recent opening of a seniors' home."

The firefighters described their hall as "the house of glass, brass, and grass" and stated that there was a lot to do to maintain the firehall, "a silent service" as one put it.

On Tuesday, November 25 council voted seven to three to give Chief Konig twenty-three more men and not to close the much-needed No. 19 Firehall.

The year ended with six fire deaths in three separate fires a week before Christmas. Two young girls lost their lives in a house fire in the East End. Then early December 20 at 0130 hours a fire was reported by a police patrol in a halfway house for parolees at 1020 Wolfe Street. The thick, black smoke rolling out the front door prevented the two policemen from affecting any rescues. Two men in the house died and four escaped the fire, believed caused by careless smoking.

The other fire that took two lives was in a co-op rooming house in Kitsilano at 2396 York Street. The fire that became a two-six alarm (second alarm) came in at about 0630 hours and was out about an hour later. Eleven people managed to escape into the snowy, cold morning air, some with little or nothing on. Cause of the fire was not known.

On March 7, 1976 retired Fire Chief Ralph Jacks died of a heart attack at the age of sixty-six years. Born in the Mt. Pleasant area of Vancouver (09-09-09), he joined the South Vancouver Fire Department on May 11, 1928, served through the ranks and retired for health reasons in February 1969.

The Norpac Fisheries Fire

At 2320 hours on March 24 a fire alarm came in for the Norpac Fisheries building on Commissioner Street, on the waterfront. Ammonia fumes hampered firefighters, but the fireboat prevented the $1 million blaze from endangering more of the buildings along the waterfront. Included in the loss of the building's contents was forty tons of herring roe being prepared for export to Japan. The four-alarm fire put about 100 employees out of work, but all were expected to be rehired at the new fish plant that

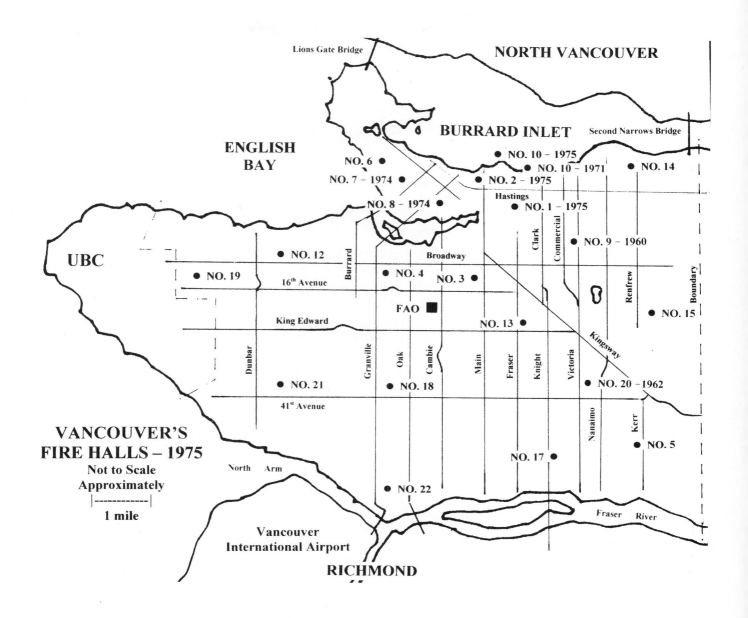

VANCOUVER'S FIRE HALLS – 1975
Not to Scale
Approximately
|------------|
1 mile

was nearing completion next door. The fire started in the compressor room below the cold storage area.

It was a miserable, cold and rainy night when the first alarm came in. On the second alarm, No. 7 Pump was tapped-out to respond to the fire and as the crew was preparing to go they were given a little razzing by the truck crew, who were remaining in quarters. En route, the pump received a medical call by radio to a skid road hotel and just as they arrived on scene the crew heard that the Norpac fire was now a third alarm and the truck from No. 7 would be responding to the fire. After No. 7 Pump finished at their call, fire dispatch changed their assignment to a fill-in at No. 14 Hall, instead of going to the fire. The pump crew had the last laugh — they stayed warm and dry while the truck crew spent the entire night at the cold and dirty fire scene until the day shift relieved them at 0800 hours.

On May 1, 1976, Vancouver police, fire and ambulance services began using the 911 emergency number.
Photo: VFD Archives

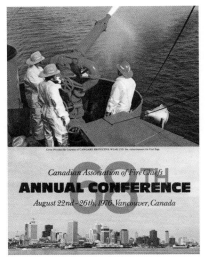

CAFC Annual Convention
Photo: VFD Archives

Two Hotel Fires

On Friday, July 16 shortly after 2300 hours, a fire at the forty-two-story Sheraton-Landmark Hotel in the West End sent twenty-five people to hospital with smoke inhalation and burns. The two-alarm fire started in a linen closet on the fifth floor and smoke spread throughout the building. The two people who suffered burns were coming down from the rooftop restaurant when the elevator suddenly stopped at the fifth floor and the doors opened to a blast of heat. A couple of people stumbled out of the elevator into the smoke, then the doors closed and the elevator continued to the ground floor.

Within minutes, firefighters had rescued the couple, a twenty-one-year-old local woman and a thirty-one-year-old visitor from Japan, and they were taken to the hospital where they were listed in stable condition. Sadly, the young woman, who had been released from hospital, was read-mitted in September, when she died. The night of the fire she had been celebrating her engagement and her twenty-first birthday. The cause of the fire was never determined, but arson was ruled out.

Then the following Monday morning at 0222 hours a call was received for the twelve-story Georgia Hotel for a fire on the third floor. The fire, confined to one room, filled the upper floors with smoke, making the evacuation of 225 people necessary. When the fire was knocked down it was discovered that the sole occupant of the room, a woman, had died in the fire. Twenty-eight people were treated in hospital, including three firefighters, who suffered from smoke inhalation. The cause of the fire appeared to have been the television set, which was destroyed.

Later that year Chief Konig hosted the Sixty-Eighth Annual Convention of the Canadian Association of Fire Chiefs (CAFC). Also around that time, Vancouver's Five-Year Plan advertisements encouraged the public to vote "yes" on the November 17 referendum to replace and relocate the old No. 14 Firehall. The ads showed a 1976 fire truck trying to fit into a 1913 firehall. Other firehalls slated for replacement were No. 19 and No. 22, and money would also be used to improve fire communications. The Five-Year Plan was passed.

VANCOUVER'S 5-YEAR PLAN

Your vote is vital on November 17th

Why is your vote so vital? Because it could decide the fate of the City's 5-Year Plan.

The firehall in the above picture is Hall #14 located at 2705 Cambridge. It was a great hall when it was built ... in 1914. Today it isn't good enough. Because it cannot handle an aerial ladder truck the north-east sector of the City is lacking in adequate rescue facilities.

Approval of the streets, street lighting and firehalls section of the 5-Year Plan would provide funds to relocate this 62-year-old hall as well as replace two other old halls, #22 (in Marpole) and #19 (West Point Grey). Funds would also be provided for the improvement of police and fire communication systems, permitting better emergency response.

This 5-Year Plan introduces a new concept for Vancouver. For the first time the plebiscite is split into four separate votes: 1. streets, street lighting and firehalls; 2. parks, recreation and libraries; 3. housing 4. Neighbourhood Improvement Program.

Each is counted separately.

Throughout the City there are essential needs. Please turn out and vote on November 17th.

For further information about the Plan, please phone City Hall Information, 873-7415.

Construction Crane Collapses

Fire and rescue companies and ambulance crews responded to a construction site near the Burrard Bridge where a construction crane collapsed about 1530 hours on April 1, 1977. Some final adjustments were being made to a mobile crane when something snapped and, within seconds, both cranes collapsed, killing one man and seriously injuring five others. Witnesses said that the counterweight and boom of the 100-foot (30-m) hammerhead (construction) crane toppled, causing the 230-foot (70-m) mobile crane to fall, crashing through a shed and several vehicles.

RESCUERS search through rubble Friday afternoon at the False Creek fishermen's wharf after 100-foot-high crane collapsed, killing one man and injuring another five.

—Glenn Baglo and George Diack Photo

Victim trapped in wrecked car is helped by rescuer.

Car lies crushed under end of crane.

Damaged tower still stands.

The Vancouver Sun

VANCOUVER, BRITISH COLUMBIA

★★★★CCC

AMID THE TANGLED RUINS . . . unidentified victim receives emergency aid from two members of the fire department's rescue squad
—Glenn Baglo Photo

Crane collapse kills one, injures five

"I heard big cracking noises and the whole thing was down in seconds. The big crane gave way and pulled the other one down with it."

Andy Kivinen was describing the collapse Friday of two cranes on a construction site at the south end of Burrard Bridge which killed one man and injured five others seriously.

Kivinen was the operator of one of the cranes and had left it at the end of his shift about 3:30 p.m., seconds before it collapsed.

Witnesses said the boom and counterweight of the hammerhead crane being erected on the site toppled, causing the 230-foot boom of a mobile crane to crash into a shed and cars on adjacent National Harbors Board property.

Dead is Thomas Rudnisky, 40, of 1305 Dominion Ave., Port Coquitlam. He was foreman on the site for Steveston Construction Co. Ltd.

Gary Hamilton, 27, of 4845 Westlawn,

Burnaby, is in poor condition in Vancouver General Hospital with broken legs and internal injuries.

John McGowan, 35, of Bolton, Ont. is in serious condition with a neck injury.

John Freeman, 25, of 1481 Berkley Ave. North Vancouver, Harold Hogan, 45, of 3964 Inverness, Port Coquitlam, and Rob McDougall, 31, of Maple Ridge, were all reported in satisfactory condition today. Freeman has a head injury, McDougall and Hogan, neck injuries.

Three of the injured were in a car and truck struck by parts of the crane booms as they crashed on to a wharf on False Creek.

The accident occurred as the 170-foot hammerhead crane was being prepared to begin work on the $6 million centre for the B.C. Credit Union League.

John Otterwell, project manager for Stevenson Construction Co. Ltd., said about 25 workers were on the job when the cranes toppled.

ANDY KIVINEN
. . . heard cracking noises

The mobile crane, supplied by Arrow Transfer Co. Ltd., had been brought to the site to erect the hammerhead crane.

Jasco Equipment Supply Ltd. was the contractor erecting the hammerhead crane.

Jack Charles, president of Arrow Transfer, said some witnesses thought a guy wire on the hammerhead crane had snapped.

The accident prevention section of the

More pictures, pages 47

Workers Compensation Board is investigating.

Kivinen, who has watched construction cranes being erected over the past 10 years, said he saw nothing unusual about the hammerhead crane's construction or the way it was being put up.

"They were just putting the finishing

"Crane" page 2

Construction crane collapse. *Photo: Vancouver Sun: Glenn Baglo*

A good friend of the firefighters was lost during April with the death of *Vancouver Sun* columnist Jack Wasserman, whose help and advice in the creation of the Firefighter's Public Relations Committee was highly regarded. He collapsed and died while giving a speech at a roast for a local businessman, with many saying that he died doing what he enjoyed in the company of his friends.

Jack Wasserman.
Photo: Vancouver Sun

Affirmative Action

Two controversial subjects that would have long-range effects on the fire department began in 1977 when the VFD was called "one of the most sexist departments in the city," by the woman who headed the city's Equal Employment Opportunities Committee on civic hiring. She claimed that, as well as women, the department was also reluctant to hire Natives and Asians. The fire chief responded saying that if a candidate met the standards they would be accepted, and that the VFD had a Chinese-Canadian member, several firefighters of East Indian ancestry, and had a long history of Native Indian members, some of whom became chiefs. One assistant chief with Native heritage took delight in saying, years earlier, "when I get to be a chief, everybody will wear a feather!" Nick Craig was an athlete, a gentleman, and a respected firefighter.

The union statement said, "The members unanimously oppose giving an advantage in hiring to any particular group, ethnic or otherwise. Giving any group special consideration would be reverse discrimination and could result in quota hiring."

Then the threats by council to remove the fireboat from service began. They had hoped that the adjacent municipalities would share the operating costs, but as one of the firefighters said in a lively firehall discussion, "why buy the cow when you can get the milk for nothing?" And a Vancouver federal cabinet minister said that it was the city's responsibility to maintain the fireboat in Burrard Inlet.

Fire alarm system public notices.
Photo: VFD Archives

In October 1977 union president Gordon Anderson retired from the department after thirty-one years' service, and in February 1978 he was appointed British Columbia's first fire commissioner, a position he would hold until February 1988. John Bunyan succeeded him as Local 18's paid president.

The tough fire bylaws regarding sprinklers and upgrading of downtown apartment houses and hotels since 1974 resulted in fewer fatal fires. In 1977 eleven people lost their lives in fires, with five lost in properties covered by the new bylaws. The fire chief commented that a poor resident living in an old, upgraded downtown hotel could be safer than some people living in some of the finest homes in the best parts of the city.

In early February 1978 city council once again wrestled with the problem of sending the fireboat to other municipalities. They couldn't bring themselves to turn their backs on a situation if the fireboat was needed, yet would have preferred to have a cost-sharing agreement. The decision was made to leave things as they were but to try and work out an equitable cost-sharing formula.

A three-alarm fire on April 9 destroyed the well-known Three Vets camping and hunting equipment store on a quiet Sunday afternoon. The first three-alarm fire in almost two years, it was difficult to fight because of the lack of windows in the building to ventilate it, and the hazard of exploding ammunition inside the store. After a two-hour fight the roof fell in, affording the firefighters better access to the fire with heavy fire streams from the two Firebirds on the scene. Damage was estimated at more than $1 million and only minor injuries were reported.

In June the city's Equal Employment Opportunities officer presented her recommendations regarding hiring on the fire department to city council. Her recommendations included advertising for firefighters so that women and minorities would have an equal chance with the "white men who apply for the jobs and who often have friends and relatives already on the job."

Members of many families have proudly served on the VFD since its early days. Chief Carlisle had three sons and a grandson on the job, and many other families can also claim up to three generations of firefighters. There are also several father/son and multiple brother groups serving.

Some people have viewed this as nepotism and unfair pull, but this charge is groundless as all candidates must meet the required standards and many firefighters' sons have been turned down for lack of qualifications. Family groups, however, serve with a certain family pride and would never bring disrepute to the family or their brother firefighters.

Chief Konig didn't want to see the standards jeopardized by changing them and, as a result, end up with a quota system, and he threatened to submit his resignation if the EEO officer's recommendations were accepted. After much discussion, council voted six to five to reject the proposal until a further study on firefighter hiring standards was completed. The fire chief had the support of the provincial fire chief's association and all local firefighter's unions. On February 6, 1979 the Equal Employment Opportunities officer's position was abolished.

In July smoke detectors began to be recognized as a good investment for private homes and a record number of them were sold in the city. Selling for between twenty and fifty dollars each, sales were reported as being brisk and dealers reported, "fire has become a big business."

The Coal Harbour Plane Crash

The worst airplane crash in the city's history occurred at 1745 hours on September 2 when an Airwest deHavilland Canada DHC-6 Twin Otter float plane crashed into Coal Harbour on a landing approach over Stanley Park. Witnesses said it appeared that the plane lost power as it was about to set down on the water, and on impact it flipped over. The pilot somehow managed to miss a line of several boathouses and probably saved many more lives by his actions. Unfortunately, he and his copilot and nine passengers died in the crash, which only had two survivors. Emergency crews were soon on the scene and VFD divers continued their search for bodies for three days following the crash. The passengers were all Japanese tourists visiting Vancouver as part of a tour group.

A twin Otter at the Coal Harbour seaplane base with West Coast Transmission Building and Sheraton Landmark in the background. *Photo: Local 18, IAFF*

On January 1, 1979 the Fire Alarm Office (FAO) once again came under the control of the fire chief and the fire department, from the city electrical department. The FAO operators now became known as "fire dispatchers."

In February 1979 a program was begun to familiarize and train fourteen City of North Vancouver and sixteen District of North Vancouver firefighters on the functions of the Vancouver fireboat as part of a cooperative arrangement among the three departments to increase firefighting capabilities on the waterfront.

The *Vancouver Sun* headline read, "*Changeling* burns fire department." When the company filming the George C. Scott movie prepared for the finale of the film, a false facade was built in front of a house and was to be burned down. Firefighters were hired to stand by to put out the false front once the scenes were shot. However, the fire load of flammables including old tires and gasoline as well as propane burners got away and soon the whole house was ablaze, resulting in a two-alarm fire. With the extra man-hours, the movie company was faced with extra billing of many thousands of dollars.

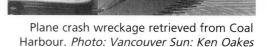

Plane crash wreckage retrieved from Coal Harbour. *Photo: Vancouver Sun: Ken Oakes*

Fire Alarm Boxes Removed

The old city fire alarm street boxes were removed from service and on July 10 the city began selling them for $10 each at the city works yard. Sales were brisk and in less than an hour they were sold out. The decision to discontinue their use was made in April 1979 because of the large num-

BOX NUMBER	1261	TYPE WB	LOCATION Hamilton Street & Hastings Street		CONTROL CARD NUMBER	1
ALARM ASSIGNMENTS	PUMPS	HOSE WAGONS	QUADS	TRUCKS	FIRE BOAT	R & S SQUAD
FIRST	1 - 2	1		1 - 2		2
SECOND	3 - 4	2				
THIRD	9 - 12					
FOURTH	14 - 20					
FIFTH	13					

CHANGE QUARTERS

FIRST						
SECOND	5X20 - 9X2 - 12X1 13X4 - 15X9 - 20X3			3X2 - 4X1 21 X 3		
THIRD	14 X 2 - 18 X 3 20 X 1 - 21 X 18					
FOURTH	13 X 1 - 17 X 3 18 X 2 - 19 X 4					
FIFTH	19 X 1					

CONTROL AREA

Refer to Card #1

E 9-MLH-70

Control card for Box 1261 at Hamilton and Hastings Streets shows a typical emergency (code 3) alarm assignment response from the first through to the fifth alarm. The bottom half of the card shows how the fire apparatus around the city are deployed to ensure that all areas are protected.
Photo: VFD Archives

ber of false alarms, the wide usage of telephones, and because they had no effect on the city's insurance rating. However, schools and government buildings would still be part of an auxiliary box system.

In August the firefighter's union ratified a two-year wage package that gave a first-class firefighter $1,907 per month, which was still 4 percent behind a policeman of equal rank.

On August 13 the new, relocated No. 14 Firehall opened at Venables and Kaslo Streets and was considered by many firefighters to be one of the best-designed new firehalls in the city.

On October 1, newly elected union president Bing Pare replaced retiring president John Bunyan who was returning to regular duties as a district chief.

October 13 saw one of the thickest fog conditions in the city in a very long time and it was said to be the reason that the fire in Hanger No. 8 at Jericho Beach was so late in being turned in. The first alarm came in 2030 hours and by 2200 hours it was a total loss, despite the efforts of the three-alarm crews on the scene. The fog was so thick that all flights in and out of Vancouver International were halted, horse races at Exhibition Park were cancelled, and massive traffic jams were caused.

No. 14 Firehall. *Photo: Author's collection*

Then on November 14 Hanger No. 7 was destroyed in another fire, ending a long controversy on the future use of the old World War II seaplane hangers that were built in 1942. Only Hanger No. 13 remained and it was used as a sailing center. Causes of the fires could not be determined.

Captain John Graham Dies in Apartment Fire

What began as a "routine" fire for No. 6 Company in the city's West End, ended in tragedy when Captain John (Jumps) Graham, age fifty-two, died in an apartment building fire.

The first alarm came in for a fire on the fifth floor of 1414 Haro Street shortly after 0100 hours on November 10. When the crew arrived, the fire could be seen on the fifth floor. The crew took the elevator, intent on going to the floor below the fire floor but accidentally ended up on the fire floor, and when the doors opened they were met by a blast of heat, smoke, and fire. The captain stumbled out of the elevator into the smoke to his

Civic funeral procession for "Jumps" Graham. *Photo: Vancouver Sun*

right and was lost. The others went to the left, found a stairway and safety but suffered burns and smoke inhalation. Attempts were made to rescue John but to no avail. The fire soon became a second alarm.

Earlier, a tenant had tried to put out a small fire in a sofa cushion by throwing some water on it. Later, the ensuing fire forced him into the hallway and by leaving his apartment door open behind him, the fire, heat, and smoke soon spread throughout the floor. One of his neighbors turned in the alarm and people began leaving the building. One tenant attempted to reach the roof of the twelve-story building from his eighth-floor apartment but collapsed in the stairway and died of a heart attack.

On Thursday, November 15 a civic funeral was held for Captain Graham at St. Andrew's Wesley United Church at Nelson and Burrard Streets, with hundreds of people and well over 1,000 firefighters in attendance. Internment was in the North Vancouver Cemetery. "Jumps" joined the VFD in 1949 and was promoted captain in 1977. He loved sports, particularly handball, and he served as president of the BC Handball Association.

Local 18's Founding Father Dies

George J. Richardson, age eighty-six, Secretary-Treasurer Emeritus of the IAFF and the Vancouver Fire Fighter's Union representative at the organization of the IAFF, died on January 5, 1980, in Long Beach, California. Born in Winchester, Massachusetts, he moved to Vancouver in 1912 and joined the VFD on March 3, 1913, at age nineteen. In 1920, he was elected secretary-treasurer of the International, a position he would hold until 1956. George loved sports and often said that it was this that helped him get on the VFD. He was a longtime fan of the NFL's Washington Redskins and manned the down markers at home games from the team's first game in 1937 until 1975.

Norm Harcus Succeeds Armand Konig as Fire Chief

On January 16, 1980 Acting Deputy Chief Norman Harcus was named to succeed retiring Fire Chief Armie Konig, effective July 1.

Also that year, council, various unions, and the surrounding municipalities discussed the fate of the fireboat at length. The decision came on May 27 (the eve of the ninety-fourth anniversary of the founding of the VFD) that city council agreed to maintain the boat until the end of the year while a search was made for other sources of funding. The finance committee wanted to get rid of the boat at the end of the year, after ten years of squabbling over its usefulness and annual costs of nearly $1 million.

The new Fire Dispatch Center at No. 1 Firehall became operational on June 23 after a new addition on the top floor was added to accommodate it. A new No. 19 Hall was built on site to replace the old, original Point Grey Firehall. Because of the way the old hall fit in architecturally with the nearby houses, one group had wanted to preserve it as a neighborhood pub. The new hall, retaining some of the design characteristics of the old building, was opened July 3, with "less glass, brass and grass."

George J. Richardson.
*Photo: from his IAFF book,
Symbol of Action*

Norman Keith Harcus

Born: Oct. 2, 1928, Vancouver, BC
Joined VFD: September 25, 1950
Appointed Chief: July 1, 1980
Retired: August 31, 1986

Also in July, the planning department of the city invited submissions from organizations interested in leasing the old No. 2 Firehall, at Gore and Cordova Streets. In October it was announced that the Vancouver Playhouse would lease the property for a theatre and an acting school. The previous tenant was the Actor's Workshop. A number of firefighter groups felt that firefighters should occupy the building with union offices, credit union, a museum, and athletic clubs but no submissions were made.

At 0800 hours on August 3 No. 2 R&S became known as No. 8 R&S.

New Fire Dispatch Center with Glen Lyon, Dave Mitchell and Dan Christie.
Photo: VFD Archives

VFD Rated "Class I" by Underwriters

On November 14, through the efforts of retired Fire Chief Konig, the Vancouver Fire Department was named Canada's first-ever, and only, Class I fire department by the Insurance Board of Canada Fire Underwriter's Survey. (Class I is the highest rating; Vancouver went from Class III to I.)

FIRE UNDERWRITERS SURVEY
A SERVICE TO INSURERS AND MUNICIPALITIES

Presented to the City of Vancouver in recognition of the establishment of fire protection and fire prevention systems for its residents which have resulted in the City being the first in Canada to achieve Class I rating for Commercial Fire Protection.

NOVEMBER 14, 1980

Insurance Bureau of Canada
Bureau d'assurance du Canada

Class I Certificate.
Photo: VFD Archives

On November 15, 1980 the Vancouver civic election was held and aldermanic hopeful Armand Konig lost in his bid to once again serve his city. He came thirteenth in a race for ten positions. Other notable losers were Philip Owen, who would be elected mayor of Vancouver in 1993, and David Schreck who, in 1991, was elected to the provincial legislature. Among the winners on the civic slate that year were Mike Harcourt, elected Vancouver's mayor (elected premier of BC in 1991, he served

ENGLESEA LODGE AFLAME ... *landmark building burns by English Bay after tenants escaped without injury Sunday*

2 explosions heard before fire

"I was just getting my breakfast. I had the radio on and someone banged on my door and said, 'There's a fire.' I just got on my coat and purse and gloves and left. The hallways were filled with smoke. I took the stairs down."

Martha MacLeod, 79, stood shivering Sunday in the early-morning cold outside Englesea Lodge as flames engulfed her home of 24 years.

She wore a coat thrown over her housedress and clutched $100 and a beautiful Italian cameo that's been in her family for generations — all she was able to save before fleeing into the damp, grey morning to stand in the mud and watch her home burn.

"Oh, it's a lovely place. I had a lovely suite on the ocean side," she said. "My goodness it went quickly."

Although she clutched her treasured cameo safe in its little red box, she feared for her "very precious pictures and paintings" still in the blazing building.

"That's the thing that makes me sick, thinking of them."

She and 13 other residents of the lodge at 2046 Beach were asking today how it all happened and why the fire alarm system didn't work.

The building was only partially occupied. Residents had moved out because there was a great deal of uncertainty over what the city would do with the lodge.

Many residents of the 69-year-old seven-storey, 45-unit building said they were awakened about 9:30 a.m. Sunday by an explosion in the basement. Others who didn't hear the blast or notice the rising smoke were alerted by a tenant who pounded on doors yelling 'fire.'

All the residents escaped unharmed.

The fire occurred only days before city council was to discuss a $1.3 million upgrading plan for the building — and only two days after park board chairman Russ Fraser wrote a letter to council advocating the 69-year-old landmark be condemned.

The board has been trying to get Englesea removed for four years in order to complete the sea wall.

Maureen Moore was in the shower of her top-floor suite when her bathroom started filling with smoke. She said she never heard an alarm.

Fellow tenant Budd Giffin said he did not hear an alarm, but the system had been "sensitive" in the past and "went off and on all the time."

Mollie Smith, a pensioner who has lived in Englesea for 27 years, said she didn't hear anything until someone pounded on the door of her fifth-floor suite.

Another tenant said two explosions in the basement about 10 seconds apart

alerted him to the fire. The second, "a big boom," lifted his chair off the floor and moved the bed in which his wife was still asleep, he said.

"Within 30 seconds I couldn't see six feet."

He said he got his wife out, then started running through the hall knocking on doors to alert other tenants before he was overcome by smoke and had to make his way outside, where firemen revived him with oxygen.

He credited three passing joggers with rousing other tenants.

Caretaker George Wright, who emerged from the building shortly after the fire broke out and was covered in soot and water, said he tried to fight the fire "until a burst of flame came back at me and I turned and went out the door."

A police spokesman said the fire was "still under investigation" and the subject of arson "has not even come up yet." He said there were no injuries, although one firefighter, Capt. William

Frederick, was taken to the hospital and treated for smoke inhalation.

Vancouver fire chief Norman Harcus said that although the building was "strong," firefighters were unable to contain the fire as it spread up ventilation shafts. He said the main blaze started in the elevator shaft.

The fire was battled by 100 firefighters and raged for close to six hours.

While the evacuated tenants were being relocated at the nearby Sylvia Hotel, hundreds of onlookers crowded the beach and watched from surrounding high-rise balconies.

In an interview outside the burning building, the director of Vancouver's civic building department said the city was planning to upgrade and renovate Englesea because "it didn't meet present-day fire standards."

Art Langley said Englesea had only a partial sprinkler system and was equipped with smoke and fire alarms which were operating when tested earlier this month.

"I know a few days ago the fire alarms were working, and a few days ago the smoke alarms were working," he said.

The building earlier had been slated for demolition by the Vancouver park board — which bought the building in 1967 with plans to complete the 11-kilometre Stanley Park seawall — but city council voted last year to improve the lodge for occupation by seniors for the next 20 years.

Fraser said Sunday he expects the Englesea property now will be turned over to the board and knocked down for completion of the seawall, but he says the park board will wait until the city makes the first move.

Englesea Lodge fire.
Photo: Vancouver Sun: Ken Oakes

until 1996); Russ Fraser, elected to the parks board (later, a provincial MLA); and school trustees included Tom Alsbury, a former mayor of Vancouver, and Kim Campbell, who later became a federal cabinet minister and the first woman prime minister of Canada. Also elected alderman was Dr. Nathan Divinsky, well-known UBC professor and ex-husband of Kim Campbell.

Firehall Security Begins

After many years, the habit of leaving firehall doors open when attending alarms came to an end because of the theft of articles while the halls were empty. All firehalls had to be secured when alarms were answered. In some districts, a nearby neighbor or "door-closer" often made sure the doors were closed; in cold weather some would put on the coffee pot as well, so the men would have a hot cup when they got back. Downtown halls sometimes had "firehall locals" who often got a free meal from the boys or a place to sleep when times were tough. After spending thousands of dollars installing switches on the overhead door tracks to shut off the furnaces when the crews left the halls, it finally occurred to someone that the money would be better spent on automatic door closing systems, like the ones that had been installed in all the new halls of the seventies.

Englesea Lodge Destroyed

An explosion in the basement of the city-owned Englesea Lodge awakened residents around 0900 hours Sunday, February 1, 1981. Within minutes the fire had spread up through light shafts and the elevator shaft of the sixty-nine-year-old, seven-story building. By 0930 hours the fire was a third alarm with the off-shift being called, and it took six hours to bring it under control.

For several years the parks board had been trying to have the building removed so that the seawall could be completed along the foreshore of English Bay. The fire occurred just days before the city council was to discuss a $1.3 million upgrading of the building. The cause of the blaze was not determined, and arson was suspected. Sixteen people were left homeless and none had insurance. On June 19 another fire was set in the building about 0300 hours and three transients were lucky to have escaped. The building was torn down in November 1981, the last remaining building in that area on the shores of English Bay.

At 0424 hours on May 2 the 120-foot (36-m) *Huntress* was taking on diesel and gasoline at the Shell Oil marine fuel barge in Coal Harbour, in preparation for a scuba diving charter to the Gulf Islands, when she exploded and sank. More than 800 gallons (3,000 liters) of diesel was on-loaded and the gasoline for the small boat engines was being loaded when the explosion and fire occurred. On board were eighteen sleeping passengers and a crew of six; two of the passengers died. Probable cause of the accident was believed to be gasoline fumes that were ignited by a refrigerator motor in the lounge. The ship carried no liability insurance.

Firefighters received a 1981 wage increase by binding arbitration and were given 19 percent of the 22.9 percent asked for, which would have given them parity with city police. With the twelve other union locals in the Greater Vancouver Regional District, negotiations had begun in October, with the same aim. When all avenues failed, strike votes were taken, with Vancouver firefighters voting 98 percent in favor of striking to get their demands, prior to the binding arbitration.

New Recruit Testing Initiated

In August aspiring firefighters underwent grueling tests during their bid to be accepted into the next recruit class. Of the almost 150 applicants, only two of the candidates of Asian descent and none of the six women showed up for the tests that included running 3,080 yards (2,820 m) in less than twelve minutes, carrying a 125-pound (56.8-kg) dummy 150 feet (48 m) in less than thirteen seconds, then traveling an obstacle (agility) course and taking a strength test. The Simon Fraser University Institute of Human Performances developed the requirements. It was also their recommendation that the height requirements not be altered, i.e. 5 ft. 9 in.–6 ft. 2 in. (175–188 cm), which the affirmative action officer had stated was discriminatory.

On October 26, twenty-seven men were selected to begin training.

A suspicious two-alarm fire destroyed the Kawasaki dealership on East Broadway, August 25.
Photo: Vancouver Sun: Rob Draper

Captain Grahame White Dies at Fire Scene

In the early morning of August 20, while helping his crew load hose after a fire, Captain Grahame White collapsed at the rear of No. 18 Pump and died, despite the efforts of his crew to revive him. Grahame was forty-nine years old and had joined the VFD on July 26, 1954.

On October 24, 1981 retired fire chief and ex-union president Hugh Bird died at the age of seventy-nine. In 1963, a year after his retirement, he was elected alderman and served on the Vancouver city council for the next twelve years.

Fire deaths for 1981 reached a total of eighteen people, with eight of those lost between October 31 and the end of the year. On Halloween night a fire started on the porch of a house claimed the life of a young woman when she ran back into the house for something forgotten. Then the next day, a man was the victim of a careless smoking in bed. Three days later, on November 4, an elderly man succumbed to smoke inhalation in an old downtown hotel fire.

On December 3 another man died in an apartment fire in the 1100-block of Bute Street and, lastly, three men and one woman perished in a rooming house fire that started in the basement of an old Shaughnessy house on Hudson Street.

Because of these fires, city council voted unanimously to begin enforcing the BC Fire Code (a provincial statute, and a higher authority governing enforcement of fire laws) rather than the city bylaw. It was deemed that implementing the provincial fire code would be more effective and give the fire warden's task force more powers to compel upgrading.

The old No. 22 Firehall.
Photo: Author's collection

The new No. 22 Firehall.
Photo: Author's collection

A second-alarm fire at 0430 hours January 1, 1982 destroyed the clubhouse of the private Shaughnessy Golf and Country Club. Believed to have started a few hours after the New Year's celebrants had gone home, the fire attracted scores of spectators who were drawn to the scene by the very large glow in the sky. Damage was estimated at $2.5 million.

At 1510 hours on June 18 the new No. 22 Firehall, a design award winner, opened at West 59th Avenue and Oak Street.

Captain Fred Fisher Collapses and Dies

The department was sad to learn that Captain Frederick Clarke (Slob) Fisher collapsed and died of a heart attack at No. 21 Firehall on June 21 shortly after coming on duty for the day shift. His crew and the paramedics worked on him at length but could not revive him. Fred had been a member of the VFD for twenty-seven years. He left a wife, a daughter and two sons.

In early 1983 advertisements were placed in the daily newspapers and on February 23 it was reported that 3,638 men and one woman requested applications for the fire department and the city's personnel department expected another 1,000 applications to fill the projected twenty-five vacant positions. By the closing date 5,050 applications were picked up.

The summer of 1983 saw over 40,000 protesters, including VFD members and the band, demonstrating in the Operation Solidarity parade against the Social Credit government's slashing of social programs. The huge rally was held at Empire Stadium.

Firefighter Ed (Garth) Stephens, on behalf of the Vancouver Firefighter's Union, presented the Vancouver General Hospital Burn Centre, a check for more than $14,000. The money was collected by donations from people who came out to see his annual Christmas light display in the front yard of his home.

On May 10 retired Captain Jack Forsgren died at the age of eighty-

Captain Fred Fisher.
Photo: Local 18, IAFF

one. Born in Sweden in 1902, he joined the VFD on September 1, 1928 and served for thirty-four years. Known by many as the "Wrestling Fireman," Jack won the Canadian championship in 1926, turned pro in 1927 and became the Canadian heavyweight champion in 1932. He had a wrestling career that spanned twenty-one years.

On June 9 the fire department held its first simulated emergency response to the new Advanced Light Rapid Transit (ALRT) "Sky Train" system on Terminal Avenue. The fire and rescue exercise was said to be a success and gave police, firefighters and ambulance personnel a feel for some of the situations they would later encounter on the automated system.

A two-alarm fire in the early hours of August 12 destroyed three downtown stores and a clothing warehouse. Also destroyed were many pieces of antique furniture owned by the clothier and several seventeenth century artifacts in an antiquarian bookstore. The fire in the 100-block West Hastings Street was started 0345 hours in the premises of a political bookstore that had been the scene of previous arson attempts. Firefighters were forced out of the building five times while trying to get an upper hand on the blaze because of the false ceilings, and it wasn't until after 0900 hours that the fire was brought under control.

A twenty-two-year-old transient from Michigan was charged with arson in connection with the fire at the Communist Party of Canada store, but charges were dismissed. Damages to the buildings and contents were almost $1 million.

At 0335 hours on November 25 fire was reported in the aging Vancouver music landmark, the Smilin' Buddha Cabaret. The fire roared through the one-story building destroying the contents, all uninsured. A former showcase for various bands, it also included performances by the late rock great, Jimi Hendrix. Rock groups said, "if they could make it at the 'Buddha' they could make it anywhere." In 1995, the Vancouver band, 54-40, paid tribute to the Buddha by putting a picture of its old neon sign on the cover of their platinum record, *Smilin' Buddha Cabaret*.

Ad from City of Vancouver personnel department.
Photo: Vancouver Sun

Sign from the Smilin' Buddha Cabaret, c. 1980.
Photo: Vancouver Sun

In December 1983 the height standards required by the department were challenged by three young men who failed to meet the requirements, from fractions of an inch to two full inches. They appealed to the BC Human Rights Branch and a demonstration was put on for a branch commissioner to illustrate the city's need to have a height standard. The city argued that the height standard was necessary for safety reasons and to reduce the incidence of injuries. The university professor who designed the battery of tests required by candidate firefighters said it would be "criminal and immoral" for the authorities to enhance the danger to firefighters by not having a uniform height among the crew. On receiving the report from the commissioner, the labor minister ruled that the height requirements were discriminatory and that the branch would listen to suggestions to correct the discrimination. The VFD did not hire the three candidates.

August 18 through 23 Chief Harcus hosted the Seventy-Sixth Canadian Association of Fire Chiefs (CAFC) conference.

Just after 0300 hours on November 28, 1984 the alarm for a house fire on the east side came in, and ended in tragedy for a young family. When firefighters arrived the building was in flames and rescues were made with ladders from the upper floor windows and roof of the old two-and-one-half-story building. Firefighter Filip Delgiglio of No. 3 Company made one rescue when he brought a five-year-old girl out of the burning building. He removed his breathing apparatus face mask and placed it over the child's face and took her out of the house. Another, a two-year-old girl, was also brought from the building but her twin sister died in the fire. Two days later both of the rescued children succumbed to the effects of the fire.

Cause of the fire was determined to be careless smoking, and it was found that the fire started in a couch following an evening of partying by the adults in the family, all of whom smoked. On further investigation it was found that the building had no smoke detectors and the fire alarm system was inoperative.

Because of Delgiglio's dramatic and courageous rescue he was recognized as the Firefighter of the Year at the annual Canadian Association of Fire Chief's Conference in Ottawa the following year. Deputy Fire Chief Pamplin said that, "he went beyond the call of duty and typified the firefighter who goes into a burning building when everyone else is running away from it." Fil modestly gave credit to his crew for being there; all would have hoped for a happier outcome.

A blast during firefighting operations injured ten people including seven firefighters, two seriously. The first alarm came in shortly after 2100 hours on April 5, 1985 for the Datchet Court Apartment, a fifty-year-old, three-story building in the West End. The fire was soon a second alarm. Then, about 2200 hours an explosion in the rear of the building blew the back half of the building apart, trapping two firefighters under the rubble in the alley. Most of the injuries were from flying glass and debris. The blast was so violent that most people thought someone would have been killed, but as one firefighter said later, "I guess it wasn't anybody's time to go. It's a miracle that everybody got out." The

Canadian Association of Fire Chiefs
76TH ANNUAL CONFERENCE
AUGUST 18TH - 23RD, 1984
VANCOUVER, CANADA

CAFC Annual Convention.
Photo: VFD Archives

last crews to leave the scene returned to quarters eight hours after the first alarm.

The cause of the fire and explosion that destroyed the building was never determined, but it was speculated that either gasoline stored in the locker room or a methane pocket in the drainpipes may have caused it.

A Towering Inferno

When the fire on a nineteenth floor room of the forty-two-story Sheraton-Landmark Hotel in downtown Vancouver is talked about, it is often referred to as Vancouver's version of *Towering Inferno*, the best disaster movie of the 1970s. The alarm came in shortly after 0700 hours on a sunny Sunday, May 26, 1985. The fire was confined to the room of origin but it looked like the top floors of the building were doomed to destruction. When the assistant chief on duty saw the fire as he came over the hill on Robson Street, he anticipated the worst and quickly put in a second alarm. Fortunately, the hotel's evacuation plan resulted in more than 600 guests getting out safely, with six minor injuries. The fire vented itself through the doors and windows leading to the balconies and did not spread. The hotel's safety measures and the fire department's high-rise evolutions (procedures) worked perfectly, and the fire was quickly extinguished. Cause was determined to be human error.

Sheraton Landmark fire.
Photo: VFD Archives

On June 7 the pump crew that had spent the previous eight months quartered on site in a portable trailer with a makeshift garage for the rig opened the new No. 21 Firehall. A week later, on June 14, No. 21 Truck returned to its new home from No. 19 Hall where it was quartered during construction.

During December 1985 all VFD firefighters were issued new Cairns & Brothers Metro Firefighter helmets, to replace the 1960s MSA-type Class D plastic helmets that had, at times, melted around the wearers' heads. This long-overdue new issue provided a stronger shell, better fit, a built-in visor for eye protection, and a lining for cold-weather comfort.

No. 21 Firehall.
Photo: Author's collection

The Centennial Year and the End of the Fireboat

(1986–1995)

A Vancouver institution and city landmark was destroyed by fire on February 7, 1986. The original White Spot restaurant at Granville and West 67th Avenue was lost along with many pieces of art and artworks, including some fifteenth century stained glass windows and over fifty years of memories of the well-known restaurant chain. The alarm came in shortly after 1430 hours when the kitchen staff noticed smoke coming from the ceiling above the kitchen. By the time the fire department arrived the fire had taken hold in the ceiling and quickly destroyed the upper floor and roof. Cause was undetermined.

White Spot Restaurant fire.
Photo: Dave Youell

The new VFD picture identification cards to replace the long-used wallet ID badges, were introduced, as they could be updated every five years, or after promotions. In the early days every member was given a badge number based upon seniority and, as his position on the seniority list changed, so did his number. Over the years the numbered badge swapping became a problem. It was decided that changing numbers would stop and when a man retired or quit, his number would be given to the man replacing him. This system worked well until the late fifties when the city developed a payroll number system. All members were given new numbers beginning with 0100, alphabetically from A to Z, to about 6000 based upon the nominal roll, with spaces left to accommodate new members. By 1999, the numbering system had to be changed again and this time each new member got a personal ID number beginning at 6001.

New uniform shoulder patches were introduced in 1986 for the centennial of the city and the fire department. A centennial logo was designed and the members of the department put together a display showing the development and progress of the VFD through the century with pictures and artifacts, as well as a video illustrating the highlights of the department's growth. Over the six months of the display at No. 1, it attracted several hundred people.

Following the May 28 opening ceremonies for the VFD display, an hour-long birthday party

Vancouver Centennial logo displayed on fire apparatus.
Photo: VFD Archives

VFD centennial display at No. 1 Firehall.
Photo: VFD Archives

Flying the flag for centennial. *Photo: Author's collection*

was held on air through the courtesy of Vancouver television station CKVU. During the year, all firehalls flew the fire department flag beneath the Canadian flag and all fire apparatus displayed the city's centennial logo.

Expo 86, Vancouver's World's Fair on transportation and communications that opened May 2, had a fire service on site made up of twenty-two volunteers from outside the Vancouver area. Over the run of the fair, the VFD responded to many alarms most of which were of a minor nature and easily handled by the fire service on the grounds.

On June 13 a celebration and parade was held at the Plaza of Nations at Expo 86 and, later, along Water Street in Gastown to remember the 100th anniversary of the Great Fire of 1886. Every elementary school student in the city received a Vancouver Day medallion, 32,000 of which were delivered to schools by city firefighters. The design had the city coat-of-arms on one side and a replica of the Volunteer Fire Brigade Badge No. 1 on the other. It was hoped at that time that the tradition of an annual Vancouver Day would be ongoing, but it was never held again. (Expo 86 ended on October 13, Thanksgiving weekend, having received a total of 22,111,578 visitors.)

Centennial Logo.
Photo: VFD Archives

Vancouver Great
Fire centennial
medallion, showing
both sides.
Photo: VFD Archives

Vancouver Fire Department Band at Expo 86. *Photo: Eric Siwik*

Donald John Pamplin

Born: July 19, 1932, Calgary, Alberta
Joined VFD: August 23, 1954
Appointed Chief: September 1, 1986
Retired: August 31, 1992

Don Pamplin New Fire Chief

On September 1 Donald J. Pamplin, age fifty-four, became the VFD's twelfth fire chief, succeeding retiring Chief Norm Harcus. The new chief's plans were to keep the department running smoothly by maintaining its high standards and his approach was "if it ain't broke, don't fix it."

On October 30 No. 12 Hall was closed after seventy-three years of service, and the pump crew was quartered in a portable trailer on the site. The truck was moved to No. 19 Hall until the new firehall was built. On November 12 the machines moved in and within hours the old hall was torn down and removed. Gone as well was the last horse trough in use in a city firehall.

In December several firehalls in the city were supplemented with tanker trucks, borrowed from city departments and outlying municipalities because of low water pressures in the city caused by damaged water mains that supplied the city from the North Shore. These tankers responded to fire calls to supply Vancouver's pumps with water in the event that the hydrants at a fire scene couldn't supply the required water.

It was revealed many months later that *Vancouver Fireboat No. 2* would have provided the only large water supply for the protection of the downtown core, the heavily populated West End, and the nine-mile (14-km) harbor waterfront, because of the failed water mains. The fire chief's report to council went on to say, "that for more than a week after the water mains broke December 11, the fire department had virtually no water downtown for fire fighting."

In February 1987 all VFD aerials and platforms (except No. 6 aerial and spare apparatus) were equipped with cellular telephones by BC Transit to be used in conjunction with any incidents involving the new Advanced Light Rapid Transit system, known as Sky Train.

Alex Betts, Centenarian

On February 25, 1987 retired firefighter Alex Betts celebrated his 100th birthday at No. 1 Firehall on Heatley Street, surrounded by family and friends. He enjoyed the day immensely and sat proudly in the seat of the restored 1912 American-LaFrance hose wagon, a rig he had seen arrive on the job many years before when he was a young fireman. In the days when Alex joined the fire department, he worked twenty-four-hour shifts, six days a week for $60 per month and his thirty-five years on the job gave him a pension of $46 per month when he retired in 1947.

He liked to tell the story about the time he and his son Martin were clearing a piece of land when he was eighty-eight years old. Not only did he fell the trees but he climbed and topped them, too. He wrote to a friend in Prince Edward Island and told him, "Martin and I are working like a good team of oxen. Unfortunately, he thinks I am both of them."

On October 30, 1987 Alex died peacefully in White Rock, BC, at 100 years, eight months and five days, having been the first-ever retired member of the Vancouver Fire Department to celebrate his 100th birthday.

Birthday celebration for Alex Betts.
Photo: Dave Youell

The Fireboat is Going...

In early May city council proposed reductions in the fire department staff and to place the fireboat out of service. On May 27, on the eve of the department's 101st anniversary, council voted seven to four to reduce the manning on all fire department pumps from five men to four and to remove the fireboat from service. The union began a campaign to solicit the help of concerned citizens and organizations to encourage the city to reconsider these cuts to the fire service, which had a cost to the people of about one cent per day or the cost of one hamburger per year. City council remained steadfast in its decision and picked January 1, 1988 as the day the fireboat would be placed out of service in spite of overwhelming public opinion.

FOR THE PRICE OF A CHEESEBURGER

VANCOUVER FIREFIGHTERS ARE STILL CONCERNED! AND SO SHOULD YOU BE! MAY 26, 1987 VANCOUVER CITY COUNCIL MADE SERVICE CUTS CONTRARY TO PUBLIC OPINION.

Results of these cuts:

1. Removed the only fireboat from Canada's busiest harbour.
2. Seriously hampered our ability to fight 75% of waterfront fires.
3. Left cruise ships and other vessels unprotected.
4. Removed the Firefighters' only alternative source of water in the event of another water-main break such as occurred December, 1987.
5. Reduced manpower on 10 pumper trucks from 5 men to 4.
6. Affected our ability to save your life and protect your property.

Manning has already dropped from 154 men minimum on duty in 1966, to 147 men minimum on duty in 1986. During this time, emergency call volume has INCREASED from 7,000 calls per year to 28,000 calls per year. These reductions dropped the minimum number of firefighters on duty to 135 men. The Vancouver Fire Department is rated by the insurance underwriters as Canada's finest! City Hall has taken the muscle out of your Fire Department for the almost insignificant cost of a cheeseburger or a penny a day! Firefighters are always there when you need them. NOW WE NEED YOU!

THESE ARE THE ALDERMEN WHO VOTED TO MAINTAIN FIRE DEPARTMENT EFFICIENCY
ALDERMAN LIBBY DAVIES
ALDERMAN DON BELLAMY
ALDERMAN BRUCE ERIKSEN
ALDERMAN JONATHAN BAKER
THE REST OF COUNCIL INCLUDING THE **MAYOR** VOTED IN FAVOUR OF SERVICE CUTS.

"VOTE" FOR A COUNCIL WHO LISTEN TO ITS "CITIZENS."

VANCOUVER FIREFIGHTERS
254-4881

Local 18 campaign to save the fireboat.
Photo: Local 18, IAFF

The Dragon Boat Team

Between June 1 and 8 the Vancouver Firefighter's Dragon Boat Team represented the Vancouver Chinese Cultural Centre and Canada in the International Dragon Boat Races in Hong Kong. Unfortunately, the team didn't win, but the competition gave its members an opportunity to visit Hong Kong's and China's local fire departments, whose members were great hosts. On the day of the final races, Hong Kong's *Fire Boat No. 6* was on site flying the Canadian and VFD flags in support of the team's efforts.

Vancouver's Dragon Boat team. *Photo: Author's collection*

Hong Kong *Fire Boat No. 6* flying the Canadian and Vancouver Fire Department flags in support of our dragon boat team.
Photo: Author's collection

At the welcome dinner each team had to sing or dance or perform a skit, and the VFD crew wowed the crowd with a song from the Blues Brothers and won the award for most original performance. Other competing teams came from China, Taiwan, Thailand, Indonesia, Singapore, Australia, Italy, England and the US, as well as the host colony.

At the celebration dinner and awards banquet, Firefighter Les White spoke on behalf of the Vancouver Dragon Boat Team. He told everyone how pleased the firefighters were to have taken part and he thanked everyone for all the courtesies extended during their stay — in Chinese! Les learned his speech phonetically and did it so well that many thought he was fluent in the language.

Some of the team's members and supporters took extended visits to Thailand and Bali, and while in Phuket, Thailand a few of the firefighters had roles as extras in the movie *Good Morning, Vietnam* with Robin Williams. As goodwill ambassadors they left everyone, everywhere, with a good impression of Canada and the VFD.

The Powell Street Fire

The warehouse fire alarm came in during the early evening of June 18 and the stubborn blaze in a five-story and adjacent three-story warehouse and storage company was difficult to put out. The fire, at 839 Powell Street, involved the same building as the Bowman Storage three-alarm warehouse fire of 1952 and was just as difficult to extinguish. At 2300 hours off-duty firefighters were still being called out to attend the blaze.

During the night, firefighters were more than a little concerned about the weakened building collapsing, and not long afterwards the front of the building crashed into the street, sending debris under two nearby rigs and causing minor damage. By dawn the next day all that remained of the building was a pile of smoldering material that had to be hosed down for days.

Six firefighters were injured, all suffering minor cuts, burns and smoke inhalation and the dollar-loss was more than $1 million. A fifty-nine-year-old man was charged with breaking and entering the premises and was believed to have started the fire.

The New No. 12 Firehall Opens

Although the pump and truck crews moved in June 23 and June 26, respectively, the official opening of No. 12 Firehall was held on July 11 with Fire Chief Don Pamplin, Mayor Gordon Campbell and VFD Chaplain, Reverend Jim Erb, officiating, along with many citizens, and the music of the Vancouver Fire Department Band. Several of the retired members of the VFD were also in attendance, and one, ninety-two-year-old retired Captain Charles L. (Buck) Buchanan, had been a fireman in the old station. An ex-member of the Royal North West Mounted Police (RNWMP), he joined the Point

Warehouse fire on Powell Street at the old Bowman Storage warehouse site.
Photo: Dave Youell

Official opening day at No. 12 Firehall was attended by active and retired firefighters, dignitaries and the VFD band.
Photo: Author's collection

Grey Fire Department on June 1, 1924.

Behind a plaque in the foyer of the new hall is a key that will open a time capsule at the city archives on the department's 150th anniversary on May 28, 2036. A box was filled with various artifacts during 1986, the centennial year, and it was decided that the occasion of No. 12's opening would be a good time to acknowledge the time capsule's location.

New Battalions Formed

At 0800 hours October 30, the designated districts in the city, numbered one through four were changed to three "battalions." District #1, covered in part by the assistant chief at No.1 Hall and the district chief at No. 2 was eliminated, but re-aligned with Battalions 1 and 2. District #2, covered by the district chief stationed at No. 7 Hall, formerly identified as Car 7 ("Car" is a designation for a district chief's vehicle), became known as Battalion 1; District #3, formerly Car 20 at No. 20 Hall, became Battalion 2, and District #4, Car 18 at No. 18 Hall, became Battalion 3. When the assistant chief at No.1 attended any alarm, he was identified as Command 1. With these changes, after almost sixty years, the rank of district chief was changed to battalion chief. Following is a list of the three new battalions and the firehalls, including cars, pumps, trucks, and rescues of which they were comprised:

> **Battalion 1:** Firehalls No. 1 (C1, 1P), No. 2 (2P & 2T), No. 6 (6P & 6T), No. 7 (B1, 7P & 7T), No. 8 (8P & 8R&S), and No. 10 (F/Boat No.2)
>
> **Battalion 2:** Firehalls No. 5 (5P & 5T), No. 9 (9P & 9T), No.13 (13P), No. 14 (14P), No. 15 (15P & 15T), No. 17 (17P & 17T), and No. 20 (B2 & 20P)
>
> **Battalion 3:** Firehalls No. 3 (3P, 3T &3R&S), No. 4 (4P, 4T & Lighting Unit 4), No. 12 (12P & 12T), No. 18 (B3, 18P & 18T), No. 19 (19P), No. 21 (21P & 21T), No. 22 (22P & 22T)
>
> Legend: B = battalion; C = command; P = pump; T = truck; R&S = rescue & safety.

During the month of November the fire department was busy with four, three-alarm fires. The first was a warehouse fire on November 1 at Station and Industrial Avenue, and on November 15 another third alarm destroyed six businesses that occupied the former Dueck Chevrolet-Oldsmobile-Cadillac dealership building in the 1300-block West Broadway. At 2010 hours on November 19 another third-alarm warehouse fire occurred in the 200-block of East 2nd Avenue and, lastly, on November 28 at 0205 hours another three-alarm in Chinatown, at 75 East Pender Street in a three-story building containing a rooming house, stores and warehouse. All were considered to be suspicious in origin.

Going . . . Going . . . Gone !

On January 1, 1988 at 0800 hours the crewmembers of *Vancouver Fireboat No. 2* ended their final shift. The City of Vancouver, with Canada's busiest port, was without a fireboat for the first time in almost sixty years. The last fireboat crew consisted of Pilot Ken Suzuki, Engineer Bill Lai and Firefighter Blair Davies. Suzuki and Lai would transfer over to the Fire Prevention Branch and Davies would return to regular firehall duties. Of the remaining displaced fireboat pilots and engineers, one retired, two quit and three took transfers to the Fire Prevention Branch as part of the

Vancouver Fireboat No. 2 shortly before being decommissioned—the end of an era. *Photo: VFD Archives*

Upgrading Task Force. Three firefighters returned to other firehall duties. There was no hoped-for eleventh-hour reprieve because of Mayor Campbell and the council's belief that the fireboat was obsolete. Lloyd Shippam was the last Chief Engineer to serve on the boat.

New Names, New Designations

February 1 saw the designation of fire department apparatus changed:

Pumps became known as engines, i.e., No. 1 Pump (1P) became Engine 1 (E1) Trucks were now known as ladders and towers, i.e., Truck 2 was now Ladder 2 (L2) and No. 7 Truck, which was a Firebird 125, was now Tower 7 (T7) R&S rigs were to be known as Rescue 3 (R3) and Rescue 8 (R8) The lighting unit at No. 4 became known as Generator 4 (Gen4) then soon after, as Gen1 when it was moved to No. 1 Hall

Fire alarm designation was also changed. The alarms known as 2-6 and 3-6, which were two- and three-alarms utilizing on-duty personnel, became known as second and third alarms, respectively. The old three-alarm assignment was now a four-alarm and the old four-alarm became a five-alarm. After that, any upgrading of the alarm gave it a new level; for example, a request for two more engines at a five-alarm fire made the new alarm a seven-alarm, and so on.

The new Computer Assisted Dispatch (CAD) system was being prepared to go on-line and the first test run, Incident #0906, was given E9 at 2008 hours March 1 to Commercial Drive and Parker Street. At 0800 hours on April 3, the CAD system was placed in operation.

Firefighter Terry O'Keeffe Dies at No. 7

Firefighter Terry O'Keeffe.
Photo: Local 18, IAFF

On April 9 Firefighter Terrance Patrick (Terry) O'Keeffe, age forty-six, collapsed at No. 7 Firehall on the company's return from a fire. Despite the efforts of the crew and paramedics, Terry died. He joined the VFD on October 16, 1967 and was survived by his wife and two daughters.

Number 6 Company moved out of No. 6 Hall to the old Vancouver automobile testing station, 1700 West Georgia Street, on April 19, while the old 1908 firehall was renovated and upgraded to meet earthquake standards. Designated a heritage building, the "new" No. 6 reopened March 18, 1989.

The First Six-Alarm Fire

Fraser Arms Hotel fire, the department's first-ever six-alarm fire. *Photo: Dave Youell*

On April 24 the first six-alarm fire in VFD history occurred at the Fraser Arms Hotel at 1450 Southwest Marine Drive. The first alarm came in at 2141 hours and was made a second-alarm eight minutes later by the first-in battalion chief, B3. Then it was upgraded to a third-alarm at 2151 hours when the fire continued its rapid spread throughout the south side of the building and through the roof. The off-shift was called out with the upgrades to fourth and fifth alarms at 2228 and 2230 respectively, then at 2237 hours it was made a six-alarm fire by Command 1, the on-duty assistant chief from No. 1.

An hour later, at 2345 hours the fire was under control and twenty-five minutes later, at 0011 hours it was struck out. Fresh manpower was brought onto the scene throughout the early morning hours to overhaul and extinguish the remaining small fires. Three firefighters received minor injuries but were checked out at the hospital and released. The fire was believed to have started under a wooden stairway at the rear of the building in a garbage container. Damage was estimated at $1 million.

The first alarm to the empty, three-story building at Clark Drive and Hastings Street was turned in by a passing

Early morning three-alarm fire at Hastings St. and Clark Drive. *Photo: Dan Goyer*

ambulance crew shortly before 0600 hours June 24, and within minutes of the fire department's arrival, it was a four-alarm fire. The rush-hour traffic on this main artery had to be diverted.

The building was used for many years as a flour mill and food warehouse, and was slated for demolition. The adjacent buildings exposed to this fire were saved and fire crews remained on scene for several hours, overhauling the area. There were no injuries and a cause wasn't determined.

For many years members of the IAFF wanted to see the International's convention held in Vancouver, but because of the lack of hotel rooms and meeting space, it was not possible. At the Thirty-Ninth IAFF Convention in Miami, August 1 to 5, the Vancouver Firefighter's Convention Committee won the right to host the convention in 1992.

The most costly house fire in the city's history occurred on Christmas Day 1988, when the dream house of local businessman

Promotional ad for Vancouver Firefighters' Union Local 18 bid for the 1992 IAFF convention. *Photo: VFD Archives*

Edgar Kaiser Jr., the grandson of industrialist Henry Kaiser, was destroyed in a spectacular three-alarm fire. The house took nearly three years to build and had been completed just two months before the fire.

About 1300 hours the alarm was turned in and firefighters from No. 19 Hall were first on scene but were hampered by hundreds of onlookers and a steep, narrow driveway leading down to the house. After about twenty minutes the building was destroyed, as was a prized art collection.

The $20-million fire was believed caused by the lights on the Christmas tree.

The Fire Fighters' Calendar

The year 1988 was the premiere for the Vancouver Fire Fighter's Union Hall of Flame Calendar, a collection of tasteful beefcake photos of firefighters of the VFD. The purpose of the calendar was to raise money and make people aware of the Vancouver General Hospital Burn Unit, and the success of this venture raised more than $105,000 for the VGH Burn Unit.

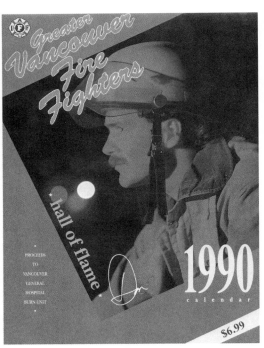

The 1990 Fire Fighters Calendar.
Photo: Local 18, IAFF

General Hospital Burn Unit

In 1990 it was expanded to become the Greater Vancouver Fire Fighters' Calendar, with proceeds going to other burn treatment units in the province of British Columbia, and later included contributions to the CKNW Orphan's Fund. By 2006 the calendar sales had exceeded $1,000,000 for the firefighters' projects.

The year 1988 was also the first since 1897 that no new men were hired.

On February 1, 1989 captains and lieutenants began wearing the new light blue shirts with trumpet collar insignia and nametags. Their fire helmets also had rank decals applied on the sides and the chief officer ranks had rank decals placed on their helmet fronts. The uniform sleeve and epaulette stripes from fire chief, deputy fire chief, assistant fire chief, and battalion chiefs were now increased to five, four, three and two respectively, from four, three, two and one, and now matched the number of trumpets on their badge insignia.

Firefighter Chuck Watt and Smokey, from the 1990 fire fighters calendar.
Photo: Local 18, IAFF

The Fireboat is Sold

On March 14, 1989, after two hours of discussion, the city council voted six to five, to accept an offer of $142,500 for the decommissioned *Vancouver Fireboat No. 2*. It was estimated that the scrap value was about $250,000 and the replacement cost would be between $5 and 6 million. The local entrepreneur who bought the boat sold it to an agent for the San Francisco Fire Department in November for more than $300,000 US (which was then almost $400,000 Canadian) where it was delivered and placed in service as the *Guardian*. The members of the SFFD couldn't believe their good fortune in acquiring such a vessel and many commented on its excellent condition.

The arguments continued pro and con regarding the need of the fireboat among firefighters, the port authorities, politicians and others, and all knew it was just a matter of time before the next big waterfront fire. The city had also lost a valuable asset that will be needed when the often talked about big earthquake strikes the area.

Program for the reopening of No. 6 Firehall. *Photo: VFD Archives*

Vancouver hosted the World Police and Fire Games III, July 29 to August 6. The previous games had been held in San Jose and San Diego, California, in 1985 and 1987.

In April 1990 it was reported that the city was planning to amend the building bylaw to include mandatory sprinkler installation in all new one- and two-family homes in the city.

On May 13, following Vancouver and nearby Richmond's lead in 1976, the remaining municipalities in the Greater Vancouver Regional District (GVRD) began using the 911 emergency telephone system for their fire, police and ambulance services. The GVRD is made up of twenty jurisdictions in an area of 1,255 square miles (3,249 square kilometers), which is larger than the State of Rhode Island.

At 0001 hours on June 15, the Records Management System (RMS) was placed in operation on the VFD computer system, for record-keeping and report purposes in all firehalls.

As in 1988, Vancouver firefighters got involved in the civic election and endorsed the candidates who believed in the restoration of the five-man crews on fire department engines. Studies had shown that the department's efficiency had fallen because of the cuts made in 1987, resulting in greater losses to property and increased injuries to firefighters.

On October 1 the *Vancouver Sun* headline read, "Plea for fireboat fleet revealed," and stated that the Vancouver Port Fire Protection Committee would urge five local municipalities to purchase four new fireboats and a fire barge for $3 million for the protection of the harbor. A nine-month secret study resulted in the proposal to equip Vancouver, Burnaby, Port Moody, and North Vancouver City and District with quick-response fireboats capable of traveling at 20 to 25 knots with a pumping capacity of 2,500 to 3,000 US gallons per minute (11,000 liters per minute).

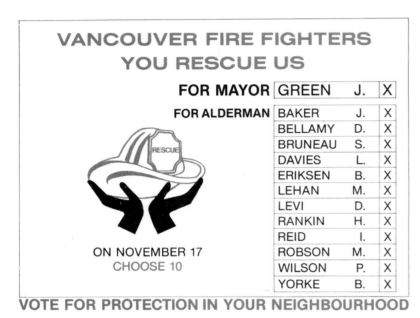

Vancouver firefighters endorse civic candidates in the upcoming election.
Photo: VFD Archives

Vancouver's city council, amid charges of deception and political opportunism, voted unanimously to endorse the proposal. Political opponents of Mayor Campbell said that they had the impression from the mayor and council that the city didn't need a fireboat and were now glad to see that fire protection on the waterfront was indeed a necessity. Campbell reiterated that the old fireboat was useless but this new fireboat system was more efficient and cost-effective. No provisions were made for any increase in personnel. David Mitchell, president of the Vancouver Firefighter's Union welcomed the report, but stated that "the city has a view that the equipment puts out fires—but without people to run it, you have nothing." The debate continued.

The New Haz-Mat Team

At the end of October the Haz-Mat team was in service after many months of training by CADRE, a Seattle-based company. The new team had more than $320,000 worth of new equipment, including a new 1990 International truck with custom-built storage compartments and a mobile communications center. No. 17 Firehall was designated as the Haz-Mat Company's home, but they still operated as a regular fire company and answered as a Haz-Mat unit, as required. The members of the team were jokingly referred to as "The Glow-worms".

No. 17 Hall's new Haz-Mat Unit.
Photo: F/F Shane MacKichan

Lieutenant Sohen Gill Injured

On December 29 at a fire inside a restaurant on West Fourth Avenue, Lieutenant Sohen Gill, age forty-nine, climbed onto the roof with a member of his truck crew. His plan was to vent the roof so that the firefighters could get at the fire easier, but he didn't plan to vent it by falling through it, which is what happened — every firefighter's nightmare. When he fell through he was in an inferno, and with more good luck than good planning, he managed to stagger to a window and to two of his crew who were able to pull him out.

He was treated and rushed to hospital where doctors weren't optimistic about his chances for survival, but his friends and colleagues knew that he was a fighter. Sohen had a long road to recovery with many skin grafts and therapy but was never able to return to work because of the fire damage to his lungs. Sohen says, "When I woke up in the hospital (where, coincidentally, he had been born) it was like being born a second time. That's why I'm (now) really only two years old," he jokingly said in 1993 when he was named Firefighter of the Year. He was recognized by the Vancouver Board of Trade for outstanding service to the community and his many years of work with sports groups, particularly the lacrosse associations, for which he served as commissioner of the Western Lacrosse Association (WLA) from 1987 to 1990.

Captain Sohen Gill.
Photo: The Province: David Clark

Of East Indian descent, Sohen was also honored at the firemen's ball for his achievements and had the audience laughing when he told them that before he fell into the fire he was white! His son, Chris, who always wanted to be a firefighter, joined the VFD in May 1993.

No. 4 Firehall Closes

On January 22, 1991 No. 4 Firehall was closed and demolished. A new firehall would be built on the site with a Vancouver Public Library branch in the same building and named the Firehall Branch. During the period of the construction, Engine 4 was relocated in the old Civil Defense bunker under the Granville Street Bridge at 1575 West 5th Avenue, and Ladder 4 went to No. 8 Hall displacing Rescue 8 that went to No. 1 Hall.

The city council passed the bylaw that made sprinkler systems mandatory in all new residential construction but the fire chief reported at a local fire chief's conference that less than 10 percent of the 700-odd buildings over seven stories had sprinkler systems. The chief wanted the same standard to apply retroactively to all high-rise buildings.

During the late 1980s and early 1990s, the city made several promotions of junior members to executive positions on the department. Most were promoted from the firefighter ranks to assistant chief, jumping ahead of 200 to 300 senior members. The removal of the fireboat in 1988 plus these promotions resulted in low morale.

The Coast Guard Dock fire.
Photo: Dave Youell

The Coast Guard Dock Fire

At about 1100 hours on Sunday, July 7 the fire that everyone dreaded would happen, actually did happen. This expected fire was the next big one on the waterfront following the disposal of the city's fireboat because of the city council's belief that the boat was obsolete and a good target for cost-cutting measures despite strong economic times.

A family of three had just launched their sixteen-foot (5-m) boat and the father had started the engine to warm it up. As the boat moved out into the water he noticed that the back of the boat was on fire, and when his attempts to put it out failed he urged his wife and daughter to jump out and swim for shore. The wind caught the burning boat and pushed it beneath the Canadian Coast Guard dock a few meters away where it struck the heavily creosoted pilings, setting them on fire. Within minutes, the fire quickly spread throughout the facility and by the time the first-in fire companies arrived on scene, three-quarters of the complex was on fire, described as having "exploded like a book of matches."

Strong westerly winds whipped the flames under the dock toward the Esso marine fuel station that was situated a few yards east of the fire scene. Beyond that were hundreds of pleasure craft, the Burrard Bridge and the always-crowded Granville Island, with people shopping, eating and enjoying the hot summer weather. Up to 48,000 gallons (220,000 liters) of various fuels were stored on the floating Esso barge, and it was only through the firefighting efforts of the Coast Guard crew on board the small vessel *Osprey,* positioned between the inferno and the fuel barge, that prevented the fuel from exploding. Had the fuel barge exploded, people would likely have been killed and there was a possibility that the Burrard Bridge, under which the fire raged, could have been badly damaged.

The city firefighters on scene could only reach the fire from shore positions as the approach ramp to the dogleg-shaped wharf was where the fire had started. It took more than eighty firefighters at the four-alarm fire all day and into the night to extinguish the fire. Crews remained on the site for three days, overhauling the fire scene. Destroyed along with the wharf were the two-story Coast Guard station and helipad, two Coast Guard vessels and a zodiac-type rubber boat and seven vehicles belonging to CCG and RCMP personnel. Damage was estimated to be about $6.5 million plus demolition costs that exceeded $1 million before a new facility could be built.

Local politicians refused to concede that the old fireboat would have done any good, saying that it would have taken too long to get to the fire scene, but with a pumping capacity nearly as much as that of Vancouver's nineteen regular fire engines, combined, it would have knocked down the fire in a few hours and it wouldn't have been necessary to have men hosing down the scene for the next three days.

In the following days, several local columnists and editorial cartoonists criticized the mayor and council members for their stand on the fireboat and pointed out how fortunate it was that people weren't killed as a result of their cost-saving measures. Many people expressed their surprise and shock to hear that such a facility, in the business of helping people, could have been so vulnerable to such total destruction.

All the men, both firefighters and coast guard members, were praised for a job well done by Alderman Don Bellamy, who stated, "the potential for a major disaster was down there if not for the tremendous work of the firefighters." The officer-in-charge of the base was convinced that the fireboat sold by the city would have helped to stop the spreading flames. He said that he had lost "his home" on that Black Sunday, as the Coast Guard crews call it, and all that was saved was "two of five boats and my suitcase," he said.

As the debate continued and the Coast Guard dock fire was still being discussed, the local newspapers were reporting that the new multimillion dollar "waterfront firefighting strategy" was underway and tenders for the new "quick response" fireboats would be put out shortly.

On September 11 the 1973 Calavar Firebird aerial platform was placed out of service and put in reserve at No. 22 Hall.

Bob Krieger's *The Province* cartoon following the Coast Guard Dock fire.

The Bon Accord Hotel Fire

At 0059 hours on February 14, 1992 the first alarm to the Bon Accord Hotel on Hornby Street came in and before it was knocked down it had progressed to a five-alarm fire. At the height of the fire the building was burning from the basement to the fourth floor, resulting in the building being completely destroyed. Several times fire crews were in the building only to be pushed out by the fire, until all the crews were advised to get out and assume a strictly defensive mode.

The Bon Accord Hotel fire.
Photo: Dave Youell

All of the tenants were able to get out, and a fire investigator who fell through a skylight during his preliminary investigations sustained the only major injury. He fell about twenty feet, suffering painful back, leg and internal injuries. He was hospitalized and took several months to recuperate and return to duty. Cause of the fire was determined to be suspicious.

The Vancouver Board of Trade named Captain Don Lee Firefighter of the Year for 1992. The thirty-year veteran and third-generation Vancouver firefighter was honored for his many years of community service. Don's father, Walt, a lacrosse hall-of-fame member, was on the VFD from 1934 to his death in 1971 when he was a district chief. His grandfather, Norman Lee, joined the department in 1904 as a hose man, after his arrival from New Zealand, and a few years later he became Fire Chief

Carlisle's secretary and administrative assistant.

In mid-April the city's engineering department reported that three earthquake-proof saltwater pumping stations would be built to give fire department engines a water supply in the event that the conventional hydrant system failed. The stations would be connected to an independent underground pipeline grid with special hydrants and have an around-the-clock, fail-safe supervision and control protocol. The first station was planned for Coal Harbour, the second on the north shore of False Creek, with the third planned for the Coast Guard site at the mouth of False Creek. Estimated cost was $30 million and was expected to take about eight years to complete. A fourth pumping station was proposed for the north foot of Victoria Drive.

On May 2 the return of the Fire Fighters' Ball, after an absence of nearly twenty years, was held at the Hotel Vancouver and attended by a sold-out crowd of 750 black-tie and uniformed guests, including Honorary Fire Chief Murray (The Pez) Pezim, well-known stockbroker and friend of the firefighters. The proceeds from the $75 per plate fundraiser were for the VGH burn unit.

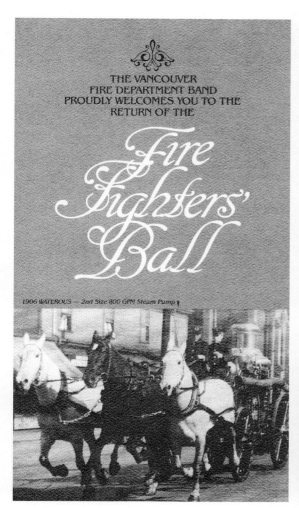

THE VANCOUVER
FIRE DEPARTMENT BAND
PROUDLY WELCOMES YOU TO THE
RETURN OF THE

Fire Fighters' Ball

1906 WATEROUS — 2nd Size 800 GPM Steam Pump

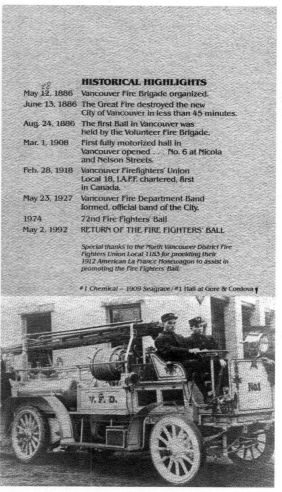

HISTORICAL HIGHLIGHTS

May 12, 1886	Vancouver Fire Brigade organized.
June 13, 1886	The Great Fire destroyed the new City of Vancouver in less than 45 minutes.
Aug. 24, 1886	The first Ball in Vancouver was held by the Volunteer Fire Brigade.
Mar. 1, 1908	First fully motorized hall in Vancouver opened . . . No. 6 at Nicola and Nelson Streets.
Feb. 28, 1918	Vancouver Firefighters' Union Local 18, I.A.F.F. chartered, first in Canada.
May 23, 1927	Vancouver Fire Department Band formed, official band of the City.
1974	72nd Fire Fighters' Ball
May 2, 1992	RETURN OF THE FIRE FIGHTERS' BALL

Special thanks to the North Vancouver District Fire Fighters Union Local 1183 for providing their 1912 American La France Hosewagon to assist in promoting the Fire Fighters' Ball.

#1 Chemical – 1909 Seagrave/#1 Hall at Gore & Cordova

Return of the Fire Fighters' Ball. *Photo: VFD Archives*

New No. 4 Hall Opens

On May 2 Engine 4 left the bunker that had served as home base during the construction and returned to the new No. 4 Firehall. On June 11 No.4 Truck came home from No. 8, and on June 27 the firehall was officially opened and placed in service. It was dedicated to Lieutenant O.B. (Bud) Swanson of the Fire Prevention Branch who collapsed and died on duty July 11, 1990. At the opening ceremony, a plaque was placed on the building in his memory.

Lieutenant Bud Swanson.
Photo: Local 18, IAFF

New No. 4 Firehall. *Photo: Author's collection*

The International Association of Fire Fighters 41st Convention met in Vancouver on August 3 to 7, after four years of planning by the Convention Committee of Local 18. With the slogan, "It's Our Turn" the union won the right to the convention in Miami in 1988.

Fire Chief Don Pamplin retired at the end of August after thirty-eight years of service. He was involved in significant changes in the department, including mandatory sprinkler systems in new-house construction and the selection of the new, fast-response fireboats. On September 1 fifty-three-year-old Deputy Fire Chief Glen Maddess was named the thirteenth fire chief of the VFD.

During 1992 retired Captain H.A. (Bert) Lowes, a long-time boxing coach and trainer, was inducted into the BC Sports Hall of Fame. He attended five Olympic Games as well as Pan-American and Commonwealth Games as a referee and judge. Bert joined the department in 1947 and died in August 2000 at age eighty-four.

On December 9 the empty buildings of the Sterling Shipyards at the north foot of Victoria Drive were the scene of a three-alarm fire. Three of the new fireboats were on the scene but not officially in service at the time as the crews were still in training. The fire gave the firefighters a chance to put the new boats through their paces, but they weren't able to save much.

In February 1993 the fifth and last of the new fireboats was delivered by the builder, Celtic Shipyards (1988) Limited, a company owned and operated by the Musqueam Indian Band of Vancouver. Fireboat 1 and Fireboat 5 were assigned to the VFD, with North Vancouver City and North Vancouver District getting Fireboat 2 and Fireboat 4 respectively,

W. GLEN MADDESS

Born: October 25, 1938, Vernon, BC
Joined VFD: April 17, 1961
Appointed Chief: September 1, 1992
Retired: May 31, 1998

Sterling Shipyards three-alarm fire.
Photo: VFD Archives

New *Fireboat 1* during pre-delivery demonstration.
Photo: VFD Archives

Cover picture from *Canadian Firefighter* magazine showing the new fleet of fireboats. *Photo: VFD Archives*

and Port Moody and Burnaby co-operating Fireboat 3. These vessels were of aluminum construction, forty feet (15 meters) in length, rated at 2,500 imperial gallons per minute (11,500 liters) with a top speed of twenty knots.

The fire chief and city medical health officer asked city council on February 3 for thirty-seven defibrillators for VFD apparatus to aid heart attack victims. It was noted that in 319 cardiac arrest cases in 1992, the fire department medical crews arrived ahead of ambulances that have these units. Until the medics arrived, VFD members had to do manual cardio-pulmonary resuscitation (CPR). The request was made to give heart patients a better chance of survival but the council voted six to five against the request stating that the defibrillators were a provincial responsibility. The cost was estimated to be $235,000 over three years with the saving of an estimated sixty lives.

Then through the efforts of Firefighter Bill Sanderson, the Vancouver City Savings Credit Union and the City of Vancouver formed a partnership and arrangements were made to purchase the necessary defibrillators. VFD members began training on them and they were subsequently put into service with dramatic results; the survival rate went from about 7 percent to around 25 percent. The success of this program resulted in similar partnerships between the Credit Union and other fire departments in Greater Vancouver.

April 3 and 4 saw the Clinton-Yeltsin Summit held in Vancouver and several members of the VFD played a small part as fire watch personnel on the floors above and below those used by the president's party at the Hyatt Hotel.

Possible loss of funding of the province's First Responder training was of great concern to local fire departments including Vancouver in April 1993. In many cases VFD crews arrive on the scene of medical emergency service alarms (MESA) before the ambulance and paramedics and, when precious minutes count, this specialized training pays off.

In 1992 it was reported that fire departments responded to 14,600 MESA calls before the arrival of the ambulance crews. The BC Ambulance Service routinely re-routes calls to the fire department when it can't respond ambulances quickly enough. One example cited was Engine 2, the city's busiest, which responded to five drug overdose patients in three hours, and of the seventeen runs it had that particular night, fifteen were emergency medical situations. The First Responders training problems were resolved giving the Greater Vancouver area one of the best emergency medical services anywhere.

On May 17 the first woman firefighter on the department started in a group of ten trainees. On completion of her basic training, Cara Goose was stationed at No. 9 Firehall, which had to have sleeping and bathing facilities altered to accommodate the needs of a female firefighter. The VFD chose to keep a low profile on her appointment to the department.

Class of 1993 recruits with the VFD's first female firefighter.
Photo: VFD Archives

Fire Chief Glen Maddess noted in the press that of 105 waterfront properties surveyed following the disastrous Coast Guard dock fire in July of 1991, not one had followed fire department recommendations for improvements. He further stated that the installation of fender logs (connected, floating logs that can prevent floating hazards from getting under wharfs), the simplest action to improve unsafe conditions, had not been undertaken in fifty-two instances and half of the marinas, wharves and piers in the harbor did not meet fire department safety standards. The problems were the responsibility of the Port Corporation and fire department notices could not be enforced. The chief also advised that he doubted that the new fireboats would be able to put out any large, creosote-fed fires that could occur under wharfs.

Fireboat 1 was officially placed in service on August 19 and berthed on the waterfront between the BC Transit Seabus Terminal and Canada Place Convention Center and cruise ship terminal. Trained engine crews from nearby Firehall Nos. 2 and 8 would man it.

The City of Vancouver amended the fire bylaw restricting the possession, use and sale of firecrackers and fireworks in the city, October 25, 1993, in an attempt to stop the damage and injuries they caused every Halloween.

Film Studio Destroyed

One of Vancouver's first movie studios in the downtown area was destroyed in a spectacular five-alarm fire on July 9. The old Cannell Films studio on Carrall Street was used in the production of the television series *Stingray, 21 Jump Street,* and *Wiseguy,* and such movies as *Look Who's Talking,* said to have been the most profitable movie made in Canada, to that time. Also destroyed was the courtroom set used in *The Accused,* for which Jodie Foster received the Academy Award for Best Actress of 1988. The cause of the fire was unknown but arson was suspected as the area was frequented by transients.

Over the years, the VFD and its members have taken part in many movies and television series filmed in and around the city. In 1975 No. 6 Firehall and its apparatus were seen in the George Segal hit, *Russian Roulette.* In the 1977 sci-fi movie, *It Happened at Lakewood Manor,* No. 6's old tiller aerial is seen in the final scenes. When the Disney movie, *The Journey of Natty Gann* was filmed here, the union's old 1937 Hayes-Anderson canteen wagon was repainted and used as a depression-era bus. The *Vancouver Fireboat No. 2,* now San Francisco's *Guardian,* was seen in *George of the Jungle,* released in 1997. Several firefighters have

Cannell Film Studios destroyed by fire.
Photo: Dave Youell

worked as extras in many movies and television series, such as *The Boy Who Could Fly*, *21 Jump Street*, *J.J. Starbuck* and *Wiseguy*, portraying civilians as well as firefighters.

A ninety-year-old Vancouver landmark was destroyed in an early morning fire on December 13. The old Arcadian Hall at 2214 Main Street was used over the years by fraternal, cultural and other social and artistic groups and was owned by the Finlandia Club. It was a popular venue for dances because it had one of only a few spring-loaded dance floors left in the city. It had been around for so long it was one of those buildings that firefighters "pre-fire planned" over the years because it was always felt that it would eventually burn down. As with every all-wooden building using old-type construction methods, it was known that it would be difficult to save. The cause was believed to be arson as a gasoline container was found under the rear stairs.

In March 1994 organizational changes resulted in the loss of five chiefs' drivers to allow the administration to hire two training officers, a lieutenant for emergency preparedness planning, a chief of inspections, a fire protection engineer and a clerk. The cuts were made eliminating driver/aides to the assistant chief and the three battalion chiefs, positions that had been in place since the early days of motorization. Union President Dave Mitchell pointed out to city council that the cuts would drop fire suppression staff to 132 men per shift, the lowest number in forty years.

The Western Boot Shop fire on East Broadway, January 6, 1993, with No. 4's Bronto Tower in the foreground.
Photo: Dave Youell

The Stanley Cup Riot

The evening of June 14 started with the night shift fighting a three-alarm apartment fire at East 14th Avenue and Main Street. As the evening progressed, a normal number of alarms were answered until about 2130 hours by which time the seventh and final game was over in New York City, with the Rangers beating the Vancouver Canucks 3 to 2 to take the coveted Stanley Cup.

Crowds of people began gathering in the downtown core and soon the postgame celebration turned ugly. That certain element in every large crowd began breaking windows, overturning cars, doing anything and everything that was both dangerous and criminal. It wasn't long before the fire alarms and MESA (medical) calls began coming in with the result that many fire crews ended up in areas full of rioters, often caught between the police and the troublemakers, on the receiving end of teargas on one side and assaults on the other. Some of the city's firehalls in the outer areas didn't know the extent of the trouble that was going on until they heard about it on the late television news.

Over the course of the evening until mid-morning the VFD responded to over 150 alarms including two- and three-alarm fires, attended by the assistant chief and each of the three battalion chiefs on duty. For the first time in the history of the Vancouver Fire Department the city had to ask for help from outside municipalities. Burnaby Fire Department helped by filling in and covering districts of Firehall Nos. 14 and 15 on the east side;

A three-alarm apartment fire at East 14th Avenue and Main Street was the beginning of a busy night for firefighters following game seven of the Stanley Cup finals.
Photo: Dave Youell

the University Fire Department covered for No. 19 Hall on the west side, with Richmond Fire Department filling in for No. 22 on the south. The Surrey Fire Department sent its air mask compressor truck into the city to assist as well, because the VFD was unable to keep up with the demand for bottles of air.

Because there had never been a need for mutual aid into Vancouver previously, there were a few problems, among them that outside departments couldn't get into the firehalls, communications were on different radio channels, and radio nomenclature and procedures were different from department to department. Outside crews didn't know how the VFD alarm system worked, and if an alarm had come in they likely wouldn't have known where to go.

Following the incident, a survey done by the union made ten recommendations to improve the handling of major emergencies:

1. Increase staff (only three extra men were hired that night).
2. Develop a plan to deal with major incidents.
3. Ensure crews get a change of clothes, bathroom and food breaks.
4. Post-incident consultation with front-line staff.
5. Better communication system.
6. Tear gas training for firefighters.
7. Improve mutual-aid system.
8. More new equipment.
9. Better reserve equipment, and
10. More reserve equipment.

The Alberta Wheat Pool Fire

At 1858 hours on August 24 a fire at the huge Alberta Wheat Pool grain elevator (now Cascadia Terminal) on the waterfront was cause for more than a little concern as any waterfront fire can develop into a major conflagration. Like the earlier Coast Guard dock fire, this fire was also under the wharf and difficult to get at, and before long the dirty, smoky fire was

The Alberta Wheat Pool four-alarm fire.
Photo: Dave Youell

made a four-alarm by Battalion Chief Joe Storoshenko. The heavy, creosote smoke was so thick that the nearby Second Narrows Bridge had to be closed from 1930 hours until 0130 hours the next morning. Also, the Pacific National Exhibition across the street was closed at 2000 hours because of the fear of a possible grain-dust explosion that would send debris flying in that direction. It was the first time in the eighty-year history of the fair that it had to be closed. Any explosion could have been devastating to the bridge as well.

There was also a fear that the fire would reach a concrete vault where electrical capacitors containing polychlorinated biphenyls (PCBs) were stored. Firefighters ensured that streams were covering that exposure to prevent any fire and exposure to the carcinogen.

With the fire burning under the deck it was once again necessary to call in help from other fire departments. Both West Vancouver and Burnaby sent zodiac boats to attack fire under the wharf, along with three supplied by the navy because the new fireboats sat too high in the water and their streams couldn't reach the fire. Three of the new fireboats were on scene but the other two were out of commission with mechanical problems.

In a statement made later, Vancouver union president David Mitchell said, "It is a technology that's unreliable. If they're all brand new they should be working and two of them are not working." He went on to say (once again), "we are relying on someone coming and bailing us out. It's our view that this is not a very good way to run a fire department. A city as large and as well off as Vancouver should have its own stand-alone fire department. If there's a disaster like an earthquake there'll be none of that shared equipment available to anybody."

It was revealed that a sprinkler system under the dock had failed, allowing the fire to come dangerously close to the PCBs. Cause of the estimated $2 million fire was attributed to a pile-driving crew working on the site.

The Vancouver Fire Fighters' Union, Local 18, IAFF, won a long battle to convince the Worker's Compensation Board (WCB) that the brain cancer death of a city firefighter was work-related and a pensionable claim. The decision, a Canadian first, was a huge victory for firefighters everywhere when it was acknowledged a link existed between firefighting and brain cancer.

On Monday, January 2, 1995 calls received by the Vancouver Fire Department headquarters reception desk were answered with, "Vancouver Fire and Rescue Services." (VF&RS) After a proud 108-year history as one of the finest fire departments anywhere, the name change reflected changing times, but to many these changes were hard to accept. The fire chief was now the "general manager" and the people served were referred to as "customers".

On January 5, 1995 four Seattle firefighters were killed while fighting a warehouse fire, the worst loss of firefighter lives in the history of the SFD. Firefighters throughout North America rallied around and paid their respects to their fallen brothers and more than 1,000 firefighters from the Greater Vancouver area attended the service, which included music by the Vancouver Fire Department Band. When the memorial service was held at the Seattle Civic Arena on January 11, flags at BC firehalls flew at half-staff in remembrance. The cause of the fire was arson and the prime suspect was later found in Brazil.

Acting Battalion Chief Paul McDonnell, one of the "Sixty-Fourers" and a man who wore many hats through his union involvements, was named the 1994 Firefighter of the Year by the Vancouver Board of Trade.

UBC Firefighters Join the VFD

On October 16 the fifty-nine firefighters on the University Endowment Lands became part of the Vancouver Fire Department and their only fire-hall was designated VFD No. 10. The University Fire Department began in the 1920s and was re-established in 1947 by the BC government as one of three provincial fire departments, the others being Riverview Hospital in Coquitlam and Tranquille Institution in Kamloops. All have now been absorbed by larger, nearby departments. With the addition of these UEL firefighters, Vancouver's manning was at the highest ever, with 840 personnel at year-end.

The new logo of the Vancouver Fire and Rescue Services. *Photo: VFD Archives*

VANCOUVER FIRE/RESCUE K-9 SEARCH PROGRAM

To the End of the Millennium

(1996–2000)

As 1996 began, one of the new pumping stations was opened downtown on False Creek as the first part of the Dedicated Fire Protection System (DFPS) that is capable of delivering 10,000 gallons per minute at 300 psi. This $30-million system will eventually include three hose tenders (only one by 2006), each capable of carrying one mile of five-inch, (125-mm) high-pressure hose and fittings. Each will have a 1,500-gpm (6,750-liters) monitor that can be used directly from the water supply. Engines will be able to lay up to three five-inch (125-mm) lines of hose from the new DFPS hydrant system in the event that the regular water mains are not useable. A second pumping station was completed in Coal Harbour in the fall of 1996 and this system was built to withstand an 8.5 earthquake. The water works department mans the pumping station on False Creek twenty-four hours a day, seven days a week.

On January 29, 1996 the *Vancouver Sun* reported that the city had come up with a newer way of choosing firefighters, ensuring a fairer competition for women and visible minorities. The physical test included tests geared toward the actual work and less toward strength and endurance. The physical tests involved carrying hose, using a sledgehammer, pulling hose lines and carrying a dummy, all completed within five minutes and thirty seconds. Bonus points would also be given for being a non-smoker and for knowledge of other languages and cultures. It further reported that the fire department now had two women firefighters, the second of whom came from the University Fire Department and was in fire prevention. Twenty years after the first accusations of sexism and nepotism by the equal employment opportunities office, the union's predictions in 1977 on quota hiring came true. Once again, the city's personnel department was criticized for discriminating against young, white male applicants for the fire department. The debate continued.

In March, Ladder 2 was moved to headquarters and became Ladder 1, once again making No. 1 Hall a two-piece firehall since opening in 1975. Rescue 8 returned to No. 2 Firehall as Rescue 2, after sixteen years as Rescue 8.

In April a survey and report done by the US-based Tri-Data Corporation was received by the city and it recommended that several fire department apparatus could be disposed of in favor of "quick response" two-man Life Support Units (LSU), because over 70 percent of the alarms received were for medical services. The report also recommended that the city adopt the Quintuple fire apparatus (a quint is a "single piece of apparatus combining a pump, hose bed, booster tank and hose, ladders and an aerial ladder"), which are a combination of the current engine and ladder companies.

Before the report was implemented, the union had long and hard discussions regarding response times and available manpower at future fires and rescue incidents. The current strength of seven men per crew may rise to eight or drop to six depending on the type of equipment required in a particular hall.

The Tri-Data report also revealed that the Class I Underwriters rating which the VFD had been so proud of, had been reduced to a "low Class II or maybe a high Class III," positions lower than it had when it earned the Class I rating in 1980 (see chapter 10).

Two views of the DFPS Hose Tender. *Photo: VFD Archives*

Vacant stores in the 1000-block Hornby Street suffered major damage in a five-alarm fire, July 11, 1996.
Photo: Dave Youell

On May 8 the city reported that earthquake preparedness was "tacky" at best with "patchy" preparations, and on a scale of from one to ten, the city rated about "five-plus." Some have said that all the simulated exercises have included a lot of simulated preparations.

It was reported, however, that plans were being made for a new communications center that would be completed by 1998. (The new center, known as E-Comm, was built near Cassiar and Hastings Streets, on the east side of the city and opened June 8, 1999. VFD fire dispatchers moved over to the facility in the fall of 2003 and after E-Comm staff learned how to handle fire dispatch, fire dispatchers either retired or returned to other fire department duties.)

When a small 5.5 tremor struck at 2105 hours on April 2, the ambulance service dispatch center was evacuated because of concerns the staff had for their safety while in the building.

Tuesday, May 28, 1996, signified 110 years since a meeting was held to form a fire brigade in the city of Vancouver. That day in 1996 the VF&RS received a total of 139 calls: 68 MESA, largely due to the high-grade heroin on the streets, and 71 fire and miscellaneous calls, all very routine and none unusual.

May 28 that year was also an election day in British Columbia. After several weeks of campaigning, the New Democratic Party (NDP) under Premier Glen Clark won by a slim margin. Ex-mayor of Vancouver, Gordon Campbell, the leader of the BC Liberal Party, was unsuccessful in his bid to become the premier when his party failed to win enough seats.

First High Angle Rescue

On August 5 the first rescue by the department's High Angle Rescue Team took place when a window washer's scaffold jammed on the

High Angle Rescue Team member at a drill.
Photo: VFD Archives

The rescued window washer and his rescuers as shown on the WCB 1996 Annual Report.
Photo: VFD Archives

twenty-fourth floor of a downtown office building. The new high-angle program was jointly administered by the WCB, the construction industry, and fire service. The team responded from No. 3 Firehall.

The BC Justice Institute took over the training of new firefighter candidates for the Greater Vancouver fire departments during the year. Each trainee paid $5,000 for training at the institute. During the year, the VFD took on twenty-two new men.

Justice Institute ad for firefighter training.
Photo: The Province

City of Vancouver personnel department ad.
Photo: The Province

By the end of the year the Vancouver Fire Department responded to 39,688 incidents and many during December were attended to in sub-freezing weather, a rarity in Vancouver's temperate climate. Though not as cold as the winter conditions endured by firefighters in many other parts of the country, it was nonetheless very cold and made everyone grateful that these conditions aren't the norm.

In another departure from the norm, Vancouver's Firefighter(s) of the Year for 1996 was a five-man group unanimously selected for their individual roles in the rescue of a man trapped in a burning basement suite. The teamwork of Acting Lieutenants Tony Neratini, Rob McCurrach, Rick Kennedy and Firefighters Mark Etheridge and Neil Smith resulted in saving the man in the early morning hours of May 6, 1996. The Vancouver Board of Trade Jaycees honored them at the Awards of Excellence banquet at the Hotel Vancouver, March 5, 1997.

The new VF&RS patches.
Photo: VFD Archives

In April 1997 firefighters lost a protracted battle when the city's budget committee adopted the Tri-Data plan to reduce fire apparatus by a third and redeploy personnel. At a union meeting in May, attendance was over 700 members, out of the 802 strength of the department, in opposition to this move. The number of engines was cut to ten from twenty, and all but four aerial ladder trucks were eliminated. They were replaced with fourteen "quints" and eight LSUs, the largest apparatus order in the history of the VFD at a cost of $14 million.

Rescue 17 LSU.
Photo: Shane MacKichan

The new No. 9 Quint.
Photo: Shane MacKichan

More Arson Fires

The total fire losses in 1997 were estimated at about $15 million with one-third of that attributed to arson fires. In one such arson fire a four-story apartment in Marpole was torched about 0330 hours and a dozen residents considered themselves lucky to have escaped. The fire was started when someone poured a flammable liquid in the hallway and lit it. No suspect was found, but residents suspected either a former tenant or someone angry with a tenant was responsible.

The arson fires continued through the spring and summer in stores, cars, carports and garages, and a suspicious house fire in August resulted in the death of one man. The alarm came in at 0645 hours on August 14 for 2076 Pendrell Street, to one of the few remaining old houses in the

West End. The victim was found in an upstairs bedroom and the death was being investigated by the major crime squad. Three men rescued two other occupants from the house before the arrival of the fire department.

In August No. 10 Firehall was designated a Haz-Mat Team station along with No. 17 Firehall and equipped with a new Haz-Mat unit.

The new No.10 Haz-Mat unit.
Photo: Shane MacKichan

The Haz-Mat team at Chess Street drill grounds. *Photo: VFD Archives*

On September 14, at 0740 hours on a quiet Sunday morning, fifteen young squatters were forced out of a vacant hotel in the 600-block Seymour Street, when one of them set his mattress on fire with a candle. When he and his friends tried to throw it out a third floor window, they only succeeded in igniting the walls of the room, and then the fire spread throughout the building. For several months squatting street kids, addicts, and the homeless had used the old boarded-up hotel that was due for demolition. Following the fire, the building was torn down.

The High Angle Rescue Team removed the operator of a construction crane from his 120-foot (36-m) perch above the street at Main and Hastings on November 23. The crane operator complained of feeling sick and was having trouble breathing when the team was called. The crew climbed up to the man, stabilized him and, because of the confined space, brought him down by rope to the waiting paramedics. When he reached the ground, the spectators cheered as the man was removed to the hospital.

High Angle Rescue Team construction crane drill. *Photo: VFD Archives*

The arson fires continued into December after a busy summer of set fires. The arsonist, referred to as the "million-dollar firebug" was believed to have set four fires in the basement of a seventy-five-unit Kitsilano apartment building on December 13. A thirty-seven-year-old drifter and convicted arsonist was arrested at the scene and taken into custody. He was released two days later. Then on December 29 two more car fires occurred in the False Creek area and police were looking for the drifter again. He targeted cars, looking for small change or other articles and, finding little or none, he then torched the vehicles. He was also known to set garbage container fires and fires in discarded furniture. The Insurance Corporation of BC (ICBC) paid out claims of over $1 million for fires in cars believed set by this man.

The city had 228 arson fires for the year, up nineteen over the previous year, for a 9.1 percent increase.

January 1998 began with continuing car fires in underground parking lots and the same man was believed responsible.

A stubborn supermarket fire in the 2900-block West Broadway was reported about 0400 hours on January 15. It was believed that the fire that started in the store's office was caused by a carelessly discarded cigarette. There were no injuries and the owners had no insurance coverage on the contents. Arson fires on Vancouver's east side during January caused more than $100,000 damage in four incidents.

The Gastown Fire

The spectacular Gastown fire.
Photo: VFD Archives

On March 2, 1998 around 2145 hours one of the heritage buildings on Water Street in Gastown was burning in a spectacular fire that lit up the area. No. 2 Company was first-in with No. 8 close behind. By 2200 hours Battalion Chief Bob Pachota had made it a third-alarm, and before midnight it was struck out. Damage was considerable but the facade of the building was saved and the 100-year-old building was restored. Many of the firefighters at first thought that the building was the site of the original firehall in Gastown because the address given was 32 Water Street, but it turned out to be a few doors further down the street.

Ray Holdgate is the New Fire Chief

Effective June 1, 1998 forty-eight-year-old Assistant Chief Ray Holdgate became Vancouver's fourteenth fire chief, succeeding Glen Maddess. He joined the department in August 1974.

Fires that cause considerable problems for area firefighters are those that occur in buildings where marijuana is being grown. The high electrical and chemical usage cause additional unsafe conditions for the firefighters attending to blazes at these grow-ops. Often the Haz-Mat team has to respond as well to properly handle the incident.

In 1998 Canada's first and only Heavy Urban Search and Rescue Team (HUSAR) became operational. The team was assembled to prepare for heavy rescue in the event of a building collapse or earthquake. In 1999 it was ready to assist in the aftermath of earthquakes in Turkey and Taiwan, but was not used because of "government indecisiveness" according to one critic. Firefighter Flynn Lamont and Search and Rescue (SAR) dog Barkley became important members of the team. Barkley received training to search and find missing persons and bodies. He was featured on the television series *Dogs with jobs*, and in 2004 Barkley's son, Cooper, joined the team.

At the end of summer of 1998 the new training center on Chess Street opened with a five-story propane burn tower, rail cars, high-rise crane, mock service station, driving course, drafting pit and auto extrication area.

SAR dog Barkley. *Photo: VFD Archives*

Raymond C. Holdgate

Born: December 17, 1949, Vancouver, BC
Joined VFD: August 12, 1974
Appointed Chief: June 1, 1998
Current Chief

Chess Street Training Center.
Photo: VFD Archives

The "Vancouver Valve" was developed by VF&RS machinist Terry Heinrich and the city engineering department for the Dedicated Fire Protection System (DFPS), a $52-million (2005) hydrant and water main system in downtown Vancouver created to supply water to firefighters should an earthquake strike the city. The valve will regulate the water pressure through the five-inch (125-mm) hose more effectively than the old Gleeson valve that was developed after the 1906 San Francisco quake. The new system was been installed with two pumping stations, one in False Creek and the other on the waterfront in Coal Harbour to supply the twenty-four-inch mains with sufficient water for firefighting from a 150,000-gallon (682,000-liter) fresh water cistern. Sea water can be also used, if necessary. The hydrants are dry-barrel types with a main spindle and each of the three five-inch (125-mm) ports has an independent shut-off.

DFPS Hydrant with "Vancouver Valve" attached.
Photo: VFD Archives

On an early October afternoon, fire in one of the city's oldest buildings left several people homeless. The Cosy Corner Grocery building at 100 East Hastings Street, built in 1888, was damaged by a fire that started in the furnace room when paint and thinners ignited. Two people had to be rescued by ladder, along with a cat that suffered smoke inhalation.

A grocery store destroyed by fire at the beginning of the year was again damaged in a three-alarm fire. At 0400 hours on December 14 The Parthenon Grocery on West Broadway was damaged in what investigators described as a suspicious fire when they found signs of a flammable liquid outside the rear door. There were no signs of entry into the building and there were no suspects. All of the stock in the store suffered smoke, water and fire damage but was covered by insurance this time.

The fire service learned of the death of Lieutenant Steve Woodworth, age 48, at home, on December 26. He laid down for a rest after complaining of fatigue and a sore back and died in his sleep. Steve had attended a big fire at work on Christmas Day.

His funeral was held on January 5, 1999 and was attended by more than 900 firefighters and friends. Everyone was reminded of the fire two years earlier where Steve rescued an elderly woman trapped in the smoke in her tenth floor apartment and how he brought her to safety, while sharing his air mask with her.

Captain Mike Gilmore was selected to be Vancouver's Firefighter of the Year for 1998 for his volunteer work in the community. A thirty-year veteran, he is the son of a Vancouver firefighter, and his brother, Pat, and son, Sean, are also on the VFD.

At the end of February the arsonist on the West Side appeared to be back at work. Three set fires were found in the Arbutus Village area and caused about $50,000 damage, which included a car, carport, garden shed and a set of garden furniture. All the fires occurred between 0415 and 0500 hours. On March 16 police arrested the known and previously convicted arsonist once again and charged him.

The city's first fatal fire in more than a year occurred when two men died in an early morning house fire in the 5800-block of St. George Street. The alarm came in at 0030 hours and by the time firefighters arrived the house was fully involved. Two neighbors tried to rescue the occupants but were forced back by the flames, heat and smoke, and were lucky not to have become victims too. They suffered smoke inhalation and minor burns.

The Point Grey Time Capsule

A time capsule was removed from the brick wall of the soon-to-be-replaced No. 18 Firehall. The Point Grey Fire Department had placed a metal container the size of a cigar box in the corner of the building on October 12, 1923 as their new firehall was being built. On June 16, 1999,

Time capsule being placed in the building.
Photo: VFD Archives

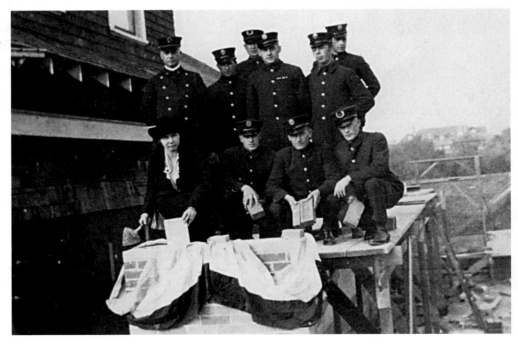

284

with the media and several officials, firefighters and friends present, Chief Ray Holdgate opened up the brick wall with a sledgehammer, exposing the box. The honor of opening the box was given to Alex Matches, the department's historian, who revealed the contents to the crowd. The box contained a list of Point Grey's firefighters, a stamped envelope containing a picture of Fire Chief Raymur, three silver coins of the day (a nickel, dime and quarter), and the October 12, 1923 editions of the *Vancouver Daily Province,* the *Sun* and the *Point Grey Gazette.*

The late retired District Chief Cecil Maddison donated some old photos to the fire department collection many years before and among them was an envelope containing some snapshots and a letter describing the placing of the box in the brickwork and how to retrieve it.

On July 21 No. 3 Firehall was vacated and slated for demolition to make way for a new, modern building. Heritage advocates were saddened to see the Mount Pleasant landmark disappear, but the old building had outlived its usefulness and was overdue for replacement. It was a drafty, cold and miserable place to work in at times, and would not have withstood an earthquake. When it opened in the summer of 1912 it had fifteen men working around the clock who, with two horse-drawn and two motorized pieces of apparatus, answered about ninety calls a year. When it closed its doors, there were three pieces of apparatus and about forty men who answered over 200 calls per month.

The time capsule was in the corner column on the left.
Photo: VFD Archives

On November 4 No. 1 and No. 9 responded at 0300 hours to a fire in an all-night market on Commercial Drive. The clerk claimed that he had been tied up by two men who then set the store on fire. He yelled for help and two men alighting from a bus heard his cries and untied him, then left him. He didn't call the fire department; he just sat on the curb awaiting their arrival. The fire went to a third-alarm and destroyed the store and two adjoining businesses. The cause was suspicious.

Commercial Drive market fire.
*Photo: The Province
Bill Cook*
(www.newsvideo.tv)

November 17 Vancouver firefighters revealed that the death of a ninety-four-year-old man in a house fire occurred when the first piece of apparatus to arrive on scene was an LSU with a two-man crew. As a rescue rig, it had no equipment to fight a fire. Two of the new quints from the closest firehalls were busy at other calls so a third had to be sent in from the next closest hall. The city now has eight Life Support Units (rescue rigs) for medical calls and the chances of rigs not capable of fighting fires arriving on scene first was becoming more likely.

On December 26 the constituency office of British Columbia Attorney-General Ujjal Dosanjh was firebombed with a Molotov cocktail, a homemade grenade, usually a gasoline- or kerosene-filled bottle with a cloth wick, that is lit and thrown. The alarm came in at 0440 hours and the first-in crew from nearby No. 20 had the fire out in eight minutes. The cause was believed to have been politically motivated. In 1985 Dosanjh was badly beaten up when he spoke out about some violence in the East Indian community. In 2000 he became the province's premier when the incumbent was forced to step down. He also served as a federal cabinet minister until 2006.

December 28 was a busy day for the fire service. An elderly man died in his West Side townhouse when a portable space heater started a fire at 0900 hours. About the same time, an exhaust fan caused a fire in a restaurant in the 3100-block of Cambie Street. Then on the East Side a first alarm came at 1030 hours for 807 East Sixth Avenue. Shortly after the arrival of first-in companies, the fire was turned into a third alarm. The large complex contained 188 units and had over 300 residents, most of whom had to be evacuated. The fire destroyed six suites on the top floor; three firefighters were injured during the incident. One received minor burns and two others were injured when a balcony collapsed onto them. One tried to pull the other out from under the rubble when his boot went through the burnt floor, causing burns to his leg. All three firefighters and one tenant were taken to Vancouver General Hospital, treated and released. The cause was undetermined.

The building was no stranger to fire calls. It had a long history of false fire alarms and drug/medical calls and on December 31, 1970, before the construction of the building was complete, a large part of it was destroyed in a three-alarm fire.

Another construction crane, high angle rescue was successfully completed about 150 feet (50 m) above Seymour Street, January 13, 2000. The thirty-eight-year-old crane operator thought he was having a heart attack. The rescue team was called, climbed up and stabilized him for removal to the ground. He was loaded into a basket stretcher and lowered to the paramedics who took him to the hospital. Doctors determined that he had had an allergic reaction. The rescue crew under Captain Bob Deeley knew exactly what to do, as a few days before the incident the crew had run a drill on that very site.

Alf Smethurst in 1918.
Photo: VFD Archives

Alf Smethurst Celebrates 100 Years

February 4 was a day of celebration at Fire/Rescue Headquarters when the fire service honored one of its own. Retired Assistant Chief Alfred Gladstone Smethurst who left the department on his sixtieth birthday, February 4, 1960, was celebrating his 100th birthday, the second VFD member to reach that milestone. (The first was Alex Betts, who reached it February 25, 1987.) Alf joined the fire department on May 17, 1918, after serving overseas during the First World War, including the attack at Vimy Ridge, long-recognized as Canada's greatest World War I victory. He credited his longevity in part to being physically fit by playing soccer, baseball and handball. He also played in the firemen's band as one of the originals. He graciously accepted gifts of a presentation fire ax and a departmental flag and listened to birthday greetings from the Queen and the prime minister, among others.

Alf Smethurst arrived at his birthday celebration on the VFD's 1912 American-LaFrance.
Photo: The Province: Nick Procaylo

Two fires in March claimed three lives. On March 3 an early morning fire claimed the life of man in an apartment fire in the 200-block East 14th Avenue. The cause was alcohol-related and due to smoking in bed. Coincidently, that afternoon in the same block a plumber's torch caused another fire, putting fifteen families on the street, but there were no injuries. Then on March 29 at 2130 hours a house on the East Side became fully involved, killing an eighty-three-year-old woman and extensively damaging the home. It is believed that her oxygen tanks may have leaked, fuelling the fire and causing it to spread to the house next door. Then, when the eighty-year-old neighbor returned home and found her house on fire, she collapsed and later died in hospital. Other neighbors said they heard an explosion before the fire.

A rash of more garage and car fires began happening on May 1, back in the neighborhood of several previous fires. Two fires, five minutes apart, were reported at 0317 and 0322 hours to the rear of West 14th and West 15th Avenue addresses, destroyed two garages, four cars and the rear of a house. Fortunately, there were no serious injuries because of the actions of quick-thinking neighbors who got the tenants out and called 911. There were no suspects in these arson fires, but they may have been linked to similar fires that happened nearby in 1997 and 1998.

New No. 18 Firehall Opens

The grand opening of the new Firehall No. 18 was held on July 22, 2000, featuring the Vancouver Fire Department Band and Honor Guard. The guests included Mayor Philip Owen, Fire Chief Ray Holdgate, IAFF Local #18 Vice-President Tony Neratini, VF&RS Chaplain Bruce Rushton and firefighters, friends and family of Captain Grahame White, in whose name the hall was dedicated. Grahame died in the line of duty fol-

The new No. 18 Firehall.
Photo: Author's collection

lowing a fire on August 20, 1981.

A report in one of Canada's women's magazines gave the VF&RS some cause for concern when an August 2000 article suggested that firehall private telephones were being used for questionable dating connections. The author was given some misinformation by a friend in Toronto and she developed a scandalous story around it. The article was an insult to all firefighters and, upon investigation of the report by the lawyer, it was concluded that there was no evidence to indicate that private firehall telephones were being used for "connections with firefighters."

"Best-In-Canada Title Won By BC Firefighter" read the headline in the morning *Province*, September 17, 2000. Firefighter Andrea Belczyk became Canada's top female firefighter when she placed first in the Canadian Firefighter Combat Challenge in New Brunswick and set a world speed record for women, winning by one second. The remainder of the four-person team included Firefighters Rich Warnock, Ken Siggers and Scott Marineau. The team came twenty-second out of a field of thirty-one teams.

The Vancouver Firefighters Credit Union moved into the new No. 3 Firehall on October 2 and began its sixtieth year in business serving the financial needs of its members. On November 29 the hall unofficially opened with the return of Engine 3 from nearby temporary quarters on Kingsway and Ladder 3 from No. 13 Firehall. The official opening was scheduled to be held early in 2001.

The year ended with 807 firefighters from twenty firehalls answering a total of 36,110 fire and medical alarms, a far cry from the 110 calls answered from two firehalls by twenty-three men in 1890, the year the city began keeping such records.

Andrea Belczyk, top female firefighter in Canada.
Photo: VFD Archives

The new VF&RS flag. *Photo: Author's collection*

Epilogue

(2001–2006)

January 1, 2001 not only marked the beginning of a new year but also the beginning of a new century and a new millennium. It was originally planned to end this book at this point, then it was decided to add a short overview of some of the events up to 2006 to bring this history up to date.

A Day in the Life of No. 2 Firehall

No. 2 Hall is located in Vancouver's Downtown Eastside, the most impoverished community in the country with the lowest annual income. It is rumored to be the busiest firehall in Canada and because it is in the poorest constituency, its regular clients are often addicts, alcoholics, psychotics, schizophrenics, and so on. The crew answers, on average, about 550 calls a month, the majority being medical emergencies that usually involve alcohol and drugs.

By noon on one particular day, they responded to a cardiac arrest, a hazardous materials spill, six assorted medical calls, a shed fire, and a garbage container fire. This was considered a normal morning with an average day being about twenty calls, with often more on "Welfare Wednesday." Their busy day can also include calls on the fireboat that they also man, if required.

The crews enjoy working at No. 2 (Engine 2 and Rescue 2) because the time goes fast and there is always the satisfaction of bringing someone back from the brink of self-destruction. They often assist EHS with drug over-dose protocols (Narcan injections that instantly revive the o/d patients — who often deny ever taking drugs). They know most of the "regulars" and "special people," and they deal with them in a professional manner; and in spite of the dangers when dealing with some of these individuals, they are always ready to go when the bell hits. The unofficial firehall motto is, "It's not the end of the world, but we can see it from here." (Paraphrased from an article by No. 2 crew for a VF&RS newsletter.)

The New No. 3 Firehall Opens

The new No. 3 Firehall officially opened on January 27, 2001, and was dedicated to the memory of Captain Lloyd Love, with family, friends, firefighters, and dignitaries in attendance, along with the Fire Department Band and Honor Guard. One of the Forty-Sixers, Lloyd collapsed and died while answering a false alarm from No. 3 on September 23, 1970. The cost of this newest firehall was $2.7 million, a far cry from the $700-plus it cost to build the first No. 1 on Water Street in 1886.

The new No. 3 Firehall. *Photo: Author's collection*

A New Tower Ladder

The new Tower 7, Spartan Gladiator with a 100-foot (30-m) Smeal aerial ladder, went into service in May 2001 at No. 7 Hall. It was the first tower ladder on the department although other departments, in particular the FDNY, has used them since the mid-1960s. Most New York firefighters would say that, along with portable radios and self-contained breathing apparatus (SCBA), the tower ladders have been the innovations that have made their jobs safer and easier.

The new Tower 7 aerial. *Photo: Shane MacKichan*

Attack on the Twin Towers

September 11, 2001. The following Group Talk message (a voice announcement to all firehalls) was sent out:

To All Members:
For your information, the Union is forwarding words of condolence to our brothers and sisters in the United States.
The tragedy of today's events cannot be captured with words. Preliminary reports from the I.A.F.F. indicate up to 500 firefighters were on scene at the time of the collapse at the World Trade Center.
I know that your thoughts and prayers will be with our brothers and sisters during these times.
Rod MacDonald
President, Local 18

The Boot Drive raised over $600,000.
Photo: Local 18, IAFF

Headquarters staff Group Talk Message read:

The staff and management of the City of Vancouver Fire and Rescue Services would like to add their sentiments to those of the Union in expressing their concern for those firefighters in peril fighting to save the lives of the many victims of this morning's terrorist attack against the United States.
The flags at all firehalls are to be lowered to half-staff immediately and until further notice.

Everyone will always remember what he or she was doing on that beautiful sunny day when the world was shaken like it had never been before by the terrible acts of terrorism. That day, almost 3,000 people in New York City, Washington, DC and Pennsylvania died, and included in that number were 343 New York firefighters. Throughout North America, firefighters rallied to do what they could to help their fallen brothers' families.

In Vancouver, by October 11 city firefighters raised more than $600,000 to assist the families of the FDNY men who died. Over the following weeks, several contingents of firefighters traveled to New York City to attend the funerals because members of the FDNY weren't able to attend all of them. As many as a dozen funerals a day were taking place in the various boroughs.

At one of the many funerals attended, *Vancouver Sun* photographer Ian Smith, who traveled to New York City with a group of Vancouver firefighters, took a picture that won him a 2001 National Newspaper Award. While attending the funeral, at St. Paul's Cathedral of Firefighter Christian Regenhard, a twenty-eight-year-old rookie of FDNY Ladder 131, he caught the emotion of Vancouver Firefighter Alex Noke-Smith as the procession passed by. The circumstances of these solemn occasions caused many of the attending firefighters to unabashedly express their feelings.

Firefighter Alex Noke-Smith at the funeral.
Photo: Vancouver Sun / Ian Smith

Vancouver and New York fire fighters paying their respects at one of the many funerals.
Photo: The Province / Arlen Redekop

Captain Dave Veljacic.
Photo: VFD Archives

Bob Spence
Photo: Vancouver Province

Gordon Anderson.
Photo: Local 18, IAFF

During the recovery operations, one of the NYC firefighters was found to be wearing a VFD T-shirt that had been given to him by one of the local firefighters during the Police/Fire Games. Some Vancouver and FDNY firefighters have had a kinship for many years, and this recovery brought the event even closer to home.

On September 16, 2001 retired Captain Dave "The Fire Chef" Veljacic died of cancer at age fifty-nine. He was recognized throughout the firefighting fraternity in the US and Canada for his barbeque and chili cooking skills, his publication of three successful cookbooks, and his own line of barbecue sauce. Despite his gruff exterior, his ever-present five o'clock shadow, and attitude, he was a warm, caring guy. In 1997 he was selected as our Firefighter of the Year for his contributions to many charities in the Vancouver area.

Broadcast journalist and good friend of Vancouver firefighters, Robert Scott (Bob) Spence died on September 25 at age fifty-three after a spirited fight against cancer. Bob became very knowledgeable about the fire department when, as a twelve-year-old kid, he began hanging around the old No. 12 Hall in the Kitsilano district following the death of his father, a Vancouver city policeman, in the early 1960s. The men at the firehall became positive role models and "adopted" him as the VFD's only two-legged mascot. His friendship and involvement with the department continued into adulthood and, as he was held in such high esteem, he was made an honorary member of the firefighters' union. He was singularly honored with the participation of the Honor Guard at his service, and in 2004 a memorial bench in his name was placed outside No. 12.

Then the members of the fire department learned of the passing of Gordon R. Anderson on September 26, 2001, at age seventy-five. Gordon's career as a firefighter began in 1946 and it included many years of service as the president of Local 18, IAFF. Later, he became the International's Sixth District vice-president and held memberships and directorships on several local boards and associations. He spent many years improving the social and economic status of firefighters in Vancouver and elsewhere, and more than thirty-five years improving the Municipal Pension Plan. In 1979 he was appointed British Columbia's first Fire Commissioner, a position he proudly held until retirement. Tributes to Gordie came from far and wide. "Here's to you; who's like you? Damn Few!"

Old No. 13 Hall Destroyed by Fire

No. 13 Firehall was closed in December of 2001 in preparation for the construction of a new firehall on site, and Engine 13 moved to temporary quarters near 41st and Fraser. At 0200 hours on January 13, 2002 a fire alarm was received to respond to the old firehall that had been set afire by persons unknown. The building was destroyed and many felt it was an undignified way for the old building to end its days. No. 13 Firehall opened on October 13, 1913.

Alf Smethurst, Vancouver's oldest retired firefighter died on April 21, 2002, at age 102 years, 2 months and 18 days. He was the longest-living retired member in the department's history. He exceeded Alex Betts, the previous oldest retiree by eighteen months.

The Wildlands Apparatus

In 2002 the fire service put two special pieces of apparatus in service in the form of off-road bush vehicles, Wildlands #5 and Wildlands #8. They are two 4 x 4 Superior-Ford F550s equipped with a compressed air foam system for fighting fires that may occur in any of Vancouver's several wooded areas, such as Crowley Park, Pacific Spirit Regional Park, and of course, the 1,000-acre (405-hectare) Stanley Park. Each rig carries 270 gallons (1,225 liters) of water and 10 gallons (45 liters) of foam.

The bush truck, Wildlands #8.
Photo: Shane MacKichan

On September 11, 2002 a parade and remembrance service was held downtown at the Vancouver Art Gallery in honor of the firefighters who died in New York City the year before. Several hundred firefighters from Greater Vancouver attended the service.

September 11, 2002 first anniversary remembrance service.
Photo: VFD Archives

New No. 13 Firehall Opens

On April 5, 2003 the new No. 13 Firehall was officially opened on East King Edward Avenue and dedicated to the memory of Captain Ozzie Howell, who died of injuries received when the squad wagon he was riding in collided with a car, June 3, 1952. The official party included Fire Chief Ray Holdgate, Reverend Bruce Rushton, Alderman Raymond Louie, representing the city, and members of the Howell family. Many retired members were also present including retired Assistant Chief Don McLeod, the last living survivor of the accident.

The new No. 13 Firehall.
Photo: Author's collection

Proposed reduction in pension benefits for Vancouver's firefighters and policemen by the Municipal Pension Board prompted almost 20 percent of them to take early retirement before the end of 2003. Some of them would have had to work many years longer to make up the shortfall of these reductions, and both departments lost dozens of good and highly qualified people.

The Pender Street Fire

Fire investigators on scene.
Photo: VFD Archives

The biggest fire of 2003 occurred July 3 at 335 West Pender Street in what some considered a very historic and important building. It was the dominant building on the block and in its heyday was the Canadian Cycle and Motor Company (CCM) auto and bicycle dealership, a dance academy, a union hall, and it had the city's first commercial bowling alley. Known as the Pender Auditorium in the 1940s through the 1960s, it also held the offices of many local unions. Its architecture epitomized the "golden years" in Vancouver, 1900 to 1910.

The fire was started when an apartment tenant, said to be making "popcorn," spilled some cooking oil on the stove and set the room on fire. She had to be rescued by ladder from the rear of the building and was treated for first-degree burns.

Highrise view at the height of the Pender Street fire. *Photo: VFD Archives*

January 6, 2004 saw one of the largest and most dramatic fires of the year when the former Mount Pleasant Baptist Church at East 10th Avenue and Quebec Street, two blocks from No. 3 Firehall, was destroyed in a four-alarm fire. At the time the weather was bitterly cold with snow on the ground, and it was learned that a transient had broken into the church basement to get warm and accidentally started the fire.

Remains of the old Baptist Church.
Photo: VFD Archives

The Blunt Brothers store three-alarm fire at 311 West Hastings Street destroyed the 1908-vintage building on April 25. Cause of the fire is unknown. The old building stood in the shadow of Vancouver's famous old Dominion Trust Building that, when built in 1908 with fifteen floors, was the tallest building in the British Empire. It is said that its height prompted the fire department to purchase a new aerial ladder, the 1909 Seagrave. With its seventy-five-foot (25 m) ladder, it would barely reach the seventh floor.

The Blunt Brothers three-alarm fire.
Photo: VFD Archives

On November 25, 2004 the firefighters union launched a cookbook appropriately titled *Pot on the Stove* featuring local and international celebrities and some of their favorite recipes. Also included were fire-hall favorites, from pastas, salads, sushi, and seafood, to cakes and desserts. The firefighters responsible for the success of the book are Brian Bogdanovich, Manny Dosange and Clint Clarke. Proceeds benefit the BC Children's Hospital.

The *Pot on the Stove* celebrity cookbook.
Photo: VFD Archives

The North Shore Mudslide

On January 19, 2005, following several days of heavy rain, a mudslide in North Vancouver destroyed a house, killing a woman and injuring others. Vancouver's Urban Search and Rescue team were called in to assist, including Firefighter Flynn Lamont and Search and Rescue dog Barkley. Miraculously, in spite of the devastation there were no other victims.

Firefighter Flynn Lamont and SAR dog Barkley on scene.
Photo: VFD Archives

Urban Search & Rescue workers at the scene of the mudslide. *Photo: VFD Archives*

Kitsilano Arson Fire

At 0330 hours on August 4, No. 12 Company responded to a fire at West 5th Avenue and MacDonald Street that turned out to be set by an arsonist. The fire, involving five older, wood-frame, multi-tenant houses, forced many people out into the street. It was soon turned into a four-alarm and had such a hold that the firefighters were unable to save very much. Many of the tenants lost everything that night. When the fire was finally struck out the firefighters, returning to No. 12 Hall found that they, too, were losers that night. While they were at the fire, someone had broken into the firehall and stolen several hundred dollars from their wallets. Firehalls are usually secure when the crews are out on a run, but it appeared that the automatic doors had somehow failed to close properly and someone had forced their way in.

The four-alarm house fires on MacDonald Street.
Photo: Graham DeNure

The rear of the house fires, likely where the fire started.
Photo: Graham DeNure

Hurricane Katrina

When the category 5 hurricane Katrina hit the Gulf Coast on August 29, Vancouver's Urban Search and Rescue Team was placed on standby, to be deployed to the area if they were called upon to assist. On August 31, the forty-six-member team responded to the call for help and went to Lafayette, Louisiana. Arriving on September 1 with ten days of supplies, they set up operations in St. Bernard Parish, east of New Orleans. Thirty-percent of the population was feared missing and the team began their long twelve-hour shifts of rescue operations. By day five of their task, it was reported by team leader Tim Armstrong, "With every square inch of real estate under water, the devastation was beyond comprehension." During their sojourn they successfully rescued 120 people.

Vancouver's Urban Search & Rescue Team on their return from Louisiana.
Photo: Vancouver Province: Ric Ernst

On September 6, after crews from the Federal Emergency Management Agency (FEMA) relieved them, the team (made up of firefighters, paramedics, police officers, and medical personnel) returned to Vancouver to a heroes' welcome. The US Ambassador to Canada praised the SAR Team for a job well done and said, "I am touched by what you did. It's an honor and a privilege to tell you face-to-face — thank you." This was the first time that the team had been used internationally and it was said to be a profound learning experience.

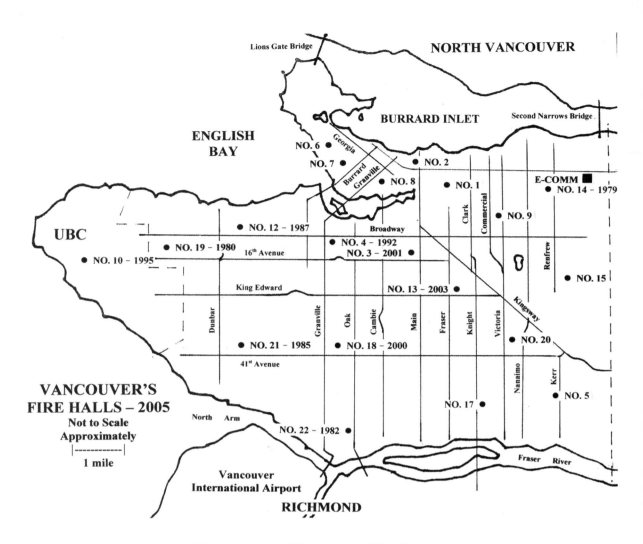

More Arson Fires and Thefts

On Christmas day 2005 at 1130 hours, firefighters from No. 7 Hall responded to a fire alarm at nearby St. Paul's Hospital. Someone had set two small fires in some cafeteria plates, cups and trays in a fourth floor storage area. The sprinkler system put the small fires out and the area suffered mostly water damage. A person was seen running from the scene. On their return to the firehall, firefighters discovered that someone had entered the building through a window and taken their lunch money that had been left on the kitchen table. Steps would be taken to make the firehalls more secure.

The year ended with a three-alarm fire, caused by a candle, at 0230 hours on December 28 in a high-rise apartment building at West 39th Avenue and Vine Street. The two tenants of the burning suite had to be rescued from their seventh-floor balcony by aerial ladder. Twenty-two people were left temporarily homeless, and three tenants suffered small cuts and smoke inhalation.

2005 ended with almost 8,700 more rescue and fire calls than 2004, with a record 46,183 calls, an increase of well over 20 percent.

Who Are Vancouver's Fire Fighters?

Through all of the 120 years since the founding in 1886, the firefighters of the Vancouver Fire Brigade/Vancouver Fire Department/Vancouver Fire and Rescue Service have come from every continent on earth. Every province and territory in Canada is represented, as well as several of the US states. Members have come from throughout the United Kingdom and Ireland and many countries in Europe, notably Sweden, Denmark, Holland, Germany, and Italy. Members represent Africa, too, from South Africa and Nigeria, and we have members from Asia, including India and China. There are two who were born in Chile and Uruguay in South America, and several from Australia, New Zealand, and the Fijian Islands.

The men and women (currently six) have come from all walks of life — carpenters, plumbers, stonemasons and bricklayers, teamsters, machinists, engineers, military ranks, farmers, miners, loggers, teachers, accountants, morticians, and policemen, too, from the famed London Bobbies to members of the mounted police, including the North West Mounted, the Royal North West Mounted, and the Royal Canadian Mounted Police. There have also been several former Vancouver police who have joined the fire department.

In the twenty-first century "Vancouver's Bravest" are truly a cosmopolitan mixture of men and women working together to preserve life, to protect property, and to prevent losses by fire.

We've come a long way since 1886!

FIREMAN'S PRAYER

When I'm called to duty, God, whenever flames may rage,
   give me strength to save some life, whatever be its age.
Help me embrace a little child before it is too late,
   or save an older person from the horror of that fate.
Enable me to be alert and hear the weakest shout,
   and quickly and efficiently to put the fire out.
I want to fill my calling, and to give the best in me,
   to guard my every neighbour and protect his property.
And if, according to my fate, I am to lose my life,
   please bless with your protecting hand my family, friends
   and wife.

Anonymous

Ri-To Fish Company, 798 Alexander St., third alarm, December 5, 1990.

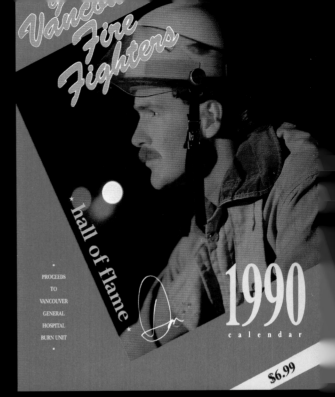

The first year of the Fire Fighters Calendar raised over $100,000 for the Vancouver General Hospital Burn Unit
Photo: Local 18, IAFF

Sterling Shipyards three-alarm fire, December 1992.
Photo: VFD Archives

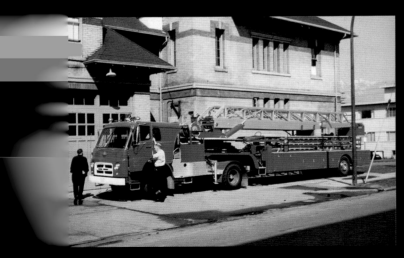

No. 6's new aerial was previewed for the crew before being

Hose Wagon No. 1 is shown in the new red and white livery, c. 197

Vancouver Fireboat No. 2 shortly before being decommissioned, 1987 — the end of an era. *Photo: VFD Archives*

The new Firebird 125 at No. 1 Firehall, 1973. Note doorway has been raised to accommodate this new large rig. *Photo: Author's collection*

Vancouver's Dragon Boat team, 1987. *Photo: Author's collection*

The spectacular Gastown fire, March 1998.
Photo: VFD Archives

Vancouver Fireboat No. 2, 1956.

Two views of the DFPS Hose Tender, c.1995 *Photo: VFD Archives*

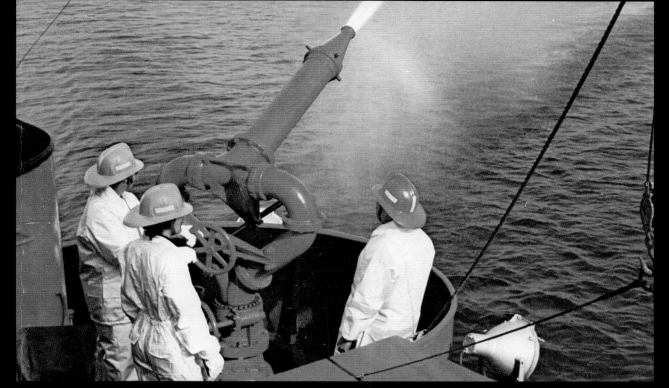

Fireboat monitor demonstration, 1976.

The 2004 *Pot on the Stove* celebrity cookbook.
Photo: VFD Archives

Fraser Arms Hotel fire, April 1988. The department's first six-alarm fire.
Photo: Dave Youell

Rescue 17 LSU in 1999. *Photo: Shane MacKichan*

The new No.10 Haz-Mat unit in 2000. *Photo: Shane MacKichan*

The Alberta Wheat Pool four-alarm fire, August 1994. *Photo: Dave Youell*

The Haz-Mat team at Chess Street drill grounds in 1997.

The BCFP lumber yard, fully involved, July 1960.

Street fire, July 2003.
Photo: VFD Archives

Sheraton Landmark fire, May 1985.
Photo: VFD Archives

The rear of the McDonald Street arson fires, 2005.
Photo: Graham DeNure

The new Tower 7 aerial, May 2001. *Photo: Shane MacKichan*

The bush truck, Wildlands #8, 2002. *Photo: Shane MacKichan*

The four-alarm arson fires on MacDonald Street, August 2005. *Photo: Graham DeNure*

Vancouver Fireboat No. 2 attacking the flaming Home
Oil Fuel barge, January 1974.
Photo: VFD Archives

BC Forest Products fire as seen from Oak Street, July 1960
Photo: VFD Archive

Early morning three-alarm
fire at Hastings Street and
Clark Drive, June 1988.
Photo: Dan Goyer

Andrea Belczyk, top female firefighter in
Canada, September 2000.
Photo: VFD Archives

White Spot Restaurant fire, February 1986.
Photo: Dave Youell

Commercial Drive market fire, November 1999.
Photo: The Province: Bill Cook

Firefighter Flynn Lamont and SAR dog Barkley on scene.
Photo: VFD Archives

Urban Search & Rescue workers at the
scene of the mudslide, January 2005.
Photo: VFD Archives

Firefighter Alex Noke-Smith at the 9/11 funeral.
Photo: Vancouver Sun/Ian Smith

Vancouver and New York firefighters paying their respects at one of the many 9/11 funerals.
Photo: The Province/Arlen Redekop

Appendix I

Vancouver Fire Department On-Duty Deaths

Driver John Smalley - No. 2 Hall
Joined VFD March 1, 1892, died June 5, 1893, age 25.
Fell from No. 2 Engine en route to a fire.

Fireman John M. McKenzie - No. 1 Hall
Joined VFD May 1, 1909, died April 12, 1912, age 25.
Fell from a ladder at a drill.

Fireman James Bryant - No. 3 Hall
Joined VFD September 24, 1912, died June 29, 1914, age 26. Collapsed at a hose drill.

Fireman Albert D. Stewart - No. 3 Hall
Joined VFD August 11, 1914, died September 11, 1914, age 24. Fell down a pole-hole at No. 3 Firehall.

Fireman Charles Milne - No. 6 Hall
Joined VFD October 11, 1913, died May 2, 1915, age 25. Thrown from No. 6 Chemical Engine in collision with No. 6 Hose Wagon.

Captain Richard S. Frost - No. 11 Hall
Joined VFD February 22, 1900, died May 10, 1918, age 44. Thrown from No. 11 Hose Wagon in collision with a streetcar.

Lieutenant Colin McKenzie - No. 11 Hall
Joined VFD January 27, 1908, died May 10, 1918, age 31. Thrown from No. 11 Hose Wagon in collision with a streetcar.

Fireman Otis Fulton - No. 11 Hall
Joined VFD November 6, 1915, died May 10, 1918, age 35. Thrown from No. 11 Hose Wagon in collision with a streetcar.

Fireman Donald Morrison - No. 11 Hall
Joined VFD August 2, 1912, died May 10, 1918, age 33. Thrown from No. 11 Hose Wagon in collision with a streetcar.

Fireman William J. Cameron - No. 12 Hall
Joined VFD September 10, 1917, died May 15, 1918, age 24. Struck by falling timbers at the Coughlan Shipyard fire.

Fireman Frederick Jenkins - No. 6 Hall
Joined VFD July 1, 1926, died June 1928, age 32. Fell from the tower of No. 6 Firehall while hanging hose.

Fireman Herbert E. Ellis - No. 2 Hall
Joined VFD March 26, 1920, died April 17, 1930, age 38. Fell from No. 2 aerial ladder during a drill at Beach Avenue drill ground.

Fireman Andrew H. Grant - No. 6 Hall
Joined VFD c. 1930, died October 16, 1933, age 25. Collapsed at No. 6 Firehall and died in hospital of a brain hemorrhage.

Fire Warden Louis B. Taylor - Fire Warden's Branch
Joined SVFD May 10, 1922, died October 20, 1942, age 54. Fell during a fire inspection at the CN Dock.

Fireman William M. Wootton - No. 18 Hall
Joined VFD December 10, 1941, died November 13, 1943, age about 27. Thrown from No. 19 Pump in collision with a police car.

Captain E. William Barnett - No. 2 Hall
Joined VFD April 18, 1917, died September 14, 1945, age 50. Collapsed at the McMaster Building fire while pulling hose.

Lieutenant James M. Hunt - No. 13 Hall
Joined VFD September 1, 1918, died September 14, 1945, age 50. Buried in collapse of floors at the McMaster Building third-alarm fire.

Fireman Reg Hill - No.2 Hall
Joined PGFD August 1, 1924, died September 14, 1945, age 44. Buried in collapse at the McMaster Building third-alarm fire.

Lieutenant Arthur E. Wilkins - No. 20 Hall
Joined VFD September 3, 1918, died May 19, 1946, age 59. Collapsed pulling hose at a fire involving four houses.

Lieutenant Eric R. Robinson - No. 17 Hall
Joined PGFD May 1, 1926, died June 26, 1947, age 49. Crushed under No. 17 Truck when it rolled following a collision with a car en route to fire.

Fireman Malcolm W. McPhatter - No. 2 Hall
Joined VFD August 1, 1946, died April 7, 1949, age 29. Fell from the drill tower at No.2 Hall during a drill.

Captain M. Oswald (Ozzie) Howell - No. 3 Hall
Joined VFD October 2, 1927, died June 14, 1952, age 46. Died of injuries suffered in collision of No. 3 R&S Squad Wagon with a car en route to rescue call.

Fireman William Jenner - No. 1 Hall
Joined VFD August 1, 1946, died September 25, 1957, age 35. Collapsed while playing handball at No. 1.

Captain George H. Layfield - No. 3 Hall
Joined VFD February 4, 1929, died May 24, 1959, age 56. Collapsed while fighting a house fire.

Captain Gordon B. Watson - No. 18 Hall
Joined VFD February 4, 1929, died January 1, 1964, age 55. Collapsed in the front seat of No. 18 Pump while returning to quarters after an alarm.

Firefighter Donald H. McCavour - No. 2 Hall
Joined VFD March 1, 1948, died March 15, 1966, age 44. Died of injuries received when thrown from No. 2 Hose Wagon following a collision with a car en route to a second-alarm fire.

Captain W. Lloyd Love - No. 3 Hall
Joined VFD August 1, 1946, died September 23, 1970, age 51. Collapsed at the scene of a false fire alarm.

Captain John F. Graham - No. 6 Hall
Joined VFD June 20, 1949, died November 10, 1979, age 52. Died of smoke inhalation in an apartment building fire.

Captain Grahame C. White - No. 18 Hall
Joined VFD July 26, 1954, died August 20, 1981, age 49. Collapsed helping his crew reload hose after a fire.

Captain Frederick C. Fisher - No. 21 Hall
Joined VFD February 7, 1955, died June 21, 1982, age 51. Collapsed at No. 21 Firehall at the start of his day shift.

Firefighter Terrance P. O'Keeffe - No. 7 Hall
Joined VFD October 16, 1967, died April 9, 1988, age 46. Collapsed at No. 7 Firehall on return from a fire alarm.

Lieutenant O.B. (Bud) Swanson - Fire Prevention Branch. Joined VFD November 1, 1967, died July 11, 1990, age 48. Collapsed while on fire prevention duties.

"When a man becomes a fireman, his act of bravery has already been accomplished."
— FIRE CHIEF EDWARD F. CROKER, FDNY, 1908

Appendix II

Vancouver Fire Fighters 1886 to 2006: Vancouver's Bravest

The Volunteers

Sam Pedgrift	George Schetky	Frank Gladwin	Thomas F. McGuigan
Robert Rutherford	J. Hoskins	Wilson McKinnon	A.L. Dunlop
James A. Moran	John A. Mateer	W.W. Griffiths	Peter Larson
A. Tyson	J.E. West	John McAllister	J. McDonald
Stephen H. Ramage	William J. Blair	Ken Smith	C. Gigier
Thomas Lillie	Frank W. Hart	W. Dawsey	Hugh Campbell
William Chadwick	Gabriel W. Thomas	William S. Cook	John Garvin
Robert Leatherdale	E. Duhamil	G.E. Upham	Ernest C. Britton
C.M. Hawley	Barnie Beckett	William McGirr	Fred Germaine
John W. Campbell	F. Curlow	G.W. West	C.M. Steinbeck
John H. Carlisle	Norman Dawsey	Arthur Clegg	C.F. Perry
David Biggar	Dr. Robert Matheson	A. McDonald	William Hamilton
John Taylor	John McKenzie	W.H. Saunders	Angus D. McKenzie
A.M. Montgomery	W.L. Hayward	George Brown	J. Kirby Douglas
C. (Chick) Heywood	Bert J. Hall	A. Laroix	A. Saint Laurent
Joseph Riox	A. Vachon	Thomas Dunn	Joseph King Davis
Andrew Hart			

Reverend Father Henry Glynne-Fynes Clinton (Father Clinton)
Police Chief John Stewart (Honourary Member)   Herbert W. Martin (waterboy)

Paid Fire Fighters

† On-Duty Death

PG Ex-Point Grey FD

† Active Member Death

SV Ex-South Vancouver FD

✠ World War I Death

UEL Ex-University FD

E.T. Morris	02/08/86	Alexander W. Cameron	15/09/86	John H. Carlisle	28/12/86
Thomas Simpson	01/05/88	William Hamilton	03/08/88	John W. Campbell †	01/09/89
Arthur Clegg	01/09/89	Hugh Campbell	01/09/89	W.A. Williamson	01/09/89
E. Scratchly	01/09/89	A. Davies	01/09/89	Thomas W. Lillie	01/09/89
C. Spence	01/09/89	W.L. Hayward	01/09/89	J. Glenn	01/09/89
G. Henderson	01/09/89	A.G. Evans	01/09/89	J. Cooper	01/09/89
James A. Moran †	01/09/90	**John Smalley †**	01/03/92	William Jordan	01/08/92
Harry Duncan †	01/10/92	George W. McDonald	01/10/92	W. Davie	01/11/92
Thomas Whiston	01/11/92	William A. McPhee	06/01/93	James A. McInnis	01/02/93
Thomas A. Tidy	01/08/93	James A. Lester	01/04/94	John Courtney †	17/06/95
Charlton W. Thompson	17/06/95	Andrew A. Gill	17/06/95	Chris H. Barker	13/08/95
Richard H. MacAuley	13/09/96	J. Edward Mitchell	01/04/98	James McMorran	01/04/98
Edward McKeating	05/04/99	W.B. Lee	01/07/99	Frederick A. Barker	01/10/99
William D. Frost	01/10/99	William E. Flood	01/10/99	James Davidson	01/10/99
Wm. M. Brownlee	01/10/99	Fred P. Murray	30/10/99	N.R. Ross	01/11/99
David Scott	01/01/00	Norman McPherson	01/01/00	**Richard S. Frost †**	22/02/00
P.T. Hartney	23/07/00	J. Sea John	01/08/00		
Robert J. Moore †	07/06/01	Robert M. Cameron †	01/09/01	D.W. McDonald	20/11/01
K. MacKenzie	01/04/02	P. Stoutenberg	08/04/02	Herbert Hyde	01/05/02
William H. Patterson	04/06/02	Robert McCurdy	10/06/02	Thomas H. Botterell	01/07/03
Ralph Ravey	01/07/03	Harry Walpole	01/12/03	Thomas (Tim) Moffatt	01/01/04
Alfred M. Clare	23/02/04	Robert Allison	29/02/04	Norman Lee	01/03/04
L. Edwards	02/03/04	James F. Loftus	21/03/04	M. Harris	17/05/04
H. Drew	01/06/04	W. Gardner	15/06/04	John Paul	21/07/04
Dudley Ritchie	21/07/04	J. McDonald	08/12/04	Wilfred B. Lutes	11/04/05
G.G. Brown	11/04/05	John A. McDonald	11/04/05	A.R. Reynolds	01/06/05
Charles Force	21/11/05	Harry Jealouse	21/11/05	William Eaton	08/02/06
John H. DeGraves	01/05/06	Sidney Burgess	07/05/06	John Warwick	10/05/06
P. Cummings	19/07/06	Neal Henderson	19/07/06	William Plumsteel	13/09/06
H. Allen	13/09/06	Charles Durrant	13/09/06	T. McInnis	08/10/06
Walter McAlpine †	20/12/06	G. Foster	20/12/06	William Tiffin	20/12/06
E.C. Thomas †	20/12/06	A. Kipp	20/12/06	Walter Garnier	20/12/06
Charles Chaplin	04/02/07	D. B. Frazer	04/02/07	F. Elliot	04/02/07
D. Johnson	04/02/07	A. A. MacDonald	29/04/07	Noel (Nemo) Layfield	29/04/07
O. Davies	27/05/07	George Dunlop	05/08/07	Archie McDiarmid	02/07/07
C.W. Cathcart	05/08/07	A.V. Johnston	16/09/07	R.H. Sowden	16/09/07
L.O. Hillier	16/09/07	John Anderson	16/09/07	A. MacDonald	16/09/07
Percy B. May ✠	16/09/07	A.J. Hughes	28/10/07	A.M. Young	28/10/07
Charles Jewett	28/10/07	N. McDonald	28/10/07	Edward Kneale	28/10/07
Albert Martin	01/12/07	Hugh H. Steen	09/12/07	Joseph Lyon	23/12/07
Robert Parr	23/12/07	George Harrop	23/12/07	J.F. McMillan	27/01/08
Colin McKenzie †	27/01/08	F.C. Lucas	28/01/08	James Mackenzie	27/01/08
T. Samworth	27/01/08	Thomas J. Burke	27/01/08	William McKechnie	28/03/08
Frank Gurney	28/03/08	Frank King	28/03/08	Gordon McGuire	28/03/08
John A. Grant	04/05/08	W. Manson	04/05/08	W.A. Mosher	04/05/08
John Fitzpatrick	04/05/08	J.W. Odlum	04/05/08	W. Kent	04/05/08
Harry T. Stephens	04/05/08	Charles C. McLennan	01/06/08	Alex McDonald ✠	01/06/08

Name	Date	Name	Date	Name	Date
G.E. Foley	01/06/08	H. McDougall	29/06/08	Alex McPherson	18/09/08
A.E. Watson	21/09/08	A.M. McLean	21/09/08	Clarence Westover	21/09/08
A.P. Reynolds	18/01/09	Allan T. Brown	01/02/09	A.R. Doe	01/02/09
John McKenzie †	01/05/09	H. Morrison	08/05/09	Peter Isbister	10/05/09
Norman F. Fox	10/05/09	William Hodson	10/05/09	A.H. Bowden	10/05/09
H. Gottschalk	10/05/09	C. Crown	20/05/09	Cy Ruddock	24/06/09
James Johnston †	01/07/09	Arthur H. Alexander	01/08/09	A. Thompson	01/08/09
William Manning	08/08/09	F.W. Johnson	16/08/09	James M. McKenzie	16/08/09
Thomas Thompson	16/08/09	Joseph A. McConnell	16/08/09	Alex Manson	16/08/09
Lawrence G. Ledwell	16/08/09	C.W. Draper	16/08/09	W.J. Simpson	16/08/09
J.A. Stewart	02/09/09	Charles R. Robinson	07/09/09	W.A. Weeks	08/10/09
C.C. Taylor	08/10/09	W.W. Elliott	11/10/09	Harry V. Sibley	11/10/09
J. Johnson	11/10/09	Claude Strang	11/10/09	N. MacDonald	28/10/09
Charles Sumner	08/11/09	Hector Morrison	08/11/09	Alex MacDonald	08/11/09
Theodore Horrobin	08/11/09	Percy Sherman	08/11/09	A.A. McKay	08/11/09
W. Harvey	08/11/09	W.M. McClellan	09/12/09	John Turner	09/12/09
Edward Roberts	18/12/09	George R. Hicks	02/01/10	Roy H. Broderick ☦	17/01/10
E.A. Burgess	08/02/10	F.A. Conway	09/02/10	B.S. Briegel	26/02/10
Harry Painter	01/03/10	Robert Sowden	09/03/10	Alex Stephens, Jr.	22/03/10
J.A. Enright	27/03/10	Lloyd E. Dunham	01/04/10	S. Jackson	04/05/10
G.A. McKenzie	03/06/10	A. Smirl	04/06/10	N.A. McMillan	01/07/10
A.D. MacDonald	04/07/10	Robert Young	07/07/10	Vernon Porter	02/08/10
George Dean	02/08/10	George A. Lefler	05/09/10	C. Hartley	22/09/10
Arthur Barrabale	06/10/10	Thomas E. Hoggarth	01/11/10	John Temple	14/12/10
A.E. Duhamel	15/12/10	T.H. Jagger	00/00/10	Ed Stratton	00/00/10
Frederick Sentell †	00/00/10				
G.W. Cadman	17/01/11	R. West	18/01/11	E.S. Palmer	21/01/11
E. Roberts	21/01/11	H.A. Hansen	24/01/11	G.C. Jones	25/01/11
C.D. Wilson	01/02/11	F.H. Taylor	01/02/11	Joseph Newton	02/02/11
E.J. Harris	02/02/11	George E. Taylor †	22/02/11	R.H. Ross	02/03/11
A. Kennedy	08/03/11	G. E. Allport	27/03/11	E. Harkelroad	03/04/11
J.D. Jardine	13/04/11	Neil MacDonald	13/04/11	H. Hepburn	02/05/11
Gordon S. Moore	04/05/11	J.W. Diamond	13/05/11	C. Beaton	29/05/11
George Jackman	01/06/11	T. W. Murphy	04/06/11	H.A. Ferguson	05/06/11
D. Murdoch	06/06/11	A.G. Sullivan	23/06/11	W.J. Black	01/07/11
A.E. Davenport	18/07/11	J.D. Muir	19/07/11	R. Henry	20/07/11
C.E. Russell	21/07/11	A. Metcalf	21/07/11	A.C. Roberts	31/07/11
Edward Nash	02/08/11	P.M. Hayward	03/08/11	Arthur Sheehan †	07/08/11
R.J. Morrison	07/08/11	A. Marshall	07/08/11	T. Fant	07/08/11
T. Inglis	07/08/11	E. Walling	07/08/11	R. Owen	08/08/11
A. Hughes	08/08/11	B.D. Moffatt	09/08/11	Edward L. Erratt	09/08/11
A.T. Edwards	09/08/11	N. Wenborn	12/08/11	L.A. Westbo	17/08/11
George Cater	28/08/11	Ambrose E. Condon	29/08/11	J.C. Orange	30/08/11
E.D. Shaw	31/08/11	P.F. Enright	01/09/11	F.M. Craig	04/09/11
H. Frazer	12/09/11	J.A. McDiarmid	12/09/11	H. Morris	19/09/11
M. Morrison	23/10/11	Torquil Campbell	23/10/11	J. Dyer	26/10/11
John Wedderburn ☦	27/10/11	M. Montgomery	02/11/11	A.M. Davidson	09/11/11
A. MacDonald	11/11/11	V. Porter	05/12/11	T.A. McLean	16/12/11
Art Johnston [PG]	19/12/11	L.B. Taylor	15/01/12	R.J. McSpadden	01/02/12
A.P. Eve	01/02/12	B. Mutch	03/02/12	J.J. Murray	07/02/12
J.W. Gillis	05/03/12	Angus McLeod	08/03/12	T. Desmond	08/03/12
H.R. McKenzie	22/04/12	H. Liddle [PG]	22/04/12	H.H. Lindsay	24/04/12
H. McDougall	01/05/12	A.F. McCaffrey	14/05/12	John (Baldy) Balderston †	02/06/12
F. Mahy	07/06/12	Harry Bower	13/06/12	Alex W. Betts	13/06/12

W. Hodgins	24/06/12	
W. Scriver	25/06/12	
D. W. Rogerson	04/07/12	
E. Frank Muir	10/07/12	
R. Stiegenberger	16/07/12	
H. Westwood	16/07/12	
A. Kelly	19/07/12	
R. Newington	26/07/12	
Donald Morrison †	02/08/12	
Stewart G. Park	05/08/12	
A. Davis	07/08/12	
C. Crum	01/09/12	
G. Williamson	24/09/12	
A. Stratton	02/10/12	
J. McLennan	30/11/12	
W.E. Clark	[SV]	00/00/12
William V. Green	01/01/13	
C. Forse	09/01/13	
O. Anderson	19/02/13	
N. McDonald	22/03/13	
A. Hull	14/06/13	
Frank Amy	01/07/13	
J. McCormick	08/07/13	
William D. Reid †	15/07/13	
G.M. Bell	16/07/13	
G. Lochead	20/08/13	
J. Bunyan	11/10/13	
J. Gallivan	13/10/13	
A. Anderson	13/10/13	
C. DeSilva	14/12/13	
A. McLennan	01/01/14	
H.G. Bowering	10/01/14	
Alexander F. Yule †	19/01/14	
J. McCormick	10/02/14	
Lorne B. Foley	09/03/14	
C.F. McCallum	09/03/14	
I. Thomas	03/04/14	
E.P. Thorne	14/05/14	
H. Otto	19/05/14	
Frank R. Mayes	20/06/14	
Arthur G. Gosse	04/07/14	
Wilfred Scott	05/08/14	
R. Hassell	07/08/14	
T. Anderson	11/08/14	
W. Hall	13/08/14	
John Horn †	17/08/14	
E. J. Lang	25/08/14	
A. Manson	12/09/14	
G.W. Morris	01/01/15	
J. Leeson	23/01/15	
H. Colling	17/02/15	
S.C. Buchanan	01/05/15	
C.W. McDonald	29/06/15	
J. Price	03/07/15	
R.E. McLaren	09/08/15	

G. Lorraine	24/06/12	
E.L. Seyfried	[PG]	26/06/12
J. Campbell	04/07/12	
H. Blower	13/07/12	
R. T. Porter	16/07/12	
L. Livingstone	17/07/12	
W. Kyle	24/07/12	
Angus Lougheed	01/08/12	
A.M. Lehman	03/08/12	
A. Stoole	05/08/12	
R.D. McRae	13/08/12	
Frank Hooper †	16/09/12	
James Bryant †	24/09/12	
John A.H. Lee	03/10/12	
J.A.G. Ross	02/12/12	
C. (Frenchie) Bouchard	01/01/13	
Sam Martin	01/01/13	
J.W. Green	14/01/13	
F. Revels	26/02/13	
C. A. Garrison	22/03/13	
P. Hatch	16/06/13	
D. A. Gillis	02/07/13	
G. Johnstone	12/07/13	
A. McKenzie	15/07/13	
H.D. (Spike) Heathorn	15/08/13	
J.A. Hamilton	22/09/13	
James Wright	11/10/13	
J. McWilliam	13/10/13	
Alexander M. Alexander †	13/10/13	
P.D. Milne	16/12/13	
C.T. Freeman	06/01/14	
M. McDonald	12/01/14	
J. L. Dundas	19/01/14	
H. Ferguson	10/02/14	
A.J. McConnell	09/03/14	
E.C. Rouse	09/03/14	
W. Arkell	06/04/14	
C. Straight	23/05/14	
E. Olsen	10/06/14	
W. F. Caldwell	20/06/14	
W. Weeks	01/08/14	
J.J. Elliott	05/08/14	
J. Schaffer	10/08/14	
J. McLeod	11/08/14	
W.G. Barker	14/08/14	
J. A. Scott	19/08/14	
E. McGeer	01/09/14	
B.C. Dinsmore	19/09/14	
Joseph Houle	21/01/15	
J. McDonald	25/01/15	
W.R. Robertson	06/13/15	
C.E. Foster	17/06/15	
Rupert Macdonald	29/06/15	
E.L. Paepke	07/08/15	
P. Christopher	25/08/15	

Thomas H. Snowden †	25/06/12
Harry Moth	04/07/12
C. A. Watson	04/07/12
W. Smirl	15/07/12
A. McDonald	16/07/12
D. Hill	19/07/12
E. Morgan	24/07/12
L. Fahey	01/08/12
M. McKay	04/08/12
D.P. Maher	06/08/12
Joseph J. Carley	21/08/12
W. Hall	24/09/12
R.C. Ellinor	02/10/12
J. Blower	22/10/12
H. Copeland	16/12/12
W.J. McDonald	01/01/13
R. Till	02/01/13
Gerald A. Paul	16/01/13
George J. Richardson	03/03/13
Percy H. Newson	01/04/13
O. Mottishaw	17/06/13
D.A. Grant	05/07/13
Allan R. Murray	14/07/13
G.L. Maitland	15/07/13
R.R. Carter	19/08/13
R.L. Manning	08/10/13
Charles Milne †	11/10/13
K. Stoddart	13/10/13
R.H. Dowling †	14/10/13
W.A. Flett	29/12/13
J. Anderson	08/01/14
Frank Carlisle	17/01/14
L. McLellan	19/01/14
A.P. Thomas	09/03/14
James Shearer	09/03/14
W. Green	10/03/14
Thomas A. Lafave	13/04/14
William Barnett	23/05/14
Charles A. Borgstrom	18/06/14
A. A. MacDonald	20/06/14
William Powell	03/08/14
T. Ainsley	06/08/14
Albert D. Stewart †	11/08/14
J.D. Matheson	12/08/14
P.D. Gerrard	17/08/14
J.O. Smith	24/08/14
E. Ravey	06/09/14
O.W. McWilliam	14/12/14
G.W. Chalmers	21/01/15
W.J. Young	28/01/15
W.S. McFarlane	30/03/15
J.F. Lee	22/06/15
A. McLellan	30/06/15
J. McDonald	08/08/15
James Smith	27/08/15

P. Russell	27/08/15	M. McKenzie [PG]	30/09/15	S. Conrod	07/10/15
W.A. Weeks	26/10/15	D. Hodges	27/10/15	**Otis Fulton** †	06/11/15
E.A. Stokes	08/11/15	F.F. Broderick	09/11/15	F.L. Hunt	23/11/15
W.B. McKinnon	01/01/16	G.A. Paul	03/01/16	J.A. Morrison	05/01/16
William A. French	18/02/16	E.C. Lewis	26/02/16	John McKenzie	28/02/16
Carl Kellett	29/02/16	Murdoch McLeod †	01/03/16	A.V. Young	17/03/16
Robert Pinkerton	25/03/16	Ronald G. Shaw	27/03/16	Henry Fraser	31/03/16
William Milne	04/04/16	N. Bradshaw	05/04/16	P. Perrett	06/04/16
P. Powell	06/04/16	Angus Beaton	06/04/16	Thomas Lawrie	06/04/16
J.J. Kent	12/04/16	George Morrison	13/04/16	James D. Laing	17/04/16
K. McKenzie	17/04/16	J.M. McLeod	18/04/16	Carmen E. Ross †	24/04/16
C. Callans	29/04/16	John R. McDonald	01/05/16	Erv McKay	01/05/16
Angus McLellan	02/05/16	O.V.B. Robinson	03/05/16	C. Woods	08/05/16
Ronald McDonald	09/05/16	M.F. Hughes	13/05/16	John McIvor †	19/05/16
J.S. Young	20/05/16	Sid Ruff	01/06/16	A. McElrea	01/06/16
M.D. Harris	03/06/16	A.W. Fullerton	05/06/16	A.H. Whipple	06/06/16
George Anderson	08/16/16	L. Wilson	16/06/16	A. Petrie	16/06/16
F. Alexander	16/06/16	E. Myron	18/06/16	A. McConaghy	24/06/16
B. Miller	24/06/16	C.M. Lee	29/06/16	F. Wait	30/06/16
John Robb ☦	01/07/16	J. E. Shaw	02/07/16	D. H. McEachern	11/07/16
J. Cline	15/07/16	W. McInnis	17/07/16	F.A. Brouillette	20/07/16
J.R. Stewart	04/08/16	W.C. Rogers	04/08/16	John McLeod	11/08/16
Edward D. Ross	16/08/16	J. McEachern	18/08/16	H. Warris	27/08/16
Richard Cloke	01/09/16	N. Dillabough	06/09/16	A. Roy West	07/09/16
Joseph McLaughlin	08/09/16	R.H. Coombs	09/09/16	Steve H. Munn	01/10/16
J. Robinson	10/10/16	E. Hall	10/10/16	A. McDonald	24/10/16
J. Magnone	27/10/16	L.A. Devos	01/11/16	Gil Martin	05/11/16
H.R. Hassell	06/11/16	C.W. Owen	29/11/16	G.A. McKenzie	04/12/16
George Johnston	13/12/16	D.H. Steves	15/12/16	T.H. Croke	16/12/16
V. Wilson	30/12/16	F. Tompkins	30/12/16	G. Moir	01/01/17
C.E. Moore	01/01/17	H. Rae	01/01/17	G.E. Coombes	06/01/17
W.S. McFarlane	08/01/17	A. Troup	08/01/17	J. Leeson	10/01/17
Samuel J. Burns †	10/01/17	A. Dagg	12/01/17	F. Elliott	12/01/17
Ralph M. Gibbons †	16/01/17	R.C. Jackson	29/01/17	C. Jensen	30/01/17
James Crutchley	06/02/17	Herbert W. Ricketts	06/02/17	D. Laird	06/02/17
J.C. Musselman	08/02/17	R. Hodson	08/02/17	A. Ferrier	15/02/17
J. Curry	22/02/17	W.M. Gutteridge	26/02/17	E.F. Smiley	01/03/17
L. (Lockie) McLeod	15/03/17	H.E. Horne	16/03/17	M. Donnachie	19/03/17
Fred E. Taylor	20/03/17	Ernest Erano	21/03/17	J. Evans	21/03/17
H.E. Mand	24/03/17	R. McGregor	27/03/17	W.N. Phinney	03/04/17
E. William Barnett †	18/04/17	John G. Jackson †	25/04/17	A. Wilson	01/05/17
F. Merash	12/05/17	R.H. Graham	07/05/17	M.M. Stewart †	08/05/17
S. Bennett	17/05/17	William Carson	18/05/17	James Montford	18/05/17
James Tyson	26/05/17	L. Eugene	20/06/17	E. Leynard	25/06/17
S. McNeil	06/07/17	Percy Trerise	09/07/17	J. Bernard	11/07/17
John Bowness	11/07/17	Syd Gillies	11/07/17	S. Hogue	11/07/17
B. Hughes	12/07/17	C. Michie	12/07/17	H.B. Frazer	14/07/17
R. Rae	14/07/17	Murray Harris	17/07/17	N. McDonald	17/07/17
William Sinclair	20/07/17	A. Shaw	24/07/17	J. Backey	28/07/17
A. Aire	30/07/17	Albert Hill	30/07/17	J.A. McIsaac	31/07/17
W.H. McPhatter	02/08/17	J. Knowles	03/08/17	P. Budd	03/08/17
Leonard Alfred Farr †	03/08/17	D. Johnston	06/08/17	R. Houston	06/08/17
M. Campbell	06/08/17	P. Trebilcock	13/08/17	W.K. Ward	14/08/17
Harry Greenslade †	15/08/17	Alfred Greenslade	15/08/17	F. Bentell †	27/08/17
E.A. Dunn	01/09/17	**William J. Cameron** †	10/09/17	M. Macauley	25/09/17

Name		Date	Name		Date	Name		Date
J. McKinnon		27/09/17	C. Daisley		01/10/17	John R. Leatherdale †		03/10/17
W. Kean		03/10/17	W.H. Gillis		03/10/17	H.K. Beamer		19/10/17
H.M. Abercrombie		23/10/17	R. Kennedy		23/10/17	C.J. Tarr		23/10/17
G. Martin		25/10/17	W. Menzies		29/10/17	H.T. Stevenson †		03/11/17
Michael Pillat		06/11/17	D.A. Moore		07/11/17	G.S. Hamilton		09/11/17
G. Smith		10/11/17	T.J. Winkell		13/11/17	S. Rand		14/11/17
W. Wilson		19/11/17	P. MacDonald		22/11/17	P. Nelson		24/11/17
W. Loader		26/11/17	C. Mathison		28/11/17	L.J. Champion		28/11/17
J.H. Galbraith		01/12/17	G. C. Coulson		02/12/17	Joseph F. Green †		03/12/17
B. Williams		04/12/17	A. Mills		05/12/17	G. Milne		05/12/17
John T. Raper		05/12/17	M. S. Wilson		05/12/17	J.W. Tompkins		17/12/17
W. Pernell		21/12/17	S. Mahood		27/12/17	S.F. Stephenson		28/12/17
D. Achison	[PG]	28/12/18	J.W. Robinson		01/01/18	H. Bullock		01/01/18
J. Murray		01/01/18	A. Pilkey		03/01/18	T.F. Hunter		05/01/18
J. Smith	[SV]	05/01/18	W. Worrall		07/01/18	F.D. Reagan		08/01/18
D. McLeod		09/01/18	H. Keefe		11/01/18	E. Callahan		11/01/18
C.A. Nord		11/01/18	C.R. McDonald		11/01/18	L. Marsh		14/01/18
F.J. Heard		17/01/18	T. Morgan		21/01/18	W.J. Kelly		22/01/18
J.J. McCarthy		25/01/18	C. W. Stark		28/01/18	W.J. Young		28/01/18
A.W. Minshull		31/01/18	J.J. Ferguson		12/02/18	A. LeClair		19/02/18
H. Bradley		23/02/18	A. Booth		01/03/18	D. Thompson		05/03/18
H. Young	[PG]	06/03/18	A. W. Lunde		07/03/18	Dan Zaklan		07/03/18
J. W. Faulkner		09/03/18	E.G. Butler		09/03/18	W. Gordon		13/03/18
C. J. Robertson		13/03/18	E. Bradley		13/03/18	G.W. Gairns		20/03/18
Walter A. Carlisle		25/03/18	W.R. Peacock		26/03/18	W.M. Rescarl		26/03/18
G.F.J. Garrett		27/03/18	W.A. Wesley		27/03/18	E. Day		31/03/18
N.A. Akins		31/03/18	G.W. Harrop		09/04/18	W.A. Henry		09/04/18
A.E. Ashcroft		27/04/18	A.J. Finlay		30/04/18	R. Strachan		01/05/18
J.D. Toderick	[SV]	05/05/18	G.W. Ford		09/05/18	G.R. Leatherdale		10/05/18
Alfred G. Smethurst		17/05/18	A.D. Bruce		20/05/18	P. Willoughby		21/05/18
R. McIntyre		21/05/18	N. Seeley		21/05/18	W. Gillis		25/05/18
P. O'Hara		27/05/18	R. Harrison		01/06/18	A.J. Pendergast		03/06/18
L.J. Shane		10/06/18	Archie King		13/06/18	P.H. Little		18/06/18
C.F. Page		18/06/18	M.H. May		19/06/18	William Tracey		19/06/18
P.D. McClure		21/06/18	J. Rose		25/06/18	R.C. Cody		25/06/18
F. Anderson		26/06/18	D. Larken		29/06/18	Peter Cole		29/06/18
A.R. Stoddard		24/06/18	J. Gawsey		06/07/18	G.R. Storry		18/07/18
G.E.A. Dailey		18/07/18	James Horie		19/07/18	A.J. Nielson		19/07/18
J.G. Hornett		19/07/18	G.G. Warnett		19/07/18	E.H. Nanthrup		19/07/18
J. McCardle		19/07/18	E. Neilson		19/07/18	J. McDonald		19/07/18
L.R. Lloyd		22/07/18	C. Thomas		23/07/18	R.S. Webb		23/07/18
F.G. Lucas		02/08/18	J.A. Jacob		07/08/18	H. Woodfield		07/08/18
C. Wood		13/08/18	Joseph Travis		20/08/18	A.E. Smith		24/08/18
James M. Hunt †		01/09/18	Frank Raymur	[PG]	03/09/18	**Arthur E. Wilkins †**		03/09/18
F.H. Taylor		07/09/18	N.K. Ward		10/09/18	J. William Jordon		25/09/18
G. Royle		25/09/18	E. Stackhouse		25/09/18	R.N. Kerr		25/09/18
C. Burnett		25/09/18	H.A. Echardt		25/09/18	G.W. Gordon		25/09/18
A. E. Dinsmore		25/09/18	A.O. Sinclair		25/09/18	H. McCarten		25/09/18
T. Collins		25/09/18	W.G. Inglis		25/09/18	T.A.G. Richardson		25/09/18
D. MacKenzie		30/09/18	Thomas Massender † [PG]		30/09/18	C. Carle		30/09/18
G.F. Mitchell		30/09/18	Eugene M. Connor		30/09/18	E. Morrissey		01/10/18
C. Smith		01/10/18	N.J. McDonald		02/10/18	R. Jacobson		03/10/18
L. Langout		04/10/18	F.J. Taylor		04/10/18	R.H. Shoesmith		05/10/18
J. McCarthey		06/10/18	W.C. Skinner		07/10/18	W.J. Wilson		07/10/18
W. Taylor		08/10/18	J.D. Corbett		08/10/18	R.A. Baker		08/10/18

A. Henderson — 11/10/18
A. Martindale — 14/10/18
H.E. Robinson — 18/10/18
Neil Henderson — 23/10/18
H. White — 31/10/18
C.A. Vereker — 09/12/18
A. Ferrier — 13/01/19
S.G. James — 09/10/19
S.W. Dubbins — 01/12/19
G. Milne — 02/04/20
R.A. Aalten — 28/05/20
A. LeClair — 08/06/20
Alex Reid — 22/07/20
A.A. Burr — 11/10/20

J.F. Green — 06/04/21
D. McKay — [SV] 01/05/21
J.F. Thom — 13/05/21
J.H. Lounsbury — 01/07/21
J. McLeod — 01/08/21
W.J. Payton — 19/09/21
F.C. Eldred — 02/11/21
W.J. Inglee — 03/04/22
Louis B. Taylor † — [SV] 10/05/22
Alex McQueen — 22/07/22
J.F. Rowland — [PG] 02/01/23
Clarence Oxley — 01/03/23
L.R. Mulligan — [PG] 07/07/23
W.O. Wadsworth — 23/05/24
H. McElvaine — 28/05/24
C.L. Buchanan — [PG] 01/06/24
J. Liddell — [PG] 01/06/24
Everett Hygh † — [PG] 01/06/24
W.G.H. Armstrong — [PG] 05/09/24
Robert J. Fleming † — [SV] 06/10/24
V. Wilson — 01/11/24
J.A. Tall — 01/04/25
W.H. (Tim) Eaton — 01/04/25
W.M. (Nick) Craig — 02/04/25
F.C. (Peggy) Duff — 22/06/25
C. Morin — [PG] 01/05/26
R. Stanton — [PG] 01/05/26
J.D. Ross — 01/07/26
V.E. Murchison — 17/10/26
J. Watters — 01/04/27
N.G. Lougheed — 20/06/27
A. Kennedy — 01/07/27
W.G.E. Upfold — 01/07/27
Elmer O. Hassard — [PG] 01/07/27
Alvin Murchison — 01/07/27
H.R. McDonald — 11/07/27
R. McDonald — 05/08/27
J.M. McWilliams — 01/10/27
J.S. Lawn — 02/10/27
M.O. (Ozzie) Howell † — 02/10/27

R. Downey — 12/10/18
P.J. Gaynor — 16/10/18
W.D. Jackson — 21/10/18
A.W. MacPherson — 24/10/18
J.G. McNulty — 30/11/18
G.P. Hearnden — 09/12/18
J. Evans — 20/03/19
S.E. Stigant — 05/11/19
E.P. Williams — 28/02/20
Fred Cooper — 19/04/20
A. McD. May — 08/06/20
William MacKenzie — 11/06/20
G.E. Taylor — 30/08/20

C.O. Ross — 06/04/21
E.A. Young — 03/05/21
H.J. Green — [SV] 01/06/21
H. Foulkes — [PG] 01/07/21
J. Murray — 02/09/21
A.J. Robertson — 22/09/21
W.J. Metcalfe — 04/03/22
W. French — 03/04/22
J. Campbell — [SV] 01/06/22
L.C. McLean — 28/08/22
V.E. Bennett — 19/02/23
E.A. Hillier — 07/05/23
P. J. Mckinlay — 08/10/23
H. Pay — [SV] 23/05/24
N. McKinnon — [PG] 01/06/24
H.C. Creelman — [PG] 01/06/24
E.G. Fralic — [PG] 01/06/24
Reginald H. Hill † — [PG] 01/08/24
Russ Docksteader — 16/09/24
J. Neale — [SV] 06/10/24
C. Goldsmith — [SV] 01/12/24
G.H. (Lefty) Kaye † — 01/04/25
J.W. Garlich — [SV] 01/04/25
W.G. McLaren — [PG] 03/04/25
E. Sewell — [SV] 22/06/25
A.J. Robertson — [PG] 01/05/26
Elmer F. Sly † — [PG] 01/05/26
Frederick Jenkins † — 01/07/26
N. Goodall — 20/01/27
W.V. Bailey — 05/06/27
F.S. Anderson — 01/07/27
F.P. Barnes — 01/07/27
Walter H. Carfrae — 01/07/27
D.A. McDonald — 01/07/27
Hugh S. Bird — 11/07/27
C. LeMoir — 25/07/27
L.N. Thompson — 13/08/27
E.J. Randall — 02/10/27
Bart T. Bean — 02/10/27
C.J. Potts — 06/10/27

J. Cottman — 12/10/18
A. Innis — 18/10/18
F.A. Dayton — 22/10/18
F. Jackson — 28/10/18
Lee Christianson — 02/12/18
A.W. Chinn — 03/01/19
T.H. Campbell — 08/05/19
T.G. Wrigglesworth — 01/12/19
Herbert E. Ellis † — 26/03/20
P.P. Johnston — 26/05/20
Leo Leavy — 08/06/20
M.K. McLennan — 22/07/20
C.R. Hodson — 24/09/20

W. Barrett — 12/04/21
William J. Middleton — 09/05/21
E.S. Vaughn — [PG] 01/07/21
J.W. Blackburn — 02/07/21
A. Wilson — 02/09/21
G.C. Putnam — 16/10/21
J. Ferguson — 03/04/22
A. Henderson — 03/05/22
W.G. McLellan — 01/06/22
C.C. Maddison — [PG] 18/12/22
Donald McDonald — 01/03/23
A. Enefer — [SV] 01/06/23
E.R. Smele — [PG] 27/11/23
J.E. Bourne — 26/05/24
J.L. Hutchinson — [PG] 01/06/24
F.E. King — [PG] 01/06/24
M. Lee — [PG] 01/06/24
W. Bradshaw — [PG] 03/08/24
J. Porter — [SV] 16/09/24
James Urquhart — [SV] 07/10/24
G.R. Hansen — 01/04/25
T.F. Warne — 01/04/25
F.A. McKelvey — 02/04/25
V. Dickson — 05/05/25
Eric R. Robinson † — [PG] 01/05/26
M.J. Gannon — [PG] 01/05/26
L B. Holden — 23/06/26
F. Delisle — 10/09/26
Thomas Baillie † — 11/02/27
Robert D. McLaren — [PG] 11/06/27
Percy R. Timmins — 01/07/27
J.A.C. Kirk — 01/07/27
T.G.D. Seggie — 01/07/27
J.R. Johnstone — 01/07/27
Charles Fitzpatrick — 11/07/27
G. Lambie — 26/07/27
D.N. Hovey — 01/09/27
A.O. Lawson — 02/10/27
Harry P. Hamilton — 02/10/27
A. Telosky — 12/10/27

J. Washbrook	20/10/27	Tom Wylie	00/00/27	R.M. MacDonald	01/02/28
J.M. Lange	02/02/28	G.H. Hawthorne [PG]	01/03/28	C.H. Hagman	09/03/28
W. A. Henry	01/05/28	J.H. Carlisle, Jr. †	05/05/28	Ralph R. Jacks [SV]	11/05/28
C.W. Creelman	16/05/28	George Sikora	20/05/28	F. Woodcock	19/07/28
K.O. Everett	24/07/28	George Black	28/07/28	Henry A. Blackwell †	01/08/28
Thomas Whitty	02/08/28	H.M. Knight	06/08/28	William E. Brereton	18/08/28
James M. Ettinger	21/08/28	R.F. Ross	23/08/28	H.R. Richardson	01/09/28
Cecil B. White [SV]	01/09/28	G. Hopper	01/09/28	J. (Jack) Forsgren	01/09/28
John A. McInnis	01/09/28	A.E. Young	01/09/28	J. Hector Wright	01/09/28
P. Jolliffe	25/09/28	Walter T. Peain	06/09/28	J.T. Gray	08/09/28
George J. McInnis	11/09/28	W.J. Agnew	18/09/28	D. Rickets	21/10/28
Norman G. Trasolini	23/10/28	G. Eric Davenport	23/11/28	S. Warren	14/12/28
G.N. Keith	14/12/28	N. McArthur	14/12/28	W. Fraser	14/12/28
John Finlayson	15/12/28	A.E. Bradshaw	15/12/28	W.E. (Sonny) Errington	15/12/28
H. Towler	15/12/28	J.L. McDonald	15/12/28	J.A. Wightman	15/12/28
R.H.N. Jones	19/01/29	Edward Hadden	21/01/29	R. McKay	25/01/29
A. Black	30/01/29	Donald R. Manuel	30/01/29	J.J. Blair	02/02/29
J. Price	02/02/29	**George H. Layfield †**	04/02/29	L.J. Wallace	04/02/29
John Telosky	04/02/29	Frank L. Perrett	04/02/29	E.G. (Duster) McCall	04/02/29
Gordon B. Watson †	04/02/29	H.C. Hunter	04/02/29	J.W. Stephens	05/02/29
J. Hutchison	09/02/29	J. Withers	09/02/29	M.K. Lee	11/02/29
Robert E. Raymond	18/03/29	H. Ausilind	01/04/29	N. Geohegan	02/04/29
R. Stanton	01/05/29	A. Phillips	16/05/29	J. Neale	20/05/29
William Child	22/05/29	T.E. Pollard	23/05/29	William J. Levett	23/05/29
Ralph E. Layfield	24/05/29	P.W. Stevenson	28/05/29	Aubrey Beaumont †	01/07/29
J.A. Morrison	01/07/29	Fred G. Hall	01/07/29	A.E. Beeton	01/07/29
A. Pratt	18/07/29	Arthur McKoski	18/07/29	H.F. Ray	08/08/29
Kenneth A. Hill	08/08/29	G.L. Clarke	29/08/29	A. Langout	18/09/29
C. Craig	02/10/29	F.E. Curtis	04/10/29	D. Johnson	05/10/29
T.K. Coughlan	23/11/29	John Wilkinson	24/11/29	H.E. Griffiths	11/12/29
C. (Charlie) Miron	12/02/30	G. Campbell	15/02/30	A. Booth	23/04/30
Patrick V. Hartney	01/05/30	R.C. Langford	16/06/30	Dan Kulai	01/07/30
M. Morrison	15/07/30	Harold Loveley	01/08/30	E.D. Ramsell	03/08/30
J.R. Johnstone	16/08/30	W. Henry	03/11/30	J. Swan	01/12/30
Andrew H. Grant †	00/00/30				
R.V. Barker	01/05/31	Robert A. Agnew	01/05/31	Sydney G. Coates †	07/06/31
D. Edward Henry	20/10/31	D.T. Brown	28/01/32	T.F. Hunter	01/04/32
L.F. Campbell	01/04/32	A.E. Brereton	01/04/32	W. Scott	01/04/32
L. Frank Gurney	01/04/32	W.G. Johnson	07/04/32	H. Walter Rand	19/07/32
C.C. Dorman	10/01/33	Charles B. Weir	02/02/33	J.E. Greenwood	20/06/33
Alex J. Gillies	04/07/33	W. Jenkinson	18/02/34	W.E. Johnstone	08/03/34
F.S. Campbell	07/06/34	Harry G. Foster †	07/06/34	R.J. Sowden	07/06/34
W.T. Moore	23/06/34	N. Thompson	14/07/34	Walter D. Lee †	07/11/34
J.W. Eaton	25/11/34	Carl Martin	05/12/34	Robert M. Ritchie	27/12/34
Albert Nassey †	30/12/34	Abe Sturrock	01/01/35	O.A. Foster	01/01/35
J. McCorkindale	01/01/35	Marvin Wortman	01/01/35	A.E. (Bert) Miller	07/01/35
D.J.W. (Doc) Bowyer	08/06/35	Roderick McIsaac	04/07/35	G.W. Brooks	30/08/35
G.R. Babcock	18/03/36	Alex P. Keay	07/04/36	George Tremblay	21/07/36
Arthur Wells	29/07/36	S.J. Bremner	12/12/36	R. Dierks	30/12/36
A.W. Moore	01/07/37	C. Davis	01/09/37	J.W. Nordlund	01/02/38
Walter A. McAlpine	15/02/38	L.V. Loughran	10/06/38	I. Duncan	26/07/38
C.A. Smith	05/08/38	H.L. Baillie	30/03/39	David L. Wheatcroft †	14/04/39
D.G. McKinnon	12/07/39	W. Percy Richards	06/08/39	G. Phillips	11/11/39
H. Stewart	17/12/39	A.J. Stephenson	25/01/40	William A. Balderston	09/04/40

Martin J. Campbell †	01/07/40	
D.D. Braun †	15/09/40	
M. Lange	07/11/40	
H. Belleck	01/02/41	
G.G. Manson	12/03/41	
J. Haydn	01/08/41	
A. Closter	06/12/41	
M.J. Fanning	10/12/41	
J. (Jack) Crouch	10/12/41	
Maurice Moye	10/01/42	
R.A. Williamson	10/01/42	
William A. McEwan	11/01/42	
C.J. Metcalfe	04/02/42	
H.A. Greenaway	23/02/42	
Arthur Winterford †	17/04/42	
H.L. Coleman	26/04/42	
E.G. Kenward	01/05/42	
James MacBeth	01/05/42	
C.J. Coleman	18/05/42	
N. McArthur	25/06/42	
James E. Tuning	16/07/42	
H.R. Roy †	03/08/42	
K.V. Olson	17/08/42	
J.A. (Al) Waine	01/09/42	
Ivan S. Hunt	09/09/42	
E.W. Bullock	11/10/42	
Roy N. Treasurer	01/11/42	
C.G. Sellars	16/12/42	
W.E. Hamm	01/01/43	
R.R. Stewart	01/03/43	
R.R. Rae	17/05/43	
M.E. Douglas	13/06/43	
A. Brook	04/10/43	
J.T. Meraw	16/10/43	
G. Smith	01/11/43	
E.T. Pollard	08/11/43	
Harold A. Lindsay †	27/01/44	
C.L. Thompson	01/03/44	
C.W. Joy	05/05/44	
G.J. Ross	17/05/44	
J. Edward Trites	23/07/44	
Archie L. Munson	01/09/44	
W.H. Churcher	01/01/45	
V.L. (Slim) Sommerville	21/02/45	
C. William Burden	13/09/45	
Karl H. Lysell	20/10/45	
Thomas F. Mitchell	25/12/45	
J.L. O'Dwyer	13/03/46	
F.E. (Ted) Barker	15/05/46	
Abe Boychuk	01/08/46	
Thomas J. Foran	01/08/46	
H.R. Hutton	01/08/46	
G. Roy Hill	01/08/46	
Arthur Trenchard †	01/08/46	

Charles Sutherland †	01/07/40
W.D. McIvor	04/10/40
H.D. Lowe	27/11/40
George F. Birnie	24/02/41
Donald A. McLeod	22/05/41
C.G. (Bud) Kellett	07/08/41
D.S. Boyce	08/12/41
D.W. Morrison	10/12/41
W.J. Buntain †	10/12/41
Bruce T. Steen	10/01/42
Percy N. Flello	10/01/42
P.L. Lloyd	29/01/42
A.G. Dahl	08/02/42
Robert Ewing	26/03/42
W.A. Philibert	22/04/42
A.M. Lewis	01/05/42
H.G. Heys	01/05/42
D. Shiell	01/05/42
John Rennie †	18/06/42
H.E. Bryant	02/07/42
James O. Campbell †	01/08/42
J.A. Sinclair	13/08/42
Harley Darnell (H.Harris)	25/08/42
E.B. Gilmore	01/09/42
Robert P. Ards †	24/09/42
J.W. Inkster	26/10/42
Thomas A. Slater	01/11/42
George Bird	23/12/42
J.A. Worrall	01/01/43
J.S. Pecjeck	11/03/43
R.A. Mather	06/06/43
Raymond L. Acheson	01/09/43
W.W. King	06/10/43
R.E. Fanning	28/10/43
O.B. Lennon	01/11/43
J.E. Morrison	14/11/43
S. Dunbrack	01/02/44
G.M. Myles	02/04/44
George M. Vervais	15/05/44
John S. Hepner	21/06/44
William S. (Heavy) Heath	30/07/44
John Forrest	06/10/44
F.W. (Curly) Brooker	22/01/45
David Patterson	11/03/45
Lester P. Cook	07/10/45
R.G. Tolhurst	01/11/45
E.A. (Ted) Bodkin	02/01/46
A. Pruden	13/03/46
Gordon P. Woodhouse	23/05/46
A.R. Owens	01/08/46
W.W. O'Hagan (Hagan)	01/08/46
Marvin E. Westcott †	01/08/46
Roy W. Riley	01/08/46
E.A. Muir	01/08/46

E.R. Thirsk	06/07/40
R. F. (Fin) Thomson †	07/11/40
David Dinsmore	05/03/41
Robert N. Middleton	10/06/41
S. Harrison	26/11/41
F.C. Wallace	10/12/41
James Rae	10/12/41
William M. Wootton †	10/12/41
G.R. Hemow	10/01/42
E.L. White	11/01/42
L.E. Desbrisay	04/02/42
J. Armstrong	16/02/42
Percy T. Nelson	13/04/42
Frank C. Bain	23/04/42
J.G. Edward	01/05/42
H.J. McColm	01/05/42
G.R. Myles	07/05/42
Donald A. Gill	25/06/42
H. Stephens	10/07/42
J. Goldie	02/08/42
A.A. Thomson	16/08/42
Donald M. Rosie	28/08/42
Harold A. Dorman	01/09/42
E.D. Fuller	28/09/42
A.W. Hewlett	30/10/42
H. Vesper	01/11/42
A. (Andy) Brand	27/12/42
W.H. Hansford	19/02/43
Arthur Coombes	11/03/43
H.W. Parkinson	10/06/43
R.D. Ellis	18/09/43
J.E. Barnott	15/10/43
C.C. Carson	31/10/43
W. Henderson	06/11/43
John Willis	23/11/43
V. McClatchey	01/03/44
W.H. Good	01/05/44
T.A. Hastings	17/05/44
F.G. Cull	01/07/44
W.M. Campbell	16/08/44
E.E. Hayes	01/11/44
J. Armstrong	07/02/45
Ian G.M. Milne †	05/08/45
Donald F. Sugden	16/10/45
Dennis P. Farina	19/11/45
C.W. (Bud) Holdgate	07/07/46
Harold O'Brien	19/03/46
T.A. Whitehead	22/06/46
G.F. Cooper	01/08/46
Donald J. Morrison †	01/08/46
R.W. Gervan	01/08/46
W.A. Easterbrook	01/08/46
J.E. Swenerton	01/08/46

G.A. Henderson	01/08/46
J.A. McGougan †	01/08/46
J.M. Walkinshaw	01/08/46
Hugh D. McDonald	01/08/46
O.R. Ballam	01/08/46
Malcolm W. McPhatter †	01/08/46
Angus McRitchie	01/08/46
V.C. Jones	01/08/46
J.A. Russell	01/08/46
A.E. McCaskill	01/08/46
C.W. Sherriff	01/08/46
C. Ellison	01/08/46
G.A. Marsh	01/08/46
D.J. Forsyth	01/08/46
C.D. Gibbons	01/08/46
A.J. Therrien	01/08/46
L.C. Newton	01/08/46
C.M. Foster	01/08/46
A.F. Herd	01/08/46
L.F. Foslien	01/08/46
N.K. Sutherland	01/08/46
C. MacKenzie	01/08/46
A.D. Sampson	01/08/46
J.W. Gillis	01/08/46
F.E. Smith	01/08/46
Stanley W. Ferris	01/08/46
John E. Bunyan	01/08/46
F.T. Franklin	01/08/46
N. Nixon	01/08/46
Edward McLachlin	01/08/46
G.A. Henderson	19/08/46
C.E. Rumball	01/09/46
L.V. Davis	15/10/46
S.W. Kirkby	01/01/47
N.J. McLeod	11/01/47
James P. Champion	26/01/47
K.R. Callaghan	05/03/47
D.R. Georgeson	31/03/47
Frank J. Ambler †	18/05/47
R.M. (Bud) Creelman	16/06/47
R.G. Ward	03/07/47
D.L. Annesley	24/08/47
James F. Miller	25/10/47
J.E. Gray	30/12/47
Donald H. McCavour †	01/03/48
P.B. Edens	01/06/48
J.E. Palmer	20/06/48
C. Bruce Milne	05/08/48
K.C. Pearson	01/10/48
E.G. Sherman	11/01/49
E.T. Shaffer	07/04/49
James E. Mullin †	07/04/49
M.M. (Max) Blair †	09/06/49
N. (Bud) Steeves	04/07/49
Edward D. Rowell †	01/08/49

John M. (Jock) Stewart	01/08/46
W. Lloyd Love †	01/08/46
R.E. (Dick) Enman	01/08/46
B.E. Anderson	01/08/46
M.N. (Mickey) Stevenson †	01/08/46
Armand (Armie) Konig	01/08/46
S. Haslam	01/08/46
John E. Brown	01/08/46
R.D. Sulivan	01/08/46
William G. Wickens	01/08/46
Thomas Urquhart †	01/08/46
E.L. McDonald	01/08/46
W.D. Thomson	01/08/46
W.R. Grennan	01/08/46
Graham Giske	01/08/46
L.W. Herder	01/08/46
L.S. Hourie	01/08/46
Thomas S. Colbeck	01/08/46
G. (Gar) Simpson †	01/08/46
A.E. Harris	01/08/46
Robert N. Rankin †	01/08/46
D.R. West	01/08/46
T.T. Hoggarth	01/08/46
Donald J. J. McLean	01/08/46
A.H. Smith	01/08/46
D.W. McLaren	01/08/46
A.A. Turkington	01/08/46
K. Clarkson	01/08/46
D.A. McLeod	01/08/46
George E. Hughes	01/08/46
G.W. Olafson	01/09/46
H.J. (Bert) Pritchard	01/10/46
Harold G. Searle	01/12/46
H.A. (Bert) Lowes	01/01/47
N.M. Edwards	18/01/47
L. Parsons	02/02/47
J.A. Hallgren	15/03/47
D.E. Bathie	19/04/47
R.L. (Dick) Kenning	01/06/47
D.R. Brown	22/06/47
W.W. Erratt	21/08/47
D.S. MacKenzie	27/08/47
L. Vincenzi	08/11/47
H. Johnson	11/01/48
W.D. McLaren	01/04/48
B.E. Ellis	08/06/48
N. Kinna	08/07/48
W.R. Sheppard	14/08/48
I. Smith	04/10/48
R.M. MacLeod	17/01/49
David H. Grieve †	07/04/49
Gordon A. Hall	09/04/49
K.S. White	14/06/49
E.J. Defoe	01/08/49
H.E. Hall	01/08/49

J.W. Sampson	01/08/46
J.J. Hayes	01/08/46
E. Lewis	01/08/46
David Ross	01/08/46
William McIntosh	01/08/46
William Jenner †	01/08/46
E.G. Lawson	01/08/46
G.F. Fletcher	01/08/46
William M. Roberts	01/08/46
D.W. Hughes	01/08/46
K. MacKenzie	01/08/46
A.A. (Sandy) Hyslop	01/08/46
Norman Lyon	01/08/46
N.A. Warren	01/08/46
W.E. Hall	01/08/46
J.E. Burns	01/08/46
F.W. Christison	01/08/46
D.Y. Stevenson	01/08/46
William A. Moore †	01/08/46
Aubrey C. (Bear) Neff	01/08/46
B.G. Mitchell	01/08/46
G.W. Kerr	01/08/46
A.W. Smith	01/08/46
R.R. Maundrell	01/08/46
William G. Parks	01/08/46
Gordon R. Anderson	01/08/46
R.B. Stiles	01/08/46
E.R. Eaton	01/08/46
Thomas Bradley	01/08/46
J. Bridge	01/08/46
E. Smith	01/09/46
James Main	01/10/46
D.A. Dann	18/12/46
J.A. Kelley	01/01/47
G.W. Belyea	22/01/47
W.P. McClure	25/02/47
C.E. (Ted) Cole	17/03/47
J.S. Farrant	17/05/47
H.N. Bonnett	01/06/47
T. Campbell	03/07/47
D.L. Stoneson	21/08/47
J.T. Coulson	12/09/47
B. Phipps	21/11/47
H.F. Keillor	23/01/48
A.J. McLeod	18/04/48
D.A. May	15/06/48
J.T. Potts	24/07/48
L.J. (Bud) Black	30/08/48
D.R. Sullivan	10/01/49
A.G. Ball	07/04/49
H.L. Armstrong	07/04/49
E. Jackson	25/05/49
John F. Graham †	20/06/49
W.J. Loutit	01/08/49
W.J. Watson	01/08/49

Name	Date	Name	Date	Name	Date
D.R. Jones	01/08/49	William T. Beatty	01/08/49	Roderick B. MacArthur †	23/08/49
R.E. Ross	29/08/49	H.K. Insley	24/10/49	T.B. Williamson	03/11/49
L.A. Bissett	07/12/49	Peter Proctor	20/12/49	J.D. Woodside	02/03/50
T.G. Williams	01/05/50	John Rae	15/05/50	D.A. Tompkins	12/06/50
W.D. Bishop	14/06/50	J.B. McGuckin	19/06/50	W.R. Leyburn	19/06/50
Charles E. Hobbs	03/07/50	T.B. Hampton	10/07/50	James R. Paterson †	10/07/50
Robert C. McLean	17/07/50	Leonard Erlendson †	24/07/50	J.D. Greig	08/08/50
A.B. Rawlings	22/08/50	D.W. (Dan) Peebles †	18/09/50	Norman K. Harcus	25/09/50
H.E. Robinson	16/10/50	E.C. Grant	23/10/50		
A. MacKay	10/01/51	R.P. Murray	14/02/51	H. Hardy	19/02/51
W. J. Huntley	19/02/51	A.M. Ross	06/03/51	J.J. Moore	02/04/51
W. T. Graham	02/04/51	K.S. Piper	02/04/51	Hugh J. Dickson †	02/04/51
E. C. (Pete) Rees	02/04/51	D. Lloyd	02/04/51	F.G. Price	02/04/51
G. R. Picchioni	02/04/51	K. York	02/04/51	R.F. West	03/04/51
K. J. Taylor	01/05/51	D.G. Penny	01/05/51	John Lewis	01/05/51
James V. Bounds †	01/05/51	Ralph Hutchinson †	01/05/51	D.C. Lamb	01/05/51
L. A. Klassen	01/05/51	A.J. Reid	01/05/51	V.R. Stone	01/05/51
A. J. Eyberson	01/05/51	B.B. Anderson	01/05/51	L.N. Paquette	01/05/51
G. E. Johnson	01/05/51	N.G. Caros	01/05/51	J.G. Dailly	01/05/51
T. G. Thompson	07/05/51	J.R. Taylor	16/05/51	J.R. Connell	16/05/51
E. C. (Bim) Clark	05/06/51	D.J. Findlay	03/07/51	M.G. (Moe) Stiles	03/07/51
D. Greenwood	03/07/51	W.R. Metcalfe	03/07/51	D.R. Morley	03/07/51
D. T. Kelly	03/07/51	W.G. MacDonald †	03/07/51	R.D. Anderson	03/07/51
Kenneth J. Fitzpatrick	03/07/51	John Morgan	03/07/51	R.N. Perry	03/07/51
Robert H. Upton	03/07/51	G.F. Betts	03/07/51	J.D. Weir	04/07/51
W.S. Chanley	04/07/51	J.D. Creamer	21/08/51	B.A. Wiffin	04/09/51
H.A. (Hal) Finch	09/10/51	A.D. Hughes	13/11/51	A.L. Pemberton	10/12/51
W.L. Johnstone	18/12/51	A.W. Rittinger	23/01/52	H. Marshall	25/02/52
A.L. Bramhall	14/03/52	L.D. Nelson	19/03/52	K.L. McLean	05/05/52
L.R. (Knobby) Clark	21/04/52	K.C. Hanke	27/05/52	A.L. (Gus) Hay	27/05/52
G.E. Lougheed	17/06/52	A.E. Weston	04/08/52	W.C. Johnson	05/09/52
W.J. Walmsley	27/10/52	E.H. Hamson	27/10/52	M.D. Johnson	23/03/53
W.W. Gurvich	17/08/53	R.E. Willis	01/09/53	D.H. Brown	08/09/53
H.A. Mindlin	14/09/53	J. W. (Doc) Cumbers	14/09/53	I.R. Hemingson	14/09/53
A.C. (Tony) Eppler	14/09/53	L.A. White	14/09/53	L.W. Allan	14/09/53
G.E. Freeborn	14/09/53	R.G. Templeton	14/09/53	S.G. Bourne	14/09/53
R.J. McLean	17/09/53	C.T. Bayntun	23/09/53	D.A. Moyes	24/10/53
L.E. Gilson	09/11/53	R.A. Fyffe	01/01/54	D.J. Kirby	01/01/54
A. Patrick	01/01/54	S.E. Brook	01/01/54	W.G. Sheanh	01/01/54
F.A. Shelby (Schmalz)	06/03/54	L.S. New	07/03/54	J.W. Bathgate	14/04/54
Earl LeBeau	21/04/54	L.M. Nielsen	21/04/54	J. Neale	21/04/54
J. Foster	21/04/54	R.M. (Tex) Bell	21/04/54	C.B. Esplen	22/04/54
Arthur Sandvik †	03/05/54	A.L. Atkins	06/05/54	W.G. Morrison	27/05/54
R.A. Wood	14/06/54	George A. Temperton	14/06/54	E.A. Durante	14/06/54
Boyd M. Coates	14/06/54	Barry F. Foster	14/06/54	G. Lloyd Shippam	26/07/54
James R. Smith	26/07/54	**Grahame C. White †**	26/07/54	Thomas Baillie	09/08/54
Donald J. Pamplin	23/08/54	G.G. Ohr	24/10/54	H.A. McCartney	16/11/54
Fred C. (Slob) Fisher †	07/02/55	L.M.R. Halverson	23/02/55	R.D. Lynch	04/04/55
Ken Pears	02/05/55	F.E. Warne	20/06/55	R.E. Fritz	01/08/55
D.W. Nicoll	15/08/55	J.R. Vollett	19/09/55	J.H. Berg	26/09/55
M.C. Turcott	07/11/55	S.J. Dunn	21/11/55	Les M. Norton	21/11/55
H.A. Stamnes	25/11/55	Derrick Ellis	00/00/55	P.M. Eades	23/01/56
Alex McLennan	12/03/56	W.A.A. Hadley	16/04/56	W.J. Fleeton	14/05/56
S.R. (Scottie) Wallace †	28/05/56	C.H. Steele	28/05/56	P.D. Ingram	28/05/56

Ronald H. Carlisle	28/05/56	P.F. Lambert	28/05/56	A.J. Hayward	28/05/56
T.A. Cowan	28/05/56	Leonard R. Peterson †	28/05/56	J. Brian Daly	28/05/56
Murdo MacIvor	01/08/56	G.W. Nordby	20/08/56	R.K. Gunderson	09/10/56
W.H. Frederick	24/10/56	M.W. McLeod	19/11/56	C.A. Heaven	21/11/56
R.D. Griswold	04/12/56	A.P. Wick	01/01/57	Douglas L. Beckett	05/02/57
G.F. Pohl	04/03/57	P.D. (Paddy) Clark	04/03/57	J. Hauck	07/03/57
K. Bramhall	18/03/57	J.E. Schollen	18/03/57	Ian F. Anderson	01/04/57
Robert G. Telosky	01/04/57	Harvey Rosenberg †	01/04/57	M. Rangeley	28/04/57
D.R. Finskars	13/05/57	J.L. Hartwick	13/05/57	I.B. Clerihue	03/06/57
J.E. Dyck	10/06/57	R.D. Farmer	10/06/57	R. Johnstone	10/06/57
W.P. (Bill) Stack	10/06/57	Roy J. Rose	02/07/57	Norman H. Combs	19/08/57
C.G. Taylor	26/09/57	N.A. Richert	07/10/57	A.A. Wager	18/12/57
D.D. (Ike) Robertson	18/12/57	K.M. Smart	18/12/57	K.N. Fox	18/12/57
Ronald H. Williams	19/12/57	J.E. Booth	19/12/57	Donald McKenzie †	19/12/57
F.B. (Wiper) McCall	19/12/57	L. Hamberg	23/12/57	Robert A. Lee	23/12/57
N.M. McCartney	23/12/57	G.S. Potter	23/12/57	R.R. Hurle	23/12/57
Joseph Pears	30/12/57	J.T. Hunter	20/01/58	R.J. Good	27/01/58
R.D. Eckersley	03/02/58	P.E. Zaharia	10/02/58	J.P.V. Morin	10/03/58
G.R. (Bud) Walton	16/06/58	P. Redkwich	07/07/58	Donald G. Danbert	14/07/58
H.R. Hassan	02/09/58	L.A. (Larry) Kraft †	08/09/58	K.C. Koch	14/10/58
L.B. Feenie	12/11/58	A.R. (Al) Bowyer	23/02/59	L.G. Thompson	09/03/59
J.W. Dale	27/04/59	R.D. Hudak	27/04/59	W.A. (Junior) Errington	19/05/59
J.W. Adams	01/06/59	T.R. Christie	22/06/59	R.Y. MacKay	29/06/59
W.C. Anderson	29/06/59	P.L. Brown	29/06/59	G.R. Findlay	13/07/59
W.R. Harris	13/07/59	George Driediger †	27/07/59	D.M. Keldsen	27/07/59
S.G. Duncan	27/07/59	E.A. (Ted) Groves	03/08/59	James (Hoss) Cartwright	10/08/59
H.H. Hansen	19/08/59	E.J. (Eddie) Bak †	01/09/59	J. Breedveld	21/09/59
Fred A. Horning †	13/10/59	D.R. McCluskie	04/01/60	T.K. McMillan	12/01/60
W.S. MacPherson	18/01/60	E.R. Bolton	18/01/60	E.A. Beketa	18/01/60
R.L. Dunne	18/01/60	A.A. Lenfesty	03/02/60	J. McVicka	07/03/60
Bernard Chew	07/04/60	A.W. (Bill) Thomson	09/05/60	A.E. (Alfie) Walker †	24/05/60
R.R. Rowland	06/06/60	R.F. Schmidt	13/06/60	H.J. Fox	13/06/60
R.F. Wilson	13/06/60	R.C. Green	18/07/60	Robert H. Smith	18/07/60
R.L. Cassidy	18/07/60	James W. Bathurst	18/07/60	J.B. Williams	22/08/60
K.G. Freelund	12/09/60	Roy E. Bird †	03/10/60	R.V. Ellis	24/10/60
F.A. Dunn	28/11/60				
A.R. (Andy) Barnes	30/01/61	R.M. Johanneson	06/02/61	K.H. Willan	27/02/61
W.G.G. Maddess	17/04/61	Gordon Ion	22/05/61	R.F. Noskin	26/06/61
Oren C. Eaton	10/07/61	L.H. Taylor	31/07/61	B.L. Bolam	07/08/61
J.G. Marleau	11/09/61	J.B. Skalle	10/10/61	D.S. Cameron	10/10/61
L.A. Hummel	17/10/61	J.A. McAughren	30/10/61	J.W. Minshull	20/11/61
G. Alex Matches	22/01/62	D.J. Salter	12/02/62	R.D. Main	12/03/62
E.W. Berda	26/03/62	P.R. Helem	16/04/62	G.C. Tack	16/04/62
D.R. King	16/04/62	L.E. Peterson (Floyd Fox)	13/08/62	R.F. Teather	13/08/62
J.H. Macht	13/08/62	D.L. Kelly	22/08/62	H.J. Jansonius	06/11/62
John B. Corker †	04/02/63	C.R. Mills	04/02/63	W. Donald Holmes †	04/02/63
J. Barry Samson †	04/02/63	A.E. (Barney) Pargee	04/02/63	S.G. Shannon (Salmon)	04/02/63
J.S. (Jack) Engler	02/04/63	J.T. Yurichuk	16/04/63	D.D. Lee	16/04/63
D.L. Corrigan	16/04/63	R.G. Loughran	16/04/63	T. McKibbon	16/04/63
W. Dave Hutton	16/04/63	J.B. Blundell	16/04/63	R.G. Babcock	16/04/63
E.A. Keen	22/04/63	R.I. Dahl	23/09/63	H.J. Gudmundson	23/09/63
J.W. Salley	23/09/63	R.D. Cowan	23/09/63	D.K. Britt	23/09/63
G.D. Rogers	23/09/63	W.J. Gillespie	18/11/63	D.G. MacPherson	18/11/63
G.G. Martin	02/03/64	B.N. Moberg	02/03/64	G.H. Pitman	04/05/64

Name	Date	Name	Date	Name	Date
D.T. Davies	04/05/64	J.L. McMillan	04/05/64	D.T. Clarke	04/05/64
J.M. (Joe) Storoshenko	04/05/64	B.K. Woodward	04/05/64	Wayne A. Hebert	04/05/64
B.F. Corrigan	04/05/64	T.E. Millar	04/05/64	H.D. Charlton	04/05/64
R.M. Bulloch	04/05/64	W.A. Long	22/06/64	P.W. Clark	22/06/64
M.P. McBride	22/06/64	C.V. Barclay	22/06/64	R.L.H. Allen	22/06/64
J.R.K. Powell	22/06/64	T.N. Henderson	22/06/64	H.R. Kerr	22/06/64
R.N. Adams	22/06/64	G.R. Tisdale	22/06/64	Victor A. Mowat †	22/06/64
Robert Pachota	22/06/64	W.M. Frith	22/06/64	A.G. Duplissie	22/06/64
N.R. (Schwartz) McIlroy	22/06/64	E.J. Murray	22/06/64	G.D. Bjarnason	22/06/64
P.B. McDonell	22/06/64	D.L. Lynge	22/06/64	R.A. Lefeaux	22/06/64
N.C. Nemrava	22/06/64	D.A. Lindvik	22/06/64	W.A. Clark	22/06/64
D.A. Richardson	22/06/64	S.L. Krewenchuk	22/06/64	R.W. Crouch	22/06/64
G.T. Mason	22/06/64	G.R. Foreman	22/06/64	I.S. Munro	22/06/64
R.C. Armstrong	22/06/64	W.E. (Ted) Burden	22/06/64	G.D. Wadsworth	22/06/64
Joseph D. Comeau †	22/06/64	M.G. Rawlins	22/06/64	J.F. Simms	22/06/64
K.C. Fadear	22/06/64	W.F. Cadwallader	22/06/64	J.M. Katanchik	22/06/64
F.F. Haller	22/06/64	P.D. Goepel	22/06/64	G. Heywood	22/06/64
S.K. Harrison	22/06/64	W.G.A. Hardy	22/06/64	J.G. Funnelle	22/06/64
L.G.J. Desautels	22/06/64	E.W. Pennell	22/06/64	R.E. Zepeski	22/06/64
R.A. Rogish	22/06/64	Sohen Gill	22/06/64	M.E. Storoshenko	22/06/64
J.D. Craigie-Halkett	22/06/64	G.J. Arden	22/06/64	W.R. Dunbar	22/06/64
L. Dickinson	22/06/64	C.B. Murray	22/06/64	F.W. Beatty	22/06/64
G.A. Sweet	22/06/64	D.A. Rosie	22/06/64	G.E. McConnell	22/06/64
K.L. Menges	22/06/64	R.J. Buntain	22/06/64	W.B. Stewart	22/06/64
L.A.F. (Eric) Geddes	22/06/64	W.G. Stevens	22/06/64	H.W. Oleson	22/06/64
W.J.M. Gebbie	22/06/64	J.L. (Beetle) Baillie	22/06/64	G.D. Buchanan	22/06/64
R.L. Callender	22/06/64	R.M. Sweeney	22/06/64	T.J. Brown	22/06/64
D.M. Young	22/06/64	W.A. Woodhall	22/06/64	Tom Walls †	22/06/64
D. W. Hatton	22/06/64	R.E. Bodner	22/06/64	R.S. Craven	22/06/64
R.J. Ewen	22/06/64	G.E. Longman	22/06/64	G.E. Shields	22/06/64
C. Rowally	22/06/64	R.G. McCullough	10/08/64	D.B. Kennedy	10/08/64
W.S. Cooksley	10/08/64	R.T. Wright	10/08/64	D.A. Clark	10/08/64
Douglas Miller †	10/08/64	R.J. Hickey	10/08/64	Alfi DeBiasio	23/11/64
R.B. Lamb	23/11/64	D.J. (Puff) Patterson	23/11/64	A.J. (Tony) Camley	23/11/64
G.B. Goodwin	23/11/64	R.W. Griffiths	23/11/64	E. Swanson	25/11/64
J.A. Treider	25/01/65	J.R. Donaldson	15/02/65	G. Jakubec	15/02/65
T.W. Taylor	15/02/65	J.J. Sulentich	17/02/65	N.E. Moodie	12/04/65
L.R. Forrest	12/04/65	L.H. (Bing) Paré	12/04/65	Stanley C. Goudie †	12/04/65
R.J. MacIsaac	12/04/65	A.A. Miron	12/04/65	W.L. Colvin	20/04/65
R.W. Rasmussen	02/09/65	K.P. Kelsch	12/10/65	R.A. Sanders	12/10/65
W.E. Lindvik	12/10/65	H.L. Funnelle	12/10/65	D.J. Choquette	12/10/65
O.W. Sallenback	22/11/65	A.W. Johnston	12/11/65	W.D. Jones	22/11/65
R.G. Nunns	22/11/65	J.S. Wick	22/11/65	T.N. Birkland	22/11/65
J.E. MacBeth	22/11/65	D.A. Beall	07/02/66	R.A. Beard	07/02/66
J.P. Quinn	07/02/66	L.M. Howell	07/02/66	W.W. Rimek	07/02/66
Donald B. Wedley	07/02/66	D.J. Brown	07/02/66	P.G. Gilmore	21/03/66
D.E. Whittome	21/03/66	R. Scott	21/03/66	J.F. Bach	21/03/66
David W. Mitchell	19/04/66	R.W. Tebbenham	12/09/66	E.E. Sauve	12/09/66
G.R. Humphrey	12/09/66	Dennis E. Kirkwood	12/09/66	G.M. Salmon	12/09/66
N.L. Gable	12/09/66	M.W.C. Wood	12/09/66	T.E. Lee	14/11/66
G.N. (Gerry) Desmarais †	14/11/66	R.E. Petersen	14/11/66	D.L. Bell	14/11/66
K.I. Campbell	14/11/66	R.G. McLeish	14/11/66	G.K. Miller	14/11/66
G.B. Hallinan	14/11/66	W.E. Dairon	23/01/67	C.S. Dearden	23/01/67
N.A. Halliday [UEL]	28/02/67	D.R. Menzies	29/05/67	D.R. Forbes	29/05/67
W.J. Widdess	29/05/67	D.G. Arsene	29/05/67	P.J. Polonio	30/05/67

C.P. Pridmore	14/08/67	
Terrance P. O'Keeffe †	16/10/67	
Greer Goldsby	16/10/67	
G.W. Miller	22/01/68	
David L. Veljacic	13/05/68	
L.E. Desbrisay	13/05/68	
R.W. McNabb	08/10/68	
W.R. Voice	04/11/68	
M.D. Gilmore	04/11/68	
C.A. Stewart	05/05/69	
I.R. Jensen	05/05/69	
K. Hurlen	06/05/69	
W.A. Ruus	23/09/69	
I.L. Monteith	22/09/69	
R.W. Gotobed	22/09/69	
A.S. Benekritis	22/09/69	
F.H. Bird	02/02/70	
B.W. Deverill	02/02/70	
K.M. Kinna	20/04/70	
D.B. Hutton	20/04/70	
K.E. Abbott	20/04/70	
S.T. Montgomery	10/08/70	
E.P. Cameron	10/08/70	
R.A. Clack	10/08/70	
L.J. Brown	10/08/70	
J.T. Enns	08/02/71	
R.K. Campsall	08/02/71	
D.A. Pears	08/02/71	
G.M. Hunter	08/02/71	
J.A. Ostermeier	[UEL]	01/05/71
K.A. Adams	[UEL]	01/05/71
D.S. Carr	09/08/71	
L.R. Morton	09/08/71	
P.G. LeLoup	09/08/71	
Kirk Lucas	09/08/71	
W.R. Pambrun	01/05/72	
G.T. Meagher	01/05/72	
B.J. MacKenzie	01/05/72	
R.C. Grennan	02/10/72	
A.K. Lanyon	02/01/73	
J.L. MacLeod	02/01/73	
M.P. Berube	02/01/73	
R.L. Mitten	[UEL]	05/04/73
R.K. Daviduk	[UEL]	07/05/73
R.A. Eliason	[UEL]	01/06/73
Lorne Mutter	09/07/73	
G.R. Tetrault	09/07/73	
J. Perrie	09/07/73	
B.J. Turkington	17/09/73	
W.J. Gilmour	[UEL]	01/11/73
D.S. Harris	07/01/74	
J.M. Stuart	22/04/74	
W.A. Hunter	22/04/74	
D.P. Sawchuk	[UEL]	01/05/74

L.J. Reisig	14/08/67	
L.B. McNeilage	16/10/67	
M.W. Graw	16/10/67	
W.C. Muchison	22/01/68	
W.P. Sly	13/05/68	
R. Goodheart	13/05/68	
K.G. Stone	04/11/68	
Robert A. Humphries †	04/11/68	
J.H. Taylor	04/11/68	
S.G. Dalziel	05/05/69	
J.E. Evans	05/05/69	
M.G. Shish	22/09/69	
W.T. Naughty	22/09/69	
J.F. Miller, Jr.	22/09/69	
D.A. Robinson	22/09/69	
W.M. Beggs	22/09/69	
R.G. Edwardes	02/02/70	
B.E. Singleton	02/02/70	
G.R. Harker	20/04/70	
D.A. Nagy	20/04/70	
I.R. Henis	20/04/70	
B.L. Bounds	10/08/70	
R.J. Sedgewick	10/08/70	
D.C. Morgan	10/08/70	
Dwight F. Kirkwood	08/02/71	
R.W. Selbee	08/02/71	
T.W. Johnstone	08/02/71	
J.B. Murchison	08/02/71	
W.J. Ferguson	[UEL]	01/05/71
A.L. Hokanson	[UEL]	01/05/71
A.L. McLeod	09/08/71	
L.V. Funnelle	09/08/71	
B.C. Powell	09/08/71	
C. Bruce Holland †	09/08/71	
R.J. Smith	01/05/72	
J.G. Morgan	01/05/72	
B.D. Smith	02/10/72	
J.W. McCloy	02/10/72	
L.F. Alderman	02/01/73	
T.J. Ovens	02/01/73	
W.A. McLeod	02/01/73	
D.A. Ward	[UEL]	07/05/73
D.R. Dennison	[UEL]	07/05/73
Peter K. Yee	09/07/73	
N. MacKinnon	09/07/73	
A.D. McRae	09/07/73	
J.G. Mitchell	17/09/73	
J.O. Dool	17/09/73	
R.J. Deeley	07/01/74	
D.M. Steele	07/01/74	
D.M. Dube	22/04/74	
R.K. Stroud	22/04/74	
J.H. Branston	12/08/74	

P.H. Janzen	14/08/67	
J.L. Michas	16/10/67	
O.B. (Bud) Swanson †	01/11/67	
D.E. Ashlee	13/05/68	
J.E. Whiting	13/05/68	
D.B. Peake	13/05/68	
J.J. Mora	04/11/68	
T.C. Leslie	04/11/68	
R.G. Salt	04/11/68	
F.W. Fischer	05/05/69	
K.D. Harris	05/05/69	
W.C. Wikene	22/09/69	
G.L. Karsen	22/09/69	
C.N. Jenkinson	22/09/69	
B.J. Flessa	22/09/69	
T.T. Chattaway	02/02/70	
David L. Mitchell	02/02/70	
H.S. Nordin	[UEL]	04/03/70
W.D. Gray	20/04/70	
B. Campbell	20/04/70	
S.G. Rogers	20/04/70	
R.E. Hutton	10/08/70	
W.T. Sanderson	10/08/70	
David Sharpe	10/08/70	
K.E. Hall	08/02/71	
B.N. Whitley	08/02/71	
R.E. Miller	08/02/71	
R.S. Ritchie	[UEL]	01/05/71
K.J. Suttill	[UEL]	01/05/71
A. Deamer	09/08/71	
L.E. Langhorn	09/08/71	
A.A. Masi	09/08/71	
R.E. Arnason	09/08/71	
Richard D. Brown †	01/05/72	
M.D. Stockwell	01/05/72	
M.J. Sapeta	01/05/72	
R.G. Bell	02/10/72	
R.D. Lawrence	02/01/73	
M.L. Holland	02/01/73	
J.E. Mramor	02/01/73	
R.J. Smith	[UEL]	27/02/73
D.W. Semler	[UEL]	07/05/73
S.E. Bawden	[UEL]	07/05/73
B.W. Graham	09/07/73	
Darrell G. Wright	09/07/73	
M.G. Kurelicz	09/07/73	
T.. Shippam	17/09/73	
F.J. Crouch	17/09/73	
T.M. Neratini	07/01/74	
Richard A. Lofgren	14/03/74	
W.H. Campbell	22/04/74	
K.M. Mullin	29/04/74	
R.A. Bronson	12/08/74	

P.S. Linder	12/08/74	A.T. Uunila	12/08/74	G.L. Lanyon	12/08/74
Raymond C. Holdgate	12/08/74	J.E. Steele	12/08/74	G.W. Munro	12/08/74
D.M. Richards	12/08/74	S.M. Webb	12/08/74	R.J. Smith	12/08/74
Aaron Feldman	12/08/74	Jeff Dighton	20/08/74	D.F. Wylie	13/09/74
U.L. (Bert) Mezzarobba	10/02/75	J.L. Olien	10/02/75	H.W. Brown	10/02/75
G.R. Loughran	10/02/75	H.A. (Sandy) Macpherson	10/02/75	D.R. Komar	10/02/75
L.C. Cooke [UEL]	01/03/75	J.R. Seeley [UEL]	01/03/75	P.M. Hayes	05/05/75
P.S. Elchuk	05/05/75	R.C. Bailey	05/05/75	T.A. O'Neill	05/05/75
B.A. Thomson	05/05/75	Stephen G. Woodworth †	05/05/75	T.J. Rose	05/05/75
G.H. Dahl	05/05/75	R.B. Kenning	05/05/75	D.R. Oshowy	05/05/75
W.T. Easterbrook	05/05/75	Rick Lacowicz	05/05/75	D.R. Stoneson	05/05/75
T.E. Dixon	07/07/75	G.J. Federal	07/07/75	W.C. Ryce	07/07/75
R.P. O'Brien	07/07/75	M.E. Woodworth	07/07/75	Harvey Olsen	07/07/75
P.J. MacRitchie	07/07/75	R.J. McCurrach	07/07/75	R.D. Berry	07/07/75
J.E. Nedelak	07/07/75	M.G. Wilson	07/07/75	W.J. McMynn	07/07/75
R.H. Newmark	07/07/75	J.A. Edgar	07/07/75	J.A. Appleby	22/09/75
Ross Styan	22/09/75	B.R. Bergum	22/09/75	D.J. Hennen	22/09/75
G. Robinson	22/09/75	A.J. Watson	22/09/75	R.E. Bogue	22/09/75
G.E. Hartney	22/09/75	G.H. Sarai	22/09/75	R.W. Clelland	22/09/75
R.E. Busch	22/09/75	F.W. Murray	22/09/75	B.W. Manning	22/09/75
G.J. Clark	22/09/75	L.P. Martin	22/09/75	D. Litzenberger	15/12/75
T.A. Jones	15/12/75	M.D. Robb	15/12/75	S.J. Edmonds	15/12/75
D.J. Ward	15/12/75	M.J. Joy	15/12/75	I.D. Parsons	15/12/75
J.B. Burnside	15/12/75	J.A. Tudge	15/12/75	J.E. Schwab	15/12/75
C.W. Thompson	15/12/75	K.R. Lewis	15/12/75	Pat Field	15/12/75
L.G. Mamoser	15/12/75	R.P. Bazaluk	15/12/75	M.R. Findley	15/12/75
D.J. McLeod	15/12/75	G.M. Miller	15/12/75	R.A. Kennedy	15/12/75
L.J. Whyte	15/12/75	C.B. Watt	15/12/75	D.H. Thibert	15/12/75
N.G. Niculeac	15/12/75	R.M. Kearney	15/12/75	B.M. Pelletier	15/12/75
Harold Erlendsen †	15/12/75	D.E. Block	03/05/76	T.S. Zweng	03/05/76
D.W. Farthing	03/05/76	Glen Lyon	03/05/76	S.H. Henderson	03/05/76
K.S. Ingram	03/05/76	D.K. Rodocker	03/05/76	A.D. Hall	03/05/76
R.M. Tapella	03/05/76	E.D. Ferguson	03/05/76	R.H. Melnick	03/05/76
G.A. Peat	03/05/76	J.S. Louden	03/05/76	Andy McNaughton	03/05/76
R.W. Chapman	03/05/76	J.D. Hudson	03/05/76	S.E. Patterson [UEL]	17/05/76
D.J. Sawatski [UEL]	17/05/76	C. Hatch [UEL]	17/05/76	W.G. Douglas [UEL]	17/05/76
J.W. Edwards [UEL]	19/07/76	A.R. Tellier	28/08/76	J.D. Hagel	04/10/76
S.N. Schwab	04/10/76	M.G. Watson	04/10/76	R.R. Durante	04/10/76
R.B. McLeod	04/10/76	A.L. Drinovz	04/10/76	R.P. Gabriel	04/10/76
David W. Ryan †	04/10/76	D.J. Schwab [UEL]	01/01/77	W.R. Storey [UEL]	01/01/77
B.M. Davidson [UEL]	01/01/77	P.E. Grassi [UEL]	01/01/77	D.J. Fell [UEL]	01/01/77
J.W. Poole	01/03/77	R.A. Young	13/06/77	E.W. Lawson	13/06/77
R.M. Allan	13/06/77	E.G. Stephens	13/06/77	C.F. Kitchen	03/06/77
M.A. Willis	13/06/77	D.C. Wright	13/06/77	B.M. Noga	13/06/77
H.L. Brown	13/06/77	S.G. Wice	03/10/77	G.A. Nicholson	03/10/77
B.A. Mathie	03/10/77	R.I. Jones-Cook	03/10/77	W.W. Gurvich, Jr.	03/10/77
J.W. Yaremy	03/10/77	M.S. McGuire	03/10/77	Peter H. Blank †	03/10/77
J.L. Platzner	03/10/77	W.F. O'Neill	01/02/78	R.H. Mattson	06/03/78
E. W. Lindsay	06/03/78	K.W. Swan	06/03/78	B.M. Crofton	06/03/78
Werner Schlechtleitner †	06/03/78	Roy S. Bissett †	06/03/78	P.A. Trenter	06/03/78
H.B. McGuire	06/03/78	R.E. Manyk	06/03/78	D. Clarke	06/03/78
G.W. Hagelund	06/03/78	M.E. Gibbons	06/03/78	M. Blair	06/03/78
A.J. Walls	06/03/78	G.J. Marshall [UEL]	01/05/78	L. Sziklai [UEL]	01/05/78
L.G. Hicks	10/07/78	P.H. MacLean	10/07/78	M.J. Sears	10/07/78
W.C. Humphry	10/07/78	R.J. Hebenton	10/07/78	T.G. Laverty	10/07/78

R.G. Dickson	10/07/78	M.D. Seggie	10/07/78	T.W. McLeod	10/07/78
D.R. Wood	01/07/78	R. Blackman †	08/08/78	R.W. Renzetti	23/04/79
R.B. Caird	23/04/79	Daniel J. Bass †	23/04/79	M.J. Peskett	23/04/79
K.R. Bryan	23/04/79	R.W. Gayton	23/04/79	K.J. Cowie	23/04/79
D.R. Philip	23/04/79	P.W. Torrance	23/04/79	D.J. Rae	23/04/79
L.W. Ewan	23/04/79	P.C. Gabriel [UEL]	01/08/79	W.J. Pye	17/08/79
W.H. Bergum	20/08/79	L. Griffiths	20/08/79	J.P. Gandolfo	20/08/79
N.B. Delmonico	20/08/79	K.C. Stevenin	20/08/79	J.B. Lamb	20/08/79
J.M. Jackson	20/08/79	Ernest Plett	20/08/79	H.C. Lehwald	20/08/79
B.W. Przednowek	20/08/79	J.R. Wilson	20/08/79	R.M. Watts	20/08/79
R.S. Jarman	20/08/79	F. Delgiglio	20/08/79	L.A. Shook	20/08/79
A.T. Roberts	20/08/79	R.W. Mitchell	20/08/79	S.D. Chila	20/08/79
R.S. (Ranny) MacDonald	20/08/79	A.C. MacGregor [UEL]	01/10/79	B.R. Harvey [UEL]	01/11/79
W.R. Longman	08/04/80	R.B. Widdess	08/04/80	G.A. Grieve	08/04/80
D.G. Bannon	08/04/80	R.B. Hatch	08/04/80	R.E. Flello	08/04/80
R.G. (Rick) Newman †	08/04/80	K.R. Steele	08/04/80	J.D. Morley	08/04/80
N.A. Leuszler	08/04/80	R.G. (Rod) MacDonald	08/04/80	M.L. Beckett	08/04/80
T.J. Seifert	08/04/80	T.V. Forss	08/04/80	A.J. Crump	08/04/80
B.T. Franklin	08/04/80	Randy Musgrave † [UEL]	14/07/80	E.E. Thompson [UEL]	14/07/80
T.C. McQuillan	21/09/80	A.O. Ortis	21/09/80	B.A. Oxenbury	21/09/80
B.G. Simonson	21/09/80	D.J. Henderson	21/09/80	B.R. Moody	21/09/80
R.L. Perrett	21/09/80	D.E. Wilson	21/09/80	M.K. Mullin	21/09/80
G.D. Miller	21/09/80	G.C. McGregor	21/09/80	M.D. Taylor	29/12/80
T.E. Lawson	29/12/80	K.B. Conway	29/12/80	R.P. Haddock	29/12/80
T.A. Sullivan	29/12/80	B.J. Brennan	29/12/80	R.C. Baglo	29/12/80
F.A. Dodich	29/12/80	J.C. Eppler	29/12/80	J.R. McKearney	29/12/80
L.D. Peskett	29/12/80	T.A. Nikolai	29/12/80	B.K. Dick	29/12/80
K.J. Abel	29/12/80				
R.E. Hall	29/06/81	Paul Flello	29/06/81	D.R. Scarr	29/06/81
N.M. Zokol	29/06/81	V.F. Rosenberg	29/06/81	A.J. Zanotto	29/06/81
R.M. Pearson	20/06/81	C.E. Thomson	29/06/81	M.E. McDonald	26/10/81
D.R. McRoberts	26/10/81	M.W. Burton	26/10/81	C.N. Smith	26/10/81
D.C. Newell	26/10/81	D.J. Laberge	26/10/81	D.G. Parson	26/10/81
C.A. Matheson	26/10/81	C.W. Scott	26/10/81	G.W. Muir	26/10/81
W.S. Hastings	26/10/81	R.M. Stewart	26/10/81	J.W. Creed	26/10/81
M.D. Lomas	26/10/81	J.V. Cowx	26/10/81	B.W. Campbell	26/10/81
J.W. Coroliuc	26/10/81	H.B. Hlushko	26/10/81	P.S. Monchamp	26/10/81
M.D. Engler	26/10/81	R.W. Pierlot	26/10/81	T.P. Armstrong	26/10/81
V.M. Minosky	26/10/81	J.D. Foster	26/10/81	A.L. Thomas	26/10/81
D.A. Saunders	26/10/81	D.L. Zaleski	26/10/81	W.Y. Lai	02/12/81
Ahmed A. Khan	15/02/82	T.R. McEwan	03/05/82	D.J. Stevens [UEL]	17/05/82
J.E. Pentland [UEL]	17/05/82	D.N. Dyer [UEL]	17/05/82	A.L. Kirk [UEL]	17/05/82
C.M. Delahunt	28/06/82	G.B. Bromley	28/06/82	W. Gilbert	28/06/82
C.G. Ball	28/06/82	D.J. Moore	28/06/82	N. Bevandick	28/06/82
G.S. Warkentin	28/06/82	S.K. Gerhardt	28/06/82	Mario Brito	28/06/82
G.I. MacKinnon	28/06/82	P.C. Bridge	28/06/82	J.M. Tougas	28/06/82
R.B. Olsen	28/06/82	B.S. Tebbutt	28/06/82	D.J. Cutler	28/06/82
M.J. Zacharuk	28/06/82	D.R. Ross	28/06/82	R.W. Jones	28/06/82
I.G. McIntyre [UEL]	12/07/82	D. Hildebrand	17/08/82	G.S. Pellegrin	17/08/82
D.W. Hilton	17/08/82	B.D. Huber	17/08/82	G.J. Kovacs	17/08/82
N.E. (Mike) Knapp	17/08/82	M.P. Gibson	17/08/82	D.P. Graf	17/08/82
D.G. Wilkinson	17/08/82	W.J. McGowan	17/08/82	N.W. Hand	17/08/82
D.J. Lanser	17/08/82	T.R. Penway	13/02/83	J.C. Cook	01/04/83
A.B. Chorney	16/05/83	S.V. Noga	16/05/83	M.J. Antunovic	16/05/83

R.G. Reiffer	16/05/83	S.P. Laleune	16/05/83	R.J. Hollier	16/05/83
D.E. Phillips	16/05/83	J.P. Wilkinson	16/05/83	G.T. Drescher	16/05/83
D.T. Christie	16/05/83	M.D. Etheridge	16/05/83	J.J. Burns	16/05/83
M.H. McGill	16/05/83	R.J. Hamilton	04/07/83	J.F. Hatch	04/07/83
Gary Ayre	04/07/83	Glen R. Taylor †	04/07/83	J.B. Boyle	04/07/83
D.G. Symington	04/07/83	S.A. Dougans	04/07/83	S.M. Letourneau	04/07/83
D.J. Friesen	04/07/83	P.A. Schnarr	04/07/83	L.A. Whitter	01/05/84
J.H. Inksater	07/05/84	D.J. Wagner	07/05/84	G.A. Hogarth	07/05/84
A.R. Orcutt	07/05/84	M.E. Anderson	07/05/84	J.P. Ridington	07/05/84
R.J. Plecas	07/05/84	G.L. Cameron	07/05/84	R.S. Garoff	07/05/84
W.H. Lealess	07/05/84	P.D. Parker	07/05/84	P.D. Kadagies	07/05/84
B.C. Gould	07/05/84	R.P. Flack	07/05/84	A.F. Faber	14/07/84
R.D. Symonds	01/10/84	J.A. Booth	01/10/84	Randy Stolp	01/10/84
R.B. Forch	01/10/84	M.R. Tookey	01/10/84	J. Reimer	01/10/84
D.K. McClelland	01/10/84	R.S. Jordan	01/10/84	J.A. Dennis	01/10/84
R.R. Critchlow	01/10/84	S.J. Duncan	01/10/84	K.J. Wilson	01/10/84
K.L. Thompson	01/10/84	K.L. Haftner	01/10/84	K.M. Suzuki	05/05/85
W.J. Goss	03/09/85	P.F. Gamble	03/09/85	B.D. Gant	03/09/85
S.A. Ross	03/09/85	M.T. Horbulyk	03/09/85	C.D. MacDonald	03/09/85
G. Crane	03/09/85	R.W. Minton	03/09/85	T.H. Lappi	03/09/85
R.A. McKinnon	03/09/85	D.B. Boufford	03/09/85	N.R. Bradley	03/09/85
R.W. Mallory	21/10/85	A.R. White	21/10/85	J.D. Buswood	21/10/85
D.A. Collins	21/10/85	R.H. Romanowski	21/10/85	D.T. Miller	21/10/85
T.D. Pierce	21/10/85	J.T. Donaldson	21/10/85	D.K. Thibodeau	21/10/85
D.A. Parno	21/10/85	R.W. Bryant	21/10/85	D.S. Aube	20/05/86
B.J. Pugsley	20/05/86	D.A. Wilson	20/05/86	D.F. Cahill	20/05/86
C.J. Van Duynhoven	20/05/86	B.R. Davies	20/05/86	M.V. Tammen	20/05/86
L.M. Achtymichuk	20/05/86	J.A. Mantei	20/05/86	R.H. Dube	20/05/86
A.S. Bains [UEL]	23/06/86	C.L. McIntyre [UEL]	01/07/86	M.B. Caton	27/10/86
T.A. Killam	27/10/86	R.C. Fuller	27/10/86	M.S. Rosychuk	27/10/86
G.G. MacFarlane	27/10/86	R.G. McDonald	27/10/86	R.C. Warnock	27/10/86
J.E. Hargreaves	11/05/87	D.S. Etheridge	11/05/87	J.T. Young	11/05/87
F.R. Lamont	11/05/87	R. Cartwright	11/05/87	T.S. Kennedy	11/05/87
R.N. Iwanson	11/05/87	D.H. Fuginski	11/05/87	J.M. Fumich	11/05/87
D.C. Murphy	11/05/87	D.F. Wallack	11/05/87	K.R. Nelson	11/05/87
G.A. Livingstone	11/05/87	M.J. Sereda	11/05/87	S.L. Errington	11/05/87
R.F. Johnson	11/05/87	J.A. Nickles [UEL]	11/07/88	P.W. Steele [UEL]	11/07/88
Donna J. Rasmussen [UEL]	03/10/88	D.C. Jackart [UEL]	17/10/88	E. Bortignon	15/05/89
P.R. Campbell	15/05/89	G.D. Carter	15/05/89	M.A. DeVito	15/05/89
K.A. Heaven	15/05/89	J.A. Kasper	15/05/89	S.T. Katanchik	15/05/89
C.D. Stanford	15/05/89	A.V. Tolusso	15/05/89	J.D. Wood	15/05/89
M.R. Conn	05/03/90	M.A. Etheridge	05/03/90	R.E. Dubbert	05/03/90
M.H. Mattu	05/03/90	J.M. Kolsrud	05/03/90	J.P. Jantunen	05/03/90
R.A. Craven	05/03/90	M.R. Paulson	05/03/90	Harry G. Foster	05/03/90
M.W. McKellar	05/03/90	A. Janzen	05/03/90	M H. McKinnon	05/03/90
G.S. Manson	05/03/90	B.S. Lander	05/03/90	R.P. Czech [UEL]	09/07/90
J.B. Dunham	15/10/90	T.T. Kennedy	15/10/90	D.P. Martin	15/10/90
M.J. Caton	15/10/90	B.J. Murton	15/10/90	T.R. Hebert	15/10/90
G.G. Roder	15/10/90	N.G. Van Laare	15/10/90	R.E. Grierson	15/10/90
G.R. Nygard	15/10/90	J.P. Zacharuk	15/10/90	T.F. Johnstone	15/10/90
D.G. Nichols	15/10/90				
E.A. Froese	06/05/91	R.J. Coulson	06/05/91	W.L. Nicol	06/05/91
E.H. Born	06/05/91	D.L. Rosenlund	06/05/91	G.A. Wace	06/05/91
T.P. Kennedy	06/05/91	P J. Hartner	06/05/91	M.J. Rostill-Huntley	06/05/91

338

G.A. Ryskie		06/05/91
B.R. Bertuzzi		06/05/91
D.M. Stroup		06/05/91
R.F. Martin		03/06/91
P.A. Matthes		09/09/91
G.S. Morrison		09/09/91
S.J. Nicholson		25/05/92
D.A. Moberg		25/05/92
S.A. Kerr		25/05/92
H.O. Groenewold		25/05/92
B.W. Singleton		25/05/92
R.M. Der		13/07/92
D.W. Pughe		19/10/92
A.D. Gregory		19/10/92
Sean Gilmore		19/10/92
D.A. Meers		04/01/93
A.N. Zawada		04/01/93
J.T. Jenkins		04/01/93
S.M. Bauer		04/01/93
D.B. Debeck		04/01/93
G.C. Clark		17/05/93
J.W. Olar		17/05/93
M.W. Stevens		17/05/93
P.H. Tithecott	[UEL]	05/07/93
G.C. Mervin		05/04/94
C.V. Cruz		05/04/94
P.J. Gibbs		05/04/94
K.W. Gemmill		05/04/94
J.A. Bryant		12/09/94
E.A. Lehwald		12/09/94
T.V. Buckingham		12/09/94
C.R. Lanthier		12/09/94
K.W. Siggers		20/02/95
B.G. Berka		20/02/95
D.A. Gill		20/02/95
T.R. Peterson		20/02/95
A.C. Sinclair		20/02/95
S.M. Ingram		20/02/95
B.J. Glover		25/04/95
M.W. Purchas		23/05/95
T.W. Howes		23/05/95
T. Jansen		23/05/95
D. Tierney		23/05/95
A. Burrero		10/06/96
W.R. Murray		10/06/96
R.A. Boruck		10/06/96
Ryan S. Cameron		10/06/96
M.N. Hickey		29/07/96
B.C. Hesse		29/07/96
D.J. Carson		29/07/96
R.D. Renning		29/07/96
G.A. Wilson		24/06/97
M.H. Persaud		24/06/97
P. Sommer		24/06/97
B.F. Haller		24/06/97
T.P. Nyhaug		06/05/91
K.S. Morishita		06/05/91
R.G. Briscoe	[UEL]	06/05/91
R.A. Manning		09/09/91
W.P. Maunsell		09/09/91
A.G. Greenwood	[UEL]	04/11/91
T.A. Brown		25/05/92
F.P. Felker		25/05/92
A.S. Sojka		25/05/92
D.A. Groves		25/05/92
Trevor Connelly		25/05/92
T.G. Heinrich		21/09/92
K.T. Suzuki		19/10/92
Walter Pereira		19/10/92
M.R. Cave	[UEL]	07/12/92
P.T. Gent		04/01/93
S.W. Young		04/01/93
R.W. Burden		04/01/93
M.M. Anzulovich		04/01/93
R. Van Acken		04/01/93
D.L. Day		17/05/93
P.E. Pfoh		17/05/93
T.M. Stuart		17/05/93
C.M. Stevens	[UEL]	13/09/93
B.J. Blundell		05/04/94
A.R. Wilson		05/04/94
P.M. Moore		05/04/94
S.M. Campbell		05/04/94
H.W. Crump		12/09/94
D.R. Messenger		12/09/94
D.A. Forster		12/09/94
G.E. Sereda		12/09/94
G.R. Wong		20/02/95
N.D. Elliott		20/02/95
B.D. Gillan		20/02/95
P.A. Peacock		20/02/95
R.G. Salas		20/02/95
J.W. Husband		20/02/95
K.J. Scott		23/05/95
C.C. Mulder		23/05/95
R.K. Gjesdal		23/05/95
B.A. Young		23/05/95
R.G. Dubbert		23/05/95
D.M. Van Horn		10/06/96
S.W. Aylett		10/06/96
D.J. Cadwallader		10/06/96
T.F. Huska		10/06/96
K. Shearer		29/07/96
C.A. Clarke		29/07/96
B.T. Bogdanovich		29/07/96
D.R. Clark		15/04/97
G.W. Ditchburn		24/06/97
C.M. Cook		24/06/97
L.T. Nichol		24/06/97
S.D. Berda		24/06/97
M.B. Breddin		06/05/91
P.J. Gobillot		06/05/91
G.D. Rodrigues	[UEL]	06/05/91
W.M. Barr		09/09/91
M.A. Scheu		09/09/91
G.J. Quilty		25/05/92
M.B. Primerano		25/05/92
A.G. Styles		25/05/92
D.P. Drezdoff		25/05/92
P.A. Morin		25/05/92
S.A. Carter		19/06/92
M.J. Smith		19/10/92
K.S. Aujla		19/10/92
B.A. Wray		19/10/92
D.E. Boone	[UEL]	07/12/92
S. Foellmer		04/01/93
B.M. Sew		04/01/93
D.A. Booth		04/01/93
T.E. Skroder		04/01/93
K.V. Lepard	[UEL]	08/02/93
Cara R. Goose		17/05/93
V.S. Skrepnik		17/05/93
C.S. Campbell		17/05/93
M.R. Evans		05/04/94
E.J. Otuomagie		05/04/94
D.R. Sampert		05/04/94
J.D. Snider		05/04/94
T.D. Moore		12/09/94
B.D. Godlonton		12/09/94
C.C. Mason		12/09/94
H.S. Ghuman		12/09/94
R.D. Ewert		12/09/94
L.O. Bright		20/02/95
D.G. Fairbairn		20/02/95
D.S. Romaniuk		20/02/95
D.L. Hansen		20/02/95
J.B. Miller		20/02/95
R.R. Rosenlund		20/02/95
J.A. Schmidt		23/05/95
S.T. Millar		23/05/95
R.J. Dulko		23/05/95
R.R. Werner		23/05/95
W.S. Abrams		23/05/95
R.P. Schenderling		10/06/96
J.P. Latta		10/06/96
B.A. Boychuk		10/06/96
D.E. Conacher		29/07/96
A.M. Zagar		29/07/96
D.C. Dickie		29/07/96
E.M. Blank		29/07/96
R.B. Hesketh		24/06/97
M.E. Von Minden		24/06/97
D.G. Shong		24/06/97
B.G. MacGillivray		24/06/97
J.W. Gray		24/06/97

C.M. Herbert	24/06/97	T.G. Johnston	24/06/97	C.E. Gelsvik	24/06/97
R.C. Peskett	24/06/97	M.A. Polonio	24/06/97	R.G. Dyck	12/08/97
S. Sovdat	12/08/97	D.R. Imrih	12/08/97	M.D. Hatcher	12/08/97
B.M. Sukul	12/08/97	R. McNutt	12/08/97	R.C. Taylor	12/08/97
R.M. Chorney	12/08/97	P.S. Sihota	12/08/97	D. Shirley	05/01/98
J. Lockey	05/01/98	K.R. Main	05/01/98	S.C. Nicol	05/01/98
V.J. Vigh	05/01/98	D.W. Kraynyk	05/01/98	S.A. Harvey	05/01/98
W. Tabata	05/01/98	R.G. Kitchener	05/01/98	J.B. Westgate	05/01/98
D.H. Higgins	05/01/98	K. Kleindienst	05/01/98	M.A. Navratil	05/01/98
M.B. Tweedie	05/01/98	A. Czeppel	05/01/98	S.J. Fraser	05/01/98
T.G. Cooper	05/01/98	T.J. Pelke	05/01/98	C. Butula	05/01/98
S. Lucks	05/01/98	W. Miller	05/01/98	D.W. Booth	22/06/98
R.S. Watson	22/06/98	R.A. Kursar	22/06/98	J.M. Keller	22/06/98
D.E. Mackie	22/06/98	K.R. Bridger	22/06/98	T.F. Dodd	22/06/98
A.D. Jeves	22/06/98	T.N. Cap	22/06/98	G.R. Penner	22/06/98
C.N. Pawlak	22/06/98	A.M. Belczyk	18/01/99	B.W. Kellner	18/01/99
A. Groenewegen	18/01/99	D.W. Bunz	18/01/99	E.T. Morrison	18/01/99
M.R. Rusticus	18/01/99	Ryan N. MacMillan †	18/01/99	B.B. Leibel	18/01/99
M.J. Blouin	18/01/99	D.E. Veuger	18/01/99	W.A. Nicholson	18/01/99
G.K. Abbott	18/01/99	C.R. Massey	18/01/99	K. Kepes	18/01/99
T.P. Dee	14/06/99	T.N. Dykes	14/06/99	M.S. Clinaz	14/06/99
A.B. Yackel	14/06/99	S.M. Bolan	14/06/99	G.S. Dick	14/06/99
J.W. Davidson	14/06/99	G.J. Gauthier	14/06/99	G.S. Marineau	14/06/99
D.R. Davidson	14/06/99	D.R. Robertson	14/06/99	J.S. Mortimer	14/06/99
B.R. Murray	14/06/99	T.J. Ballard	14/06/99	J.S. Pace	14/06/99
S.A. Kitt	14/06/99	J.J. Shandera	14/06/99	E.J. Pickett	14/06/99
R.K. Stewart	04/01/00	J.C. McQuarrie	04/01/00	M.G. McCash	04/01/00
C.S. Potter	04/01/00	J. Evans	04/01/00	J. Kiem	04/01/00
R. Culbert	04/01/00	M. Steer	04/01/00	M.J. Schaap	04/01/00
B.J. Call	04/01/00	C.W. Thompson	04/01/00	B.C. Wold	04/01/00
D. Ferris	04/01/00	D. Cochrane	04/01/00	R. Coy	04/01/00
J.T. Blount	04/01/00	C.J. Breure	04/01/00	A.S. Meyer	04/01/00
J.I. MacInnes	23/05/00	R.J. Hayes	23/05/00	E.G. Reisen	23/05/00
L.L. Daminato	23/05/00	M.J. Gillis	23/05/00	C.A. Brodziak	23/05/00
D.R. Lyons	23/05/00	R.A. DeAlbuquerque	23/05/00	M.R. Ogden	23/05/00
G.W. Frost	23/05/00	L.M. Richards	23/05/00	R.G. Hutton	23/05/00
J.E. Dawkins	23/05/00	S.N. Dighton	23/05/00	P.J. McKeown	23/05/00
R.S. Hegedus	23/05/00	M.A. Miller	23/05/00	C. Ball	23/05/00
C. Kampmeinert	23/05/00	A.C. Noke-Smith	23/05/00	A.J. Ryan	23/05/00
R.D. Veer	23/05/00	D.G. Pighin	23/05/00	K.K. Sahota	23/05/00
C.A. Swant	02/10/00	D.E. Granchar	02/10/00	T.J. Felts	02/10/00
L.P. Weibe	02/10/00	R.P. Wilson	02/10/00	D. Gaos	02/10/00
D.R. Green	02/10/00	S.D. Turner	02/10/00	S.P. Hendrickson	02/10/00
D.W. Oppenlander	02/10/00	B.W. Hutchinson	02/10/00	S.A. MacKichan	02/10/00
L.G. Brown	02/10/00	M.D. Green	02/10/00	K.T. Zoppa	02/10/00
C.M. Macaulay	02/10/00	T.C. Bourne	02/10/00	J.A. Shanahan	02/10/00
S.W. Humenny	02/10/00	B.G. Randall	02/10/00		
G.J. Smith	05/02/01	D.E. Johnson	05/02/01	S.G. Dick	05/02/01
K.A. Lees	05/02/01	A.D. Nelson	05/02/01	D.R.Dickerson	07/05/01
R.R. Renville	07/05/01	G.J. Hiebert	07/05/01	T.R. Yarych	07/05/01
G.P. Wood	07/05/01	S.J. Lessard	07/05/01	R. Anderson	07/05/01
B.A. Dales	07/05/01	S.L. Howell	07/05/01	D.J. Kirincic	07/05/01
R.A. Klein	07/05/01	F.V. Schmidt	01/10/01	P.J. Nicholson	01/10/01
M. Baker	01/10/01	P.M. Sheasby	01/10/01	M.W. Guns	01/10/01

340

Name	Date	Name	Date	Name	Date
M.D. Canaday	01/10/01	J.C. Short	01/10/01	J.R. Perodie	01/10/01
M.F. Kane	04/02/02	K.C. Friesen	04/02/02	D.B. Korstrom	04/02/02
G.S. Sidhu	04/02/02	T.R. Kielan	04/02/02	J.E. Aitken	04/02/02
K.T. Sears	04/02/02	S.A. Korry	04/04/02	J.G. Lay	04/02/02
M.S. Dosange	04/02/02	M.F. Lee	04/02/02	D.H. Murru	18/03/02
M.O. Greissel	27/05/02	D.O. Bourdeaud'huy	27/05/02	J.M. Jakubec	27/05/02
D.A. Robertson	27/05/02	S.W. Rivet	27/05/02	J.A. Nicholson	27/05/02
S.T. Ward	27/05/02	C.W. Clarke	27/05/02	K.B. Tomyk	05/05/03
W.W. Boscher	05/05/03	K.R. Heywood	05/05/03	C.M. Merrill	05/05/03
M. Chen	05/05/03	R.P. Essinger	05/05/03	R.S. Aver	05/05/03
B.T. McDonnell	05/05/03	D. Hayre	05/05/03	W.K. Yeung	06/08/03
K.J. Ekman	08/09/03	J.D. Greenwood	08/09/03	D.W. Baldwin	08/09/03
J.P. Neale	08/09/03	R.L. McKinnon	08/09/03	D.P. Phillips	08/09/03
A. Friesen	08/09/03	J.E. Palmer	08/09/03	H.G. Petticrew	08/09/03
J.S. Gormick	08/09/03	R.S. Naughty	08/09/03	C.W. Won	08/09/03
C.W. Stadnek	08/09/03	R.W. Landles	08/09/03	J.M. Shalist	08/09/03
S.D. Klassen	08/09/03	J.V. Chamberland	06/10/03	B.J. Walker	06/10/03
M.R. Heslop	06/10/03	M.C. Savage	06/10/03	T.J. Latimer	06/10/03
T.C. Struthers	06/10/03	A.J. Eidher	06/10/03	S.A. Barzen	06/10/03
D.N. Pereira	06/10/03	A.R. Brown	06/10/03	M. Cerantola	06/10/03
I.M. Pickett	06/10/03	D.E. Robinson	06/10/03	D.V. Klassen	06/10/03
D.J. Loverin	06/10/03	C.M. Duncan	06/10/03	G.S. Motokado	06/10/03
A.P. Scott	06/10/03	T.L. Boldt	02/02/04	C.D. Pidcock	02/02/04
R.E. Peskett	02/02/04	K.H. Hastings	02/02/04	M.D. Martin	02/02/04
R.C. Deans	02/02/04	L.S. Hobbis	02/02/04	C. Lee	02/02/04
A.J. Begg	02/02/04	T.J. Heppner	02/02/04	B.A. Quennell	02/02/04
T.J. Lucas	02/02/04	C.W. Newman	17/05/04	M.T. Frew	17/05/04
T.M. MacLeod	17/05/04	M.D. Gibson	17/05/04	J.P. Dyck	17/05/04
R.G. Lauzon	17/05/04	M.W. Zupan	17/05/04	C.J. Coleman	17/05/04
R.M. Deacon	17/05/04	J.M. LaGreca	17/05/04	M.P. Duifhuis	17/05/04
R.D. Araujo	17/05/04	K.A. Trott	17/05/04	G.G. Nielson	17/05/04
P.C. Essinger	17/05/04	D.D. Leopold	17/05/04	N.S. Austin	04/10/04
D.P. Hallgren	04/10/04	B.O. Meyer	04/10/04	T.D. Girard	04/10/04
D.P. Baxter	04/10/04	P.R. Jeffery	04/10/04	R.P. Gonzaga	04/10/04
G.D. Murphy	04/10/04	B.L. Cooke	04/01/05	M.F. Yates	31/01/05
S.A. Stoneson	31/01/05	M.J. Whincup	31/01/05	S.S. Chera	31/01/05
A.R. Johnston	31/01/05	J.T. Sheppard	31/01/05	M.A. Lemire	31/01/05
J.G. Malcom	09/05/05	R.C. Powell	09/05/05	E.L. Barron	09/05/05
A.P. Starritt	09/05/05	R.C. Weeks	09/05/05	D.E. McKay	09/05/05
P.A. Neal	09/05/05	D.D. O'Sullivan	09/05/05	K.J. Sandberg	09/05/05
Keith M. Oliver †	09/05/05	R.N. Mulligan	08/08/05	L.D. Kuva	10/03/05
S.D. Doggett	10/03/05	D.C. Mafferi	10/02/05	A.S. Litell	10/03/05
T.E. Wellington	10/03/05	C.J. McArthur	10/03/05	N.H. Gorseth	10/03/05
M.R. Chevrefils	08/05/06	R.M. Delaurier	08/05/06	C.A. Edge	08/05/06
M.G. Goulet	08/05/06	J.M. Mulcahy	08/05/06	C.S. Sandhu	08/05/06
R.D. Steele	08/05/06	M.A. Varga	08/05/06	S.J. Wood	08/05/06

Appendix III

City of Vancouver Firehalls 1886 to 2006

No. 1

• No. 1 Firehall, 14 Water Street. Opened about October 14, 1886. Closed March 15, 1906, torn down in 1927.

• No. 1 Firehall, Gore and Cordova Streets. Opened March 15, 1906. Designated No. 2 Firehall, March 7, 1951, when No. 1, 729 Hamilton Street, opened. Closed August 8, 1975. Used as an actor's workshop, later became the Firehall Theater. Designated a civic heritage building, July 1980.

• No. 1 Firehall, 729 Hamilton Street. Opened March 7, 1951. Official opening ceremony March 16, 1951. Closed June 16, 1975. Torn down July 1975.

• No. 1 Firehall, 900 Heatley Street. Opened June 16, 1975 with administration offices and machine shops; as a fire company with No. 1 Pump, August 8, 1975.

No. 2

• No. 2 Firehall, 724 Seymour Street. Opened August 3, 1888. Closed end of April 1903. Torn down May 13, 1903. Fire company moved to rented quarters at 845 Granville Street until new hall completed.

• No. 2 Firehall, 754 Seymour Street. Officially opened February 1, 1904. Closed March 7, 1951. Used by the British Columbia Telephone Company until torn down for BC Tel head office expansion, late 1950s.

• No. 2 Firehall, 296 East Cordova at Gore Street. From March 7, 1951 until August 8, 1975. Originally No. 1, opened 1906.

• No. 2 Firehall, 199 Main Street at Powell. Opened August 8, 1975. (Built as the new No.8.)

No. 3

• No. 3 Firehall, Ninth Avenue, west of Westminster Ave. (later, 133 East Broadway). Opened November 1, 1892. Closed June 24, 1912. Torn down 1917.

• No. 3 Firehall, 96 East 12th Avenue. Opened June 24, 1912. Closed July, 21, 1999 and torn down. Company to temporary quarters at No.13 Hall and 2700-block Kingsway.

• No. 3 Firehall, New address, 2801 Quebec Street. Officially opened January 27, 2001 and dedicated to Captain Lloyd Love.

No. 4

• No. 4 Firehall, Broadway and Granville Streets (southeast corner). Opened January 1, 1904. Closed February 2, 1911.

• No. 4 Firehall, 1475 West 10th Avenue. Opened February 2, 1911. Closed January 22, 1991. E4 to old Civil Defense bunker under Granville Bridge; L4 to No.8 Hall during construction.

• No. 4 Firehall, 1475 West 10th Avenue. Opened May 11, 1992; official opening, June 27, 1992. Dedicated to Lieutenant O.B. (Bud) Swanson. The Firehall shares building space with the Vancouver Public Library's Firehall Branch.

No. 5

• No. 5 Firehall, Vernon and Keefer Streets. Opened February 1, 1905. Closed April 4, 1918. Used as a neighborhood/service club clubhouse it was torn down and rebuilt in the late 1970s.

• No. 5 Firehall, ("Fort Apache") 3090 East 54th Avenue at Kerr Street. Opened April 2, 1952.

No. 6

• No. 6 Firehall, 1500 Nelson Street at Nicola. Opened March 1, 1908. Addition at rear built 1927-28. Company moved to empty City of Vancouver automobile testing station, Bidwell and Georgia Streets during remodeling and upgrading. Designated a civic heritage building, September 1988. Hall re-opened March 18, 1989. New address, 1001 Nicola Street. Dedicated to Captain John Graham.

No. 7

• No. 7 Firehall, West 5th Avenue and Yew Street. Opened January 1, 1908. Closed May 1, 1933. Used by the Olympic Pie Company as a bakery. Torn down in the 1950s.

• No. 7 Firehall, 1090 Haro Street at Thurlow. Opened December 5, 1974, when No.1 Hose Wagon became No. 7 Pump. No. 1 Truck, the Calavar Firebird became No. 7 Truck, March 10, 1976. (Built as new No. 2.)

No. 8

• No. 8 Firehall, Victoria Drive and Pandora Street, opened January 1, 1908. Closed May 1, 1933, and later became part of the Terminal City Iron Works foundry.

• No. 8 Firehall, 893 Hamilton Street at Smithe. Opened December 5, 1974, with No.2 R&S in service. (Built as new No.7.)

No. 9

• No. 9 Firehall, Salsbury and Charles Streets. Opened October 11, 1909. Closed February 12, 1960.

• No. 9 Firehall, 1805 Victoria Drive at East 2nd Avenue. Opened February 12, 1960. Official civic opening ceremony, March 11, 1960.

No. 10

• No. 10 Firehall, West 13th Avenue and Heather Street. Opened July 1, 1909. Closed April 4, 1918. Used as storage for reserve apparatus' and residence for chief's drivers. In

1950s while occupied by ambulance service was destroyed by fire.

• No. 10 Firehall, Immigration Building, north foot of Burrard Street. Fire barge and *Vancouver Fireboat No. 2* station from January 14, 1946 until closed November 30, 1971.

• No. 10 Firehall, north foot of Campbell Avenue on Burrard Inlet. Fireboat station, portable trailer construction. Opened November 30, 1971. Closed in 1975.

• No. 10 Firehall, north foot of Dunlevy Avenue on Burrard Inlet. Fireboat station, portable trailer construction. Opened 1975, closed January 1, 1988 when the fireboat was put out of service.

• No. 10 Firehall, 2992 Westbrook Mall, University Endowment Lands (UEL). Ex-UBC Fire Department HQ Firehall. Amalgamated with Vancouver Fire Dept. October 16, 1995.

No. 11

• No. 11 Firehall, East 12th Avenue and St. Catherines Street. Opened January 16, 1911, closed October 25, 1935. Used as a service club boys' club. Torn down 1970s.

No. 12

• No. 12 Firehall, West 8th Avenue and Balaclava Street. Opened October 13, 1913. Closed October 30, 1986 and torn down. Pump and crew remained in portables on site, truck and crew to No. 19 Hall during construction.

• No. 12 Firehall, 2460 Balaclava Street at West 8th Avenue. Occupied June 23rd by No. 12 Pump and June 26th, No. 12 Truck; officially opened July 11, 1987.

No. 13

• No. 13 Firehall, 790 East 24th Avenue at Prince Albert Street. Opened October 13, 1913. Closed June 8, 1917, and re-opened June 1, 1927. Closed December 2001 to be rebuilt on site. Destroyed by arson fire January 13, 2002. Temporary site at 5616 Fraser Street during construction. Officially opened April 4, 2003, and dedicated to Captain Ozzie Howell. New address, 4013 Prince Albert Street.

No. 14

• No. 14 Firehall, 2705 Cambridge Street, at Slocan. Opened July 11, 1913. Closed August 13, 1979, and torn down.

• No. 14 Firehall, 2804 Venables at Kaslo Street. Opened August 13, 1979.

No. 15

• No. 15 Firehall, 3003 East 22nd Avenue at Nootka Street. Opened September 15, 1914. Closed June 8, 1917, and re-opened July 1, 1927. In 2006, city council voted to save it. The last remaining VFD hall to have used horses.

No. 16

• No. 16 Firehall, south foot of Drake Street alignment, north shore of False Creek. Fireboat *J.H. Carlisle* station.

Opened September 1, 1928. Also used as a fire barge station, 1946 to 1951. Closed August 5, 1971, when the fireboat was put out of service.

No. 17

• No. 17 Firehall, 600-block East 41st Avenue at Draper Street. Originally SVFD No. 3 Firehall and FD HQ, it became part of the VFD on amalgamation January 1, 1929. Closed June 3, 1955. Torn down in 1955 for John Oliver High School expansion.

• No. 17 Firehall, 7070 Knight Street at East 55th Avenue. Opened June 3, 1955.

No. 18

• No. 18 Firehall, 1375 West 38th Avenue at Cartier Street. It was the second PGFD No. 1 Firehall and HQ Hall and opened April 30, 1924, becoming part of the VFD, January 1, 1929. Closed June, 1999, and torn down June 17, 1999.

• No. 18 Firehall, 1375 West 38th Avenue at Cartier Street. Opened July 22, 2000. Dedicated to Captain Grahame White.

No. 19

• No. 19 Firehall, West 12th Avenue and Trimble Street. Originally PGFD No. 2 Firehall, it opened September 1, 1922, and joined Vancouver on amalgamation, January 1, 1929. Closed and torn down in January 1979. No. 19 Pump and crew were quartered in trailers on site during construction of new hall.

• No. 19 Firehall, 4396 West 12th Avenue at Trimble Street. Opened July 3, 1980.

No. 20

• No. 20 Firehall, Wales Road and Kingsway. Originally SVFD No.2 Firehall and became part of VFD on amalgamation. Closed November 9, 1962, and torn down.

• No. 20 Firehall, 5402 Victoria Drive at East 38th Avenue. Opened November 9, 1962.

No. 21

• No. 21 Firehall, West 38th Avenue and Carnarvon Street. Originally PGFD No. 3 Firehall, it became part of the VFD on amalgamation. Opened May 1, 1926, it was closed in October of 1984 and torn down. The pump crew stayed on the site in trailers during construction and the truck crew went to No. 19 Hall.

• No. 21 Firehall, 5425 Carnarvon Street at West 38th Avenue. Opened June 7, 1985.

No. 22

• No. 22 Firehall, West 70th Avenue and Hudson Street. Originally PGFD No.4, it opened "sometime after October 30, 1922," when the final payment of $2,250 was made to the contractors. It closed June 18, 1982, and became a senior citizen's drop-in center.

• No. 22 Firehall, 1005 West 59th Avenue at Oak Street. Opened June 18, 1982.

South Vancouver Firehalls 1910–1929

At the time of amalgamation, the South Vancouver Fire Department only had two firehalls* which joined the city, but at its peak in 1916, the department had five firehalls.

- No. 1 Firehall, Joyce Road and Kingsway. Opened 1910, closed 1920s.
- No. 2 Firehall, Miller Street and Kingsway. Opened c. 1910, closed c. 1917.
- No. 2 Firehall, Wales Road and Kingsway. Opened 1925. Became VFD No. 20.*
- No. 3 Firehall, Wilson Road (later East 41st Avenue) and Draper Street. The SVFD HQ Firehall opened c. 1912 and became VFD No.17.*
- No. 4 Firehall, 20 West 26th Avenue at Main Street. Opened 1914, closed 1920.
- No. 5 Firehall, Fraser Street and East 59th Avenue (later 57th Avenue). Opened in 1914 and closed c. 1917.

Point Grey Firehalls 1912–1929

The Point Grey Fire Department joined Vancouver with four firehalls:*

- No. 1 Firehall, West 38th Avenue and Cartier Street. Opened c. 1912. Closed April 30, 1924 and torn down.
- No. 1 Firehall, 38th Avenue and Cartier Street. Opened April 30, 1924. PGFD Headquarters. Became VFD No. 18 Firehall on amalgamation, 1929.*
- No. 2 Firehall, West 11th Avenue and Trimble Street. A small garage, it opened c. 1914, closed c. 1916.
- No. 2 Firehall, West 10th Avenue and Sasamat Street. A rented property, it opened c. 1916. Given notice by owner to vacate, June 6, 1922.
- No. 2 Firehall, West 12th Avenue and Trimble Street. Opened September 1, 1922. Became VFD No. 19.*
- No. 3 Firehall, West 38th Avenue and Carnarvon Street. Opened May 1, 1926. Became VFD No. 21.*
- No. 4 Firehall, West 70th Avenue and Hudson Street. Opened 1922. Became VFD No. 22.*

In 1996, all city firehalls were given names that identified the general neighborhood of the city in which they were located. They were named as follows:

No. 1 Firehall, 900 Heatley Street . . . . .Strathcona
No. 2 Firehall, 199 Main Street . . . . .Downtown Eastside
No. 3 Firehall, 2801 Quebec Street . . . . .Mt. Pleasant
No. 4 Firehall, 1475 West 10th Avenue . . . . .Fairview
No. 5 Firehall, 3090 East 54th Avenue . . . . .Champlain
No. 6 Firehall, 1001 Nicola Street . . . . .West End
No. 7 Firehall, 1090 Haro Street . . . . .Downtown
No. 8 Firehall, 893 Hamilton Street . . . . .Yaletown
No. 9 Firehall, 1805 Victoria Drive . . . . .Grandview/Woodland
No.10 Firehall, 2992 Westbrook Mall . . . . .University
No.12 Firehall, 2460 Balaclava Street . . . . .Kitsilano
No.13 Firehall, 795 East King Edward . . . . .Riley Park
No.14 Firehall, 2804 Venables Street . . . . .Hastings/Sunrise
No.15 Firehall, 3003 East 22nd Avenue . . . . .Renfrew
No.17 Firehall, 7070 Knight Street . . . . .Fraserview
No.18 Firehall, 1375 West 38th Avenue . . . . .Shaughnessy
No.19 Firehall, 4396 West 12th Avenue . . . . .West Point Grey
No.20 Firehall, 5402 Victoria Drive . . . . .Victoria
No.21 Firehall, 5425 Carnarvon Street . . . . .Kerrisdale
No.22 Firehall, 1005 West 59th Avenue . . . . .Marpole

Also during the year, to show the department's new name, the signage on all firehalls was changed to read VF&RS Firehall No. XX.

Appendix IV

Fire Apparatus of the Vancouver Fire Department 1886 to 2006

1886 Ronald 600 gpm 3rd-size steam pump, Engine No.1, hand-drawn, later two-horse. In service No.1, August 1886.

1886 Silsby hose reels (four), hand-drawn. In service No.1, August 1886.

1887 Hayes village hook & ladder truck, hand-drawn. In service, No.1, June 1887.

1888 Silsby hose reels (two), hand-drawn. In service No.2 and City Hall, May 1888.

1888 Ronald 600-gpm 3rd-size steam pump, Engine No.2, hand-drawn, later two-horse. In service, No.2, August 1888.

1889 Silsby hose wagon, Hose Wagon No.1, two-horse. In service, No.1, September 1889.

1892 Winch hose wagon, Hose Wagon No.2, two-horse. In service, No.2, May 1892.

1892 Morrison 100-gallon chemical engine, two-horse. In service, No.2, December 1892.

1893 St. Charles hose wagon, Hose Wagon No.3, two-horse. In service, No.3, October 1893.

1899 Waterous 1,000-gpm 1st-size steam pump, Engine No.3, three-horse. in service, No.2. September 1899.

1899 Hayes 75-ft. aerial ladder truck, two-horse. In service, No.2, October 1899.

1899 Champion 120-gallon chemical engine, two-horse. In service, No.2, October 1899.

1890s Silsby 1,000-gpm steam pump, on loan when Engine No.3 sent to Dawson City. In service, September 1900, to March 1901.

1901 Waterous 1,000-gpm 1st-size steam pump, new Engine No.3, three-horse. In service at No.2, March 1901.

1904 (make unknown) combination hose/chemical, Hose Wagon No.4, two-horse. In service, No.4, January 1904.

1905 (make unknown) Hose Wagon No.5, two-horse. In service, No.5, February 1905.

1905 Waterous 600-gpm 3rd-size steam pump, Engine No.4, two-horse. In service, No.3, August 1905.

1906 Waterous 800-gpm 2nd-size steam pump, Engine No.5, three-horse. In service, No.5, February 1906.

1907 (make unknown, ex-express wagon), Hose Wagon No.9, two-horse. In service, No.9, October 1909.

1907 Seagrave Model 9A hook & ladder truck, two-horse. In service, No.1, March 1907.

1907 Canadian 600-gpm 3rd-size steam pump (no engine number), two-horse. In service, No.5, September 1907.

1907 Seagrave AC-53 hose wagon, Registration No.2382. Shop No.27. In service, No.5, January 1908.

1907 Seagrave AC-53 hose wagon, Reg. No.2383. Shop No.28. In service, No.6, March 1908.

1907 Seagrave AC-53 120-gallon chemical wagon, Reg. No.2384. Shop No.29. In service, No.6, March 1908.

1908 Amoskeag (Built 1906) 1,200-gpm self-propelled steam pump, Reg. No.789. In service, No.1, April 1908.

1908 Waterous 800-gpm 2nd-size steam pump, Engine No.6, two-horse. in service, No.3, November 1908.

1909 Seagrave AC-53 120-gallon chemical wagon, Reg. No.3301. Shop No.30. In service, No.2, August 1909.

1909 Seagrave AC-53 hose wagon, Reg. No.3302. Shop No.31. In service, No.1, August 1909.

1909 Seagrave AC-90 75-ft. tractor aerial, Reg. No.3586. Shop No.50. In service, No.1, September 1909.

1909 Seagrave AC-40 hose wagon, Reg. No.3587. Shop No.32. In service, No.2, December 1909.

1910 Seagrave (Canada) AC-53 120-gallon chemical wagon, Reg. No. unknown, Shop No.33. In service, No.2, January 1911.

1910 Seagrave (Canada) AC-53 hose wagon, Reg. No. unknown, Shop No.34. In service, No.3, June 1911.

1910 Seagrave (Canada) AC-53 120-gallon chemical wagon, Reg. No. unknown, Shop No.35. In service, No.3, June 1911.

1910 Seagrave (Canada) AC-53 120-gallon chemical wagon, Reg. No. unknown, Shop No.36. In service, No.5, September 1911.

1911 Seagrave (Canada) AC-80 hose wagon, Reg. No. unknown Shop No.37. In service, No.4, October 1911.

1912 Webb 85-ft. Couple-Gear electric aerial. Shop No.51. In service, No.2, July 1912.

1912 American-LaFrance Type 10, combination hose/chemical wagon. Reg. No. 206. Shop No.95. In service, PGFD No.1, Jan. 1913.

1912 American-LaFrance Type 10, hose wagon, Reg. No.208. Shop No.53. In service, No.6, January 1913.

1912 American-LaFrance Type 10, hose wagon, Reg. No.209. Shop No.52. In service, No.1, December 1912.

1912 American-LaFrance Type 10, combination hose/chemical wagon, Reg. No.210. Shop No.54. In service, No.7, January 1913.

1913 American-LaFrance Type 14, hose wagon, Reg. No.375. Shop No.55. In service, No.5, January 1914.

1913 American-LaFrance Type 14, city service ladder truck, Reg. No.347. Shop No.56. In service, No.9, January 1914.

1914 Seagrave (Canada) WC-80 city service ladder truck, Reg. No. unknown, Shop No.57. In service, No.4, April 1914.

1914 American-LaFrance Type 14, city service ladder truck, Reg. No.609. Shop No.92. In service, PGFD No.1, July 1914.

1914 American-LaFrance Type 10, combination hose/chemical, Reg. No 642. (Used) Shop No.63. In service, No.1, October 1921.

1914 American-LaFrance Type 15, 1,250-gpm pump, Reg. No.710. Shop No.58. In service, No.2, January 1915.

1914 American-LaFrance Type 12, 625-gpm pump, Reg. No.718. Shop No.93. In service, PGFD No.1, December 1914.

1914 Seagrave (Canada) WC-80 combination hose/chemical. Reg. No.402. Shop No.85. In service, SVFD No.3, June 1914.

1914 Seagrave (Canada) WC-80 combination hose/chemical. Reg. No.403. Shop No.84. In service, SVFD No.2, June 1914.

1916 Oldsmobile hose wagon (1912 model; shop-built in 1916). Shop No.38; rebuilt as Shop No.61. In service, No.11, June 1916.

1917 Stearns combination hose/squad wagon. (Used for extra manning at fires during meal hours.) In service, No.3, 1917.

1917 Oldsmobile 120-gal. chemical wagon (1912 model, shop-built 1917, using Seagrave tanks). Shop No.59. In service, No.2, June 1917.

1918 American-LaFrance (Canada) Type 45, 840-gpm pump, Reg. No.1929. Shop No.60. In service, No.4, June 1918.

1919 American-LaFrance (Canada) Type 45, 840-gpm pump, Reg. No.2633. Shop No.62. In service, No.8, February 1920.

1922 White 120-gallon chemical wagon (shop-built, using Seagrave tanks). Shop No.65. In service, No.2, January 1922.

1922 Packard Twin Six 625-gpm pump. Shop No.94. In service, PGFD No.1, October 1922.

1924 American-LaFrance (Canada) Type 45, 840-gpm pump, Reg. No.4433. Shop No.69. In service, No.3, August 1924.

1925 American-LaFrance (Canada) Type 12, 840-gpm pump, Reg. No.5269. Shop No.71. In service, No.6, October 1925.

1926 American-LaFrance (Canada) Type 12, 840-gpm pump, Reg. No.5408. Shop No.72. In service, No.7, September 1926.

1926 American-LaFrance (Canada) Type 12, 625-gpm pump, Reg. No.5726. Shop 88. In service, PGFD No.1. November 1926.

1926 American-LaFrance (Canada) Type 14, city service truck, Reg. No.5725. Shop 89. In service, PGFD No.1, November 1926.

1926 Dodge Brothers-Graham bush car, Shop 90. In service, PGFD No.1, June 1926.

1927 Studebaker city service ladder truck. Shop No.86. In service, SVFD No.3, May 1927.

1927 Studebaker combination hose/chemical wagon, 35-gallon (shop-built). Shop No.74. In service, No.15, June 1927.

1927 American-LaFrance (Canada) Type 17, 85-ft. tractor-aerial, Reg. No.6025. Shop No.75. In service, No.2, August 1927.

1928 LaFrance Type 145, Metropolitan 840-gpm pump, Reg. No.6065. Shop No.76. In service, No.9, May 1928.

1928 LaFrance Type 145, Metropolitan 840-gpm pump, Reg. No.6078. Shop No.77. In service, No.11, May 1928.

1928 LaFrance Type 112, combination hose/chemical wagon, 40 gallon, Reg. No.6302. Shop No.78. In service, No.12, May 1928.

1928 LaFrance Type 31, 85-ft. front wheel drive aerial, Reg. No.6301. Shop No.80. In service, No.6, July 1928.

1928 LaFrance Type T-70-6 city service ladder truck, Reg. No.6300. Shop No.79. In service, No.7, July 1928.

1928 Fireboat *J.H. Carlisle* 4,550 gpm. Hull No.115, Burrard Dry Dock. Shop No.96. In service, September 1, 1928.

1929 LaFrance Type 112, combination hose/chemical wagon, 40 gallon, Reg. No.6663. Shop 98. In service, No.11, October 1929.

1929 Bickle Type 700, Canadian 840-gpm pump, Reg. No.9002. Shop No.99. In service, No.6, November 1929.

1930 LaFrance Type 114, city service ladder truck, Reg. No.6527. Shop No.57. In service, No.3, June 1930.

1937 LaFrance Type 412-RB, 1,000-gpm pump, Reg. No.7786. Shop No.49. In service, No.4, September 1937.

1937 Hayes-Anderson Model PCT-32 Bus (Canteen wagon). Shop No.32. In service, No.2, November 1951.

1938 LaFrance Type 512, 85-ft. tractor aerial, Reg. No.8014. Shop No.48. In service, No.1, November 1938.

1939 LaFrance Type 512-BO, hose wagon, Reg. No.8043. Shop No.55. In service, No.2, March 1939.

1939–1945 Converted panel delivery trucks of the Fire Auxiliary.

1934 Ford ½-ton panel	1937 International ½-ton panel	1939 Dodge ½-ton panel
1935 Ford ½-ton panel	1938 Chevrolet ½-ton panel	1939 Fargo ½-ton panel
1935 Reo ½-ton panel	1938 Ford ½-ton panel	1939 Ford ½-ton panel
1935 International ½-ton panel	1938 Ford 1-ton panel	1939 International ½-ton panel
1936 International ½-ton panel (3)	1939 Chevrolet 1-ton panel	1942 Chevrolet 1-ton panel

Each panel was equipped with a portable trailer pump, nine of which were Bickle-Seagrave, with 420-gpm capacity.

1941 LaFrance Type B-57-5 DO, city service ladder truck, Reg. No.L-1498. Shop No.51. In service, No.18, October 1941.

1943 Bickle-Seagrave Type 66-E, Underwriter 625-gpm pump, Reg. No.6036. Shop 63. In service, No.2, February 1944.

1945 LaFrance Type JOX M5-100, 100-ft. Service aerial, Reg. No.L-2175. Shop No.59. In service, No.2, November 1945.

1946 Fire Barges No.1 and No.2, 3,000-gpm each. Shop Nos. 84 and 85. In service, No.10, January 1946.

1947 LaFrance Spartan 1,000-gpm pump, Reg. No.L-3005. Shop No.47. In service, No.2, October 1947.

1947 LaFrance Spartan 1,000-gpm pump, Reg. No.L-3006. Shop No.61. In service, No.19, October 1947.

1948 LaFrance Type 712, PEO 1,200-gpm pump, Reg. No.L-9086. Shop No.82. In service, No.3, September 1948.

1948 LaFrance-Dodge Type DE-6, 500-gpm pump, Reg. No.B-1839. Shop No.54. In service, No.20, July 1948.

1948 LaFrance-Dodge Type DE-6, 500-gpm pump, Reg. No.B-1840. Shop No.53. In service, No.15, July 1948.

1948 LaFrance-Dodge Type DE-6, 500-gpm pump, Reg. No.B-1841. Shop No.81. In service, No.14, July 1948.

1949 LaFrance Type 7-100 AEO, 100-ft. service aerial, Reg. No.9123. Shop No.52. In service, No.3, March 1949.

1949 LaFrance Type 7-100 AEO, 100-ft. service aerial, Reg. No.9135. Shop No.90. In service, No 12, October 1949

1950 Bickle-Seagrave Model 12JA, 85-ft. service aerial, Reg. No.E8825. Shop No.100. In service, No.18, July 1950.

1951 Bickle-Seagrave Model 900, city service ladder truck, Reg. No.F4165. Shop No.67. In service, No.9, October 1951.

1951 Bickle-Seagrave Model 900, 1,200-gpm pump, Reg. No.F4405. Shop No.64. In service, No.17, November 1951.

1951 LaFrance Type 715 PEO, Dominion 1,200-gpm pump, Reg. No.9228. Shop No.38. In service, No.2, June 1951.

1951 LaFrance Type 715 PEO, Dominion 1,200-gpm pump, Reg. No.9250. Shop No.58. In service, No.6, June 1951.

1951 *Vancouver Fireboat No.2,* 20,000-gpm. Shop No.39. In service at No.10, April 1951.

1952 Kenworth city service ladder truck. Shop No.34. In service, No.12, May 1952.

1955 LaFrance Type 715 PEO, Dominion 1,250-gpm pump, Reg. No.9366. Shop No.56. In service, No.22, February 1955.

1955 LaFrance Type 715 PEO, Dominion 1,250-gpm pump, Reg. No.9367. Shop No.60. In service, No.4, February 1955.

1955 LaFrance Type 715 PEO, Dominion 1,250-gpm pump, Reg. No.9368. Shop No.93. In service, No.1, February 1955.

1955 LaFrance Type 700 BSEO, Veteran city service ladder truck, Reg. No.9378. Shop 72. In service, No.22, March 1955.

1956 LaFrance Type 715 PMO, Dominion 1,250-gpm pump, Reg. No.N-96. Shop No.69. In service, No.9, September 1956.

1956 LaFrance Type 715 PMO, Dominion 1,250-gpm pump, Reg. No.N-97. Shop No.84. In service, No.1, September 1956.

1956 LaFrance Type 715 QMO, Dominion 1,250-gpm quad., Reg. No.N-121. Shop No.75. In service, No.17, January 1957.

1956 LaFrance Type 715 QMO, Dominion 1,250 gpm quad., Reg. No.N-122. Shop No.62. In service, No.12, January 1957.

1959 Mercury generator/lighting unit, 15,000 watts (shop-built) Shop No.33. In service at No.14, May 1973.

1960 LaFrance Type 4-100M, 100-ft. service aerial, Reg. No.N-769. Shop No.48. In service No.1, March 1960.

1961 LaFrance Type 915 PMO, Dominion 1,250-gpm pump, Reg. No.N-61-9800A. Shop 71. In service, No.1, April 1961.

1961 LaFrance Type 915 PMO, Dominion 1,250-gpm pump, Reg. No.N-61-9800B. Shop 98. In service, No.2, April 1961.

1961 LaFrance Type 915 PMO, Dominion 1,250-gpm pump, Reg. No.N-61-9800C. Shop 78. In service, No.6, May 1961.

1961 LaFrance Type 915 QMO, Dominion 1,250-gpm quad., Reg. No.N-61-9900. Shop 79. In service, No.17, June 1961.

1964 Hub-International 1,000-gpm pump, Reg. No.225. Shop No.88. In service, No.20, August 1964.

1964 Hub-International 1,000-gpm pump, Reg. No.220. Shop No.99. In service, No.2, August 1964.

1966 Hub-International 100-ft. tractor aerial, Reg. No. unknown. Shop No.57. In service, No.6, March 1966.

1966 Hub-International 1,000-gpm pump, Reg. No.285. Shop No.77. In service, No.18, August 1966.

1967 Hub-International 1,000-gpm pump, Reg. No.315. Shop No.49. In service, No.1, January 1968.

1967 Hub-International 1,000-gpm pump, Reg. No.320. Shop No.55. In service, No.3, February 1968.

1970 LaFrance-Ford 1,050-gpm pump, Reg. No.F-70-155. Shop No.53. In service, No.6, November 1970.

1970 LaFrance-Ford 1,050-gpm pump, Reg. No.F-70-156. Shop No.64. In service, No.9, December 1970.

1970 LaFrance-Ford 100-ft. service aerial, Reg. No.F-70-157. Shop No.59. In service, No.2, February 1971.

1971 King-Seagrave-Ford 1,500-gpm pump, Reg. No.700015. Shop No.54. In service, No.4, April 1971.

1971 King-Seagrave-Ford 1,500-gpm pump, Reg. No.700016. Shop No.81. In service, No.1, April 1971.

1971 King-Seagrave-Ford 1,500-gpm pump, Reg. No.700045. Shop No.63. In service, No.3, July 1971.

1971 King-Seagrave-Ford 1,500-gpm pump, Reg. No.700046. Shop No.47. In service, No.2, July 1971.

1971 Thibault-Ford 100-ft. service aerial, Reg. No.555. Shop No 80. In service, No.1, February 1972.

1973 Calavar Firebird 125 125-ft.aerial platform. Shop No.100. In service at No.1, June 1973.

1973 Pierreville-Ford 1,050-gpm pump, Reg. No.PFT 303. Shop No.58. In service, No.1, August 1973.

1974 Pierreville-Ford 1,050-gpm pump, Reg. No.PFT 369. Shop No.61 became Shop No.A9161.** In service, No.1, April 1974.

1974 Pierreville-Ford 1,050-gpm pump, Reg. No.PFT 370. Shop No.A9182. In service, No.2, April 1974.

1974 Pierreville-Imperial Custom 1,050-gpm pump, Reg. No.PFT 456. Shop No.A9138. In service, No.6, October 1974.

1974 Pierreville-Imperial Custom 1,050-gpm pump, Reg. No.PFT 455. Shop No.A9184. In service, No.9, October 1974.

1975 Thibault-White 100-ft. service aerial, Reg. No.PT 270. Shop No.A9167. In service, No.2, January 1975.

1975 Pierreville-Imperial Custom 100-ft. service aerial, Reg. No.PFT 621. Shop No.A9119. In service, No.9, May 1975.

1975 Pierreville-Imperial Custom 100-ft. rearmount aerial, Reg. No.PFT 622. Shop No.A9134. In service, No.6, June 1975.

1976 Pierreville-Imperial Custom 1,050-gpm pump, Reg. No.PFT 605. Shop No.A9160. In service, No.17, January 1976.

1976 Pierreville-Imperial Custom 1,050-gpm pump, Reg. No.PFT 604. Shop No.A9156. In service, No.22, February 1976.

1976 Calavar Firebird 125 125-ft. aerial platform. Shop No.A9151. In service, No.3, June 1976.

1976 Pierreville-Ford 100-ft. rearmount aerial, Reg. No.unk. Shop No.A9157. In service, No.4, July 1976.

1978 Pierreville-Scot 100-ft. rearmount aerial, Reg. No.PFT 669, Shop No.A9152. In service, No.2, April 1978.

1978 Pierreville-Scot 1,050-gpm pump, Reg. No.PFT 894. Shop No.A9169. In service, No.2, November 1978.

1979 Pierreville-Scot 625-gpm, 100-ft. rearmount quint, Reg. No.PFT 895. Shop No.A9179. In service, No.12, June 1979.

1979 Pierreville-Scot 625-gpm, 100-ft. rearmount quint, Reg. No.PFT 882. Shop No.A9162. In service, No.15, June 1979.

1979 Pierreville-Scot 100-ft. rearmount aerial. Reg. No.PFT 954. Shop No.A9111. In service, UBC, June 1979.

1979 Pierreville-International 1,050-gpm pump, Reg. No. unk. Shop No.A9171. In service, No.13, July 1979.

1979 Pierreville-Scot 625-gpm, 100-ft. rearmount quint, Reg. No.PFT 998. Shop No.A9187. In service, No.17, February 1980.

1980 Pierreville-Scot 100-ft. rearmount aerial, Reg. No.PFT 999. Shop No.A9146. In service, No.9, June 1980.

1980 Pierreville-International 1,050-gpm pump, Reg. No.PFT 987. Shop No.A9114. In service, No.14, August 1980.

1980 Pierreville-International 1,050-gpm pump, Reg. No.PFT 988. Shop No.A9141. In service, No.18, August 1980.

1981 Pierreville-International 1,050-gpm pump, Reg. No.PFT 1108. Shop No.A9193. In service, No.5, June 1981.

1981 Pierreville-International 1,050-gpm pump, Reg. No.PFT 991. Shop No.A9198. In service, No.6, July 1981.

1982 Pierreville-Pemfab Custom 100-ft. rearmount aerial, Reg. No.PFT 1023. Shop No.A9178. In service, No.5, April, 1982.

1982 Superior-International 1,050 gpm pump, Reg. No.SE 465. Shop No.A9188. In service, No.12, October 1982.

1982 Superior-International 1,050 gpm pump, Reg. No.SE 464. Shop No.A9199. In service, No.9, November 1982.

1983 Pierreville-Pemfab Custom 100-ft. rearmount aerial, Reg. No.PFT 1256. Shop No.A9175. In service, No.9, June, 1983.

1986 GMC generator/lighting unit (GEN-1). Shop No.A9133. In service, No.1, December 1986.

1987 Anderson-International 1,050-gpm pump, Reg. No.104. Shop No.A9155. In service, No.15, September 1987.

1987 Anderson-Freightliner 1,500-gpm pump, Reg. No.109. Shop No.A9149. In service, No.1, November 1987.

1987 Anderson-Freightliner 1,500-gpm pump, Reg. No.110. Shop No.A9177. In service, No.3, February 1988.

1988 Hub-Mack 1,250 gpm pump, Reg. No.1345. In service UBC, June 1988. Became VFD Shop No.A9110. In service, No.10, Oct. 1996.

1989 Hub-Freightliner 1,050-gpm engine,* Reg. No.1610. Shop No.A9197. In service, No.17, October 1989.

1989 Hub-Freightliner 1,050-gpm engine, Reg. No.1630. Shop No.A9166. In service, No.7, November 1989.

1989 Anderson-Pacific-Bronto 110-ft. tower.* Reg. No.175. Shop No.A9189. In service, No.2, December 1989.

1989 Hub-Freightliner 1,050-gpm engine, Reg. No.1620. Shop No.A9126. In service, No.20, January 1990.

1990 Hub-Freightliner 1,050-gpm engine, Reg. No.1640. Shop No.A9109. In service, No.9, March 1990.

1991 International Haz-Mat Response Team Unit. (shop-built). Shop No.A9205. In service, No.17, August 1991.

1991 Superior-Pierce 105-ft. rearmount ladder,* Reg. No.SE 1155. Shop No.A9122. In service, No.12, December 1991.

1991 Superior-Pierce 105-ft. rearmount ladder, Reg. No.SE 1156. Shop No.A9120. In service, No.17, March 1992.

1991 Anderson-Simon Duplex 1,500-gpm engine, Reg. No.194. Shop No.A9108. In service, No.8, September 1991.

1991 Anderson-Simon Duplex 1,500-gpm engine, Reg. No.193. Shop No.A9104. In service, No.4, January 1992.

1991 Chevrolet/Grumman Van 30, Shop No.A9233. Technical Rescue Team Unit; X-UBC Haz-Mat, now Haz-Mat 10.

1992 *Fireboat No.1,* 2,500-gpm. Shop No.A9001. In service, Burrard Inlet, September 1992.

1992 *Fireboat No.5,* 2,500-gpm. Shop No.A9005. In service, False Creek, March 1993.

1994 Anderson-Simon Duplex 1,500-gpm engine, Reg. No.590. Shop No.B9160. In service, No.2, February 1994.

1994 Anderson-Simon Duplex 1,500-gpm engine, Reg. No.600. Shop No.B9158. In service, No.6, March 1994.

1994 Anderson-Simon Duplex 1,500-gpm engine, Reg. No.595. Shop No.B9184. In service, No.9, April 1994.

1994 Anderson-Simon Duplex 1,500-gpm 110-ft. rearmount quint, Reg. No.605. Shop No.B9159. In service, No.7, August 1994.

1994 Anderson-Simon Duplex 1,500-gpm 110-ft. rearmount quint, Reg. No.610. Shop No.B9167. In service, No.9, November 1994.

1995 Anderson-Freightliner Hi-Pressure Unit. Reg. No.835. Shop No.A9128. In reserve, No.1

1996 Anderson-Simon Duplex 1,500-gpm 106-ft. quint, Reg. No.870. Shop No B9151. In service, No.10, July 1996.

1997 Anderson/Int'l./ Collins Haz-Mat tender. Shop No.A9207. In service, No.10, August 1997.

1998 Spartan 7,000l/m, 23m rearmount quint, Shop No.B9119. In service, No.1, October 1998.

1998 Spartan 7,000l/m, 23m rearmount quint, Shop No.B9142. In service, No.3, December 1998.

1998 Spartan 7,000l/m, 23m rearmount quint, Shop No.B9138. In service, No.5, February 1999.

1998 Spartan 7,000l/m, 23m rearmount quint, Shop No.B9156. In service, No.15, March 1999

1998 Spartan 7,000l/m, 23m rearmount quint, Shop No.B9146. In service, No.18, March 1999.

1998 Spartan 7,000l/m, 23m rearmount quint, Shop No.B9157. In service, No.21, April 1999.

1998 Spartan 7,000l/m, 23m rearmount quint, Shop No.B9161. In service, No.22, April 1999.

1999 Spartan 7,000l/m, 23m rearmount quint, Shop No.B9187. In Service, No.2, February 1999.

1999 Freightliner 1,150l/m life support unit (LSU), Shop No.B9101. In service, No.3, February 1999.

1999 Freightliner 1,150l/m LSU, Shop No.B9102. In service, No.2, February 1999.

1999 Spartan 7,000l/m, 23m rearmount quint, Shop No.B9198. In service, No.8, February 1999.

1999 Freightliner 1,150l/m LSU, Shop No.B9105. In service, No.17, February 1999.

1999 Freightliner 1,150l/m LSU, Shop No.B9112. In service, No.22, February 1999.

1999 Freightliner 1,150l/m LSU, Shop No.B9103. In service, No.8, March 1999.

1999 Freightliner 1,150l/m LSU, Shop No.B9106. In service, No.9, March 1999.

1999 Spartan 7,000l/m, 23m rearmount quint, Shop No.B9162. In service, No.14, April 1999.

1999 Spartan 7,000l/m, 23m rearmount quint, Shop No.B9169. In service, No.13, April 1999.

1999 Spartan 7,000l/m, 23m rearmount quint, Shop No.B9179. In service, No.19, April 1999.

1999 Freightliner 1,150l/m LSU, Shop No.B9107. In service, No.21, April 1999.

1999 Freightliner 1,150l/m LSU, Shop No.B9115. In service, No 1, (spare) April 1999.

1999 Spartan 7,000l/m, 23m rearmount quint, Shop No.B9171. In service, No.4, May 1999.

1999 Spartan 7,000l/m, 23m rearmount quint, Shop No.B9182. In service, No.20, September 1999.

2000 Ford Haz-Mat tender, Shop No.B9245. In service, No.10, October 2000.

2001 Spartan Gladiator 30m Tower, Shop No.B9189. In service, No.7, May 2001.

2002 Superior-Ford F550 4x4 Wildlands #5 (Bush truck), Shop No.A9237. In service, No.5, 2002.

2002 Superior-Ford F550 4x4 Wildlands #8 (Bush truck), Shop No.A9238. In service, No.8, 2002.

2006 An order was placed for 13 Spartan engines, two Spartan-Smeal aerials, and two new Haz-Mat rigs for delivery in 2007.

*   Denotes nomenclature change in 1989; pumps to engines; aerials/trucks to ladders/towers, etc.

** Shop Nos. were changed in 1990; i.e., Shop No.01 changed to Shop No.A9101, then future replacements will be B9101, C9101, etc.

    For the Civil Defense program during the 1960s, each shop number was given the prefix "VA" (Vancouver Area).

Appendix V

Fire Alarms 1890 to 2006

This list includes all Western Union, B.C. District Telegraph (BCDT), box, telephone, silent, and verbal alarms.

YEAR	ALARMS	HALLS	MEN	YEAR	ALARMS	HALLS	MEN	YEAR	ALARMS	HALLS	MEN
1890	110	2	23*	1929	1,896	20	336	1968	8,996	19	
1891	60	2		1930	2,309	20	368	1969	9,066	19	
1892	46	3	30	1931	1,795	20		1970	9,607	19	
1893	47	3		1932	1,554	20		1971	9,550	18	
1894	58	3		1933	1,670	18		1972	10,968	18	
1895	97	3		1934	1,839	18		1973	11,833	18	
1896	64	3		1935	2,466	17		1974	12,862	20	762
1897	62	3		1936	2,270	17		1975	12,659	20	834
1898	131	3		1937	2,129	17	369	1976	12,573	20	820
1899	127	3		1938	3,119	17		1977	15,032	20	
1900	108	3		1939	2,413	17	348	1978	18,188	20	806
1901	145	3	28	1940	2,547	17	368	1979	18,971	20	
1902	157	3		1941	2,519	17		1980	19,754	20	
1903	164	3		1942	2816	17	378	1981	22,017	20	
1904	167	4	39	1943	3,356	17	381	1982	19,760	20	
1905	132	5	45	1944	3,400	17		1983	15,312	20	
1906	216	5	46	1945	3,954	17	396	1984	19,023	20	
1907	229	5		1946	3,668	18	537	1985	21,473	20	
1908	276	8	72	1947	4,319	18	539	1986	23,353	20	
1909	359	10	114	1948	4,501	18	542	1987	23,041	20	
1910	453	10		1949	4,688	18	554	1988	23,849	19	
1911	546	11	145	1950	4,638	18	558	1989	21,073	19	
1912	583	11	156	1951	5,372	18	598	1990	23,589	19	
1913	552	14	182	1952	5,064	19		1991	30,122	19	
1914	642	15	201	1953	4,924	19	610	1992	36,655	19	
1915	672	15	171	1954	5,193	19	626	1993	37,053	19	
1916	540	15	166	1955	6,055	19	621	1994	37,104	19	
1917	608	13	159	1956	6,369	19	632	1995	37,887	20	840
1918	772	11	150	1957	5,847	19	651	1996	39,688	20	
1919	969	11	187	1958	6,072	19		1997	39,670	20	812
1920	1,109	11		1959	5,826	19		1998	37,878	20	
1921	1,032	11	191	1960	6,221	19	657	1999	35,838	20	
1922	1,055	11		1961	6,659	19		2000	36,110	20	807
1923	772	11	192	1962	6,685	19		2001	35,305	20	
1924	1,022	11	194	1963	7,297	19		2002	33,400	20	
1925	1,080	11	206	1964	7,480	19	751	2003	31,444	20	
1926	1,122	11		1965	7,914	19	758	2004	37,490	20	805
1927	1,141	13	228	1966	7,944	19		2005	46,183	20	
1928	1,258	14	256	1967	8,911	19		2006	42,984	20	811

* Also includes call-out men

Appendix VI

Vancouver Fire Fighters Union Presidents

City Firemen's Union, Local 15363, American Federation of Labor (AF of L)
CHARLES A. WATSON   1916–1918

Vancouver Local No.18, International Association of Fire Fighters (IAFF)
CHARLES A. WATSON   1918–1920
PERCY TRERISE   1920–1922
NEIL MacDONALD   1922–1925
JOHN ANDERSON   1925–1927
NEIL MacDONALD   1927–1936
ERNEST A. YOUNG   1936

City Fire Fighters' Union Local No. 1
ERNEST A. YOUNG   1936–1938
ELMER R. SLY   1938–1941
F.G. LUCAS   1941–1942
HUGH S. BIRD   1942–1947

Vancouver Fire Fighters, Local No.18, IAFF

HUGH S. BIRD   1947–1949
J. HECTOR WRIGHT   1949–1952
HUGH S. BIRD   1952 January–April
J. HECTOR WRIGHT   1952–1959
A. (HARRY) LANGOUT   1959–1960
GORDON R. ANDERSON   1960–1969
J. BARRY WILLIAMS   1969–1971
GORDON R. ANDERSON   1972–1977
JOHN E. BUNYAN   1977–1979
L.H. (BING) PARÉ   1979–1981
WILLIAM D. (BILL) JONES   1981–1983
WILLIAM C. (BILL) ANDERSON   1983–1987
L.H. (BING) PARÉ   1987–1989
DAVID W. MITCHELL   1989–1996
R.G. (ROD) MacDONALD   1996–1999
JEFF DIGHTON   1999–2001
R.G. (ROD) MacDONALD   2001–

Local 18's Sixth District IAFF International Vice-Presidents
GEORGE J. RICHARDSON 1918–1920
CHARLES A. WATSON   1920–1921
NEIL MacDONALD   1928–1936
HECTOR WRIGHT   1954–1959
HARRY LANGOUT   1959–1962
GORDON R. ANDERSON   1965–1977

1911 Vancouver firemen organized a union with the Trades and Labor Council.

December 1916 The firemen joined the AF of L, as Local 15363.

February 28, 1918 Vancouver firemen became charter members of the IAFF as Local No.18. Local numbers were designated by seniority in the AF of L.

June 1, 1928 Members of South Vancouver Local 259, IAFF and Point Grey Local 260, IAFF, disbanded and joined Vancouver Local 18.

1936 to 1947 Vancouver's IAFF charter was revoked because of a strike vote. The union then became the City Fire Fighters' Union, Local No.1.

May 21, 1947 The union was reinstated as Vancouver Fire Fighters, Local No. 18, IAFF.

January 1, 1972 Gordon R. Anderson became Local 18's first full-time paid president.

Index

People

Buildings & Locations

* Page number in **bold** indicates photograph or illustration.